Biodiversity Conservation (Volume I)

Biodiversity Conservation (Volume I)

Bradley Hunt

Biodiversity Conservation (Volume I)
Bradley Hunt
ISBN: 979-8-88836-050-7 (Hardback)

© 2023 Larsen & Keller

Published by Larsen and Keller Education,
5 Penn Plaza,
19th Floor,
New York, NY 10001, USA

Cataloging-in-Publication Data

Biodiversity conservation (Volume I) / Bradley Hunt.
   p. cm.
Includes bibliographical references and index.
ISBN 979-8-88836-050-7
1. Biodiversity conservation. 2. Biodiversity. 3. Ecosystem management.
I. Hunt, Bradley.
QH75 .B56 2023
333.951 6--dc23

This book contains information obtained from authentic and highly regarded sources. All chapters are published with permission under the Creative Commons Attribution Share Alike License or equivalent. A wide variety of references are listed. Permissions and sources are indicated; for detailed attributions, please refer to the permissions page. Reasonable efforts have been made to publish reliable data and information, but the authors, editors and publisher cannot assume any responsibility for the validity of all materials or the consequences of their use.

Trademark Notice: All trademarks used herein are the property of their respective owners. The use of any trademark in this text does not vest in the author or publisher any trademark ownership rights in such trademarks, nor does the use of such trademarks imply any affiliation with or endorsement of this book by such owners.

For more information regarding Larsen and Keller Education and its products, please visit the publisher's website www.larsen-keller.com

Table of Contents

	Preface	**VII**
Part I	**Background and Context**	**1**
Chapter 1	**A Modern Synthesis of Biodiversity in Angola: Approach and Purpose**	**3**
	Brian J. Huntley and Nuno Ferrand	
Chapter 2	**Physical Geography and Biodiversity Characteristics of Angola**	**14**
	Brian J. Huntley	
Chapter 3	**Biogeography and Conservation of Angolan Coastal and Marine Systems**	**42**
	Stephen P. Kirkman and Kumbi Kilongo Nsingi	
Chapter 4	**Angola and its Alpha Paleobiodiversity**	**52**
	Octávio Mateus, Pedro M. Callapez, Michael J. Polcyn, Anne S. Schulp, António Olímpio Gonçalves and Louis L. Jacobs	
Part II	**Flora and Vegetation: Patterns and Evolutionary Processes**	**75**
Chapter 5	**Botanical Activity in Angola: Current and Historical Overview**	**77**
	David J. Goyder and Francisco Maiato P. Gonçalves	
Chapter 6	**Surveying, Mapping and Classifying Angolan Vegetation**	**95**
	Rasmus Revermann and Manfred Finckh	

VI Contents

Chapter 7 **Geoxylic Suffrutices and Suffrutex-Grasslands in Angola** 106
Paulina Zigelski, Amândio Gomes and Manfred Finckh

Chapter 8 **Changes in Vegetation, Soils and Water Quality at Landscape Scales** 118
John M. Mendelsohn

Part III **Environmental Indicators and Diversity among Dragonflies and Butterflies** 133

Chapter 9 **Odonata of Angola: Researches and Biogeography** 135
Jens Kipping, Viola Clausnitzer, Sara R. F. Fernandes Elizalde and Klaas-Douwe B. Dijkstra

Chapter 10 **A Checklist of Angolan Lepidoptera and Papilionoidea** 160
Luís F. Mendes, A. Bivar-de-Sousa and Mark C. Williams

Part IV **Diversity and Distribution of the Vertebrate Taxa** 197

Chapter 11 **Angolan Freshwater Fishes: A Biogeographic Model** 199
Paul H. Skelton

Chapter 12 **Research on Angolan Amphibians: Past and Present** 234
Ninda Baptista, Werner Conradie, Pedro Vaz Pinto and William R. Branch

Permissions

Index

Preface

The main aim of this book is to educate learners and enhance their research focus by presenting diverse topics covering this vast field. This is an advanced book which compiles significant studies by distinguished experts in the area of analysis. This book addresses successive solutions to the challenges arising in the area of application, along with it; the book provides scope for future developments.

Biodiversity conservation refers to the preservation, enhancement and management of biodiversity. It aims to provide sustainable benefits to current and future generations. Ecological modeling refers to the creation and study of mathematical models of ecological processes, including integrated biophysical and purely biological models. These models are used for assisting in the creation of strategies to manage biodiversity. Ecological modeling allows the simulation of those ecological processes, which in reality might take centuries but can be done within minutes on a computer model. This book provides comprehensive insights on biodiversity conservation. The topics included herein are of utmost significance and bound to provide incredible insights to readers. Coherent flow of topics, student-friendly language and extensive use of examples make this book an invaluable source of knowledge.

It was a great honour to edit this book, though there were challenges, as it involved a lot of communication and networking between me and the editorial team. However, the end result was this all-inclusive book covering diverse themes in the field.

Finally, it is important to acknowledge the efforts of the contributors for their excellent chapters, through which a wide variety of issues have been addressed. I would also like to thank my colleagues for their valuable feedback during the making of this book.

Bradley Hunt

Part I
Background and Context

A Modern Synthesis of Biodiversity in Angola: Approach and Purpose

Brian J. Huntley and Nuno Ferrand

Abstract Angola possesses an unusually rich diversity of ecosystems and species, but this natural wealth is poorly documented when compared with other countries in the region. Both colonial history and extended wars challenged progress in biodiversity research and conservation, but since peace was achieved in 2002 a rapidly increasing level of collaboration between Angolan and visiting scientists and institutions has seen a blossoming of biodiversity research. The absence of comprehensive reviews and syntheses of existing knowledge, often published in extinct journals and inaccessible official reports, necessitates a modern synthesis. This volume brings together the existing body of scientific results from studies on Angola's landscapes, ecosystems, flora and fauna, and presents an outline of opportunities for biodiversity discovery, understanding and conservation as well as collaborative research.

Keywords Africa · Biomes · Collaborative research · Conservation · Ecoregions

Background and Context

Angola is a country of unusually rich physiographic, climatic and biological diversity. It occupies only 4% of the terrestrial area of Africa, yet it possesses the highest diversity of biomes and is second only to mega-diverse South Africa in terms of the number of ecoregions found within its borders. However, scientific literature on its

---

B. J. Huntley (✉)
CIBIO-InBIO, Centro de Investigação em Biodiversidade e Recursos Genéticos, Universidade do Porto, Vairão, Portugal
e-mail: brianjhuntley@gmail.com

N. Ferrand
CIBIO-InBIO, Centro de Investigação em Biodiversidade e Recursos Genéticos, Laboratório Associado, Campus de Vairão, Universidade do Porto, Vairão, Portugal

Departamento de Biologia, Faculdade de Ciências, Universidade do Porto, Porto, Portugal

Department of Zoology, Auckland Park, University of Johannesburg, Johannesburg, South Africa
e-mail: nferrand@cibio.up.pt

biodiversity is extremely limited when compared with most African countries. Much of that which has been published is difficult to access or out of print. This volume seeks to redress this situation.

Here we present a review of what is known about Angola's biodiversity. Much of the existing literature dates from the nineteenth and early to mid-twentieth centuries. Following independence in 1975, field studies were curtailed by the instabilities of an extended civil war. It was not until after the peace settlement of 2002 that a new wave of research has been possible. Initial attempts to establish collaborative field expeditions were frustrated by visa and permit restrictions, but these challenges were gradually overcome and by the 2010s a vibrant programme of joint projects has evolved. Today many foreign specialists work in partnership with Angolan researchers and institutions, producing a new flow of scientific results of which many are presented in this volume.

For any comprehensive synthesis, both temporal depth and spatial breadth is necessary. An historical perspective is presented in each chapter. Angolan indigenous knowledge has contributed to the insights and materials that have informed visiting researchers from the eighteenth century to the present day. The pioneering studies and exhaustive botanical collections of the Austrian botanist, Friedrich Welwitsch (1806–1872), the zoological collections of the indefatigable Portuguese naturalist José Anchieta (1832–1897) and the Swiss botanist John Gossweiler (1873–1952) set benchmarks for later work (Swinscow 1972; de Andrade 1985). Each succeeding student of Angola has added to the description of its biological diversity. While botanists such as Romero Monteiro (1970) and zoologists such as Crawford-Cabral (1983) have summarised available biogeographic information within a national context, no comprehensive synthesis of studies on Angola's fauna, flora and ecosystems has yet been undertaken. The need for an integrated account has become evident in the past decade, as increasing numbers of expeditions and collaborative projects have evolved as part of the country's 'peace dividend'.

Approach and Purpose of This Synthesis

A modern synthesis is not easily achieved. Much of the early literature on Angola's biodiversity resides in publications and reports that are difficult to source. This review attempts to reference these important but sometimes elusive accounts, in order to provide students with access to what information is available. While focusing on papers in peer-reviewed journals, some topics need to draw on unpublished reports filed in government departments. It also seeks to bring together the findings of recent, post-independence studies, many of which are still in progress or in press. It is intended to serve the new generation of Angolan students by providing a comprehensive but focused synopsis of what is known on the biomes, landscapes, flora and fauna of Angola. It should also bring Angola to the attention of researchers across Africa and beyond, revealing the great diversity of life, and the multiple questions on the structure and functioning of Angola's biodiversity that await exploration, examination and explanation.

In structuring this book, this introduction leads through synopses on the country's terrestrial and marine biogeography, paleontological record, recent landscape evolution and land transformation, to chapters on its flora and vegetation. The main body of this volume is devoted to accounts of its fauna – selected invertebrate groups that have promise as indicators of environmental stress, and all vertebrate groups. In each treatment, the need for increased conservation measures for threatened taxa and habitats is a recurrent theme, while research opportunities are highlighted. While general inventories and checklists are progressing well, the state of ecological knowledge remains rudimentary. Topics as fundamental as ecological processes such as the flows of energy, water and nutrients; the ecological impacts of phenomena such as fire, invasive species, herbivory, droughts and frosts; community structure, plant-animal interactions and the impacts of land-transformation and of climate change are yet to be researched in Angola. This volume's content is limited by the availability of information. It is therefore opportunistic, covering those taxonomic groups and those features and processes for which a critical mass of information is available. The focus is primarily on the terrestrial ecosystems and biota of Angola, but the importance of the marine environment is described in accounts on marine biodiversity and ocean dynamics, and on the richness of the whale, dolphin and marine turtle faunas of Angolan waters.

In comparison with similar reviews for other African countries with long and strong traditions of research into their biodiversity and ecology, and for which comprehensive syntheses of the state of knowledge are available (e.g. Namibia: Barnard 1998; Southern Africa: Davis 1964; Werger and van Bruggen 1978; Huntley 1989; Tanzania: Sinclair 2012), this account reveals both the strengths and weaknesses of the research agenda of the colonial era, and the challenges of the recent past. While institutions such as the *Instituto de Investigação Científica de Angola* and the *Instituto de Investigação Agronómica de Angola* undertook very important studies on many taxa, and on vegetation, soils and agronomy, and the *Museu do Dundo* amassed and distributed a vast series of collections of the animal species of the Lundas, the coverage of disciplines and of the remote regions of Angola was weak.

Biodiversity Surveys: Historical Synopsis

The history of scientific exploration and biological collection in Angola is relatively modest. Whereas South Africa, by 1975, had over three million herbarium specimens collected by some 2500 botanists since the late eighteenth century (Gunn and Codd 1981), Angola had less than 300,000 specimens collected by just 300 botanists during the same period (Figueiredo and Smith 2008). Despite the relatively limited coverage of Angolan collections, the great botanist Francisco Mendonça was occasioned to state in his preface to Gossweiler and Mendonça (1939):

> We are happily able to confirm that the flora of Angola is the best known in tropical Africa, due to the attention given by the state towards the botanical exploration of the colony, and the great interest and zeal of scientists in its study.

The Swiss zoologist, Monard (1935), had been less sanguine:

> A regrettable fact about the Natural History of Angola is the scarcity of concrete information about the nature, distribution, and habits of the large game. The Boers ... never communicated their observations. The Portuguese hunters ... did not write reports on their hunts, or if they did, did so in newspapers or magazines that never enter the scientific literature. The observations remain, in this way, lost to the naturalist who is not able to locate such work.

In truth, during the colonial era, investment in research on the country's biodiversity was limited. The achievements of early pioneers such as Friedrich Welwitsch, José Anchieta and John Gossweiler were quite remarkable, and those of more recent agronomists, botanists and zoologists such as Castanheira Diniz, Romero Monteiro, Grandvaux Barbosa, Brito Teixeira, Crawford-Cabral, Rosa Pinto, Barros Machado, etc., were equally laudable, indeed amazing.

The war years from 1975 to 2002 saw very few researchers venturing into the field. Most activities were limited to brief searches for remnant populations of giant sable (Estes 1982), marine turtles (Carr and Carr 1991), birds (Günther and Feiler 1986a, b; Hawkins 1993), and a countrywide assessment of the state of wildlife populations (Huntley and Matos 1992). The Southern African Botanical Diversity Network (SABONET) project attempted to stimulate botanical studies in Angola from the mid-1990s (Huntley et al. 2006), while the Kissama Foundation funded a vegetation survey of the northern extreme of Quiçama (Jeffrey 1996) and introduced a mixed assemblage of antelope and ostriches into the park in 2000 (Walker 2004). The last decades of the twentieth century were aptly described as a period of *confusão* – confusion (Maier 2007). In brief, from Angola's independence in 1975, until the twenty-first century, cooperative field research over most of the country was challenged by the impacts of war. But dramatic and positive change came with the dawn of the new millennium.

Research Collaboration in the Twenty-First Century

From 2000, especially after the peace agreement of April 2002, field activities expanded rapidly. Most notably, Vaz Pinto has focused on a long-term study on giant sable in Cangandala (Walker 2004; Vaz Pinto 2018), Morais (2017) has led surveys of marine turtles along the Angolan coast, and Mills (2010, 2018) has undertaken field studies on birds across the country.

International support for environmental conservation and research strengthened from 2001, when the Global Environment Facility, through the United Nations Development Programme, initiated a multidisciplinary project to develop a transboundary diagnostic analysis of hydro-environment threats within the Okavango River Basin, known as the Environmental Protection and Sustainable Management of the Okavango River Basin Project (EPSMO). The project's aim was to facilitate the protection of the Basin's aquatic ecosystems and biological diversity (OKACOM 2009, 2011). The project included participation from Angola, Botswana and

Namibia and provided a strong impetus to future multi-national projects in the Basin. A further initiative, the Integrated River Management Project, was funded by USAID/Southern Africa between 2004 and 2009 and provided both institutional and management planning support to the national partners (OKACOM 2009). The EPSMO project was succeeded by the SAREP project described below.

The OKACOM projects were focused on major water management needs and did not embrace detailed biodiversity surveys. Indeed, until 2009, biodiversity research activities in Angola had been essentially individual efforts, with limited funding. Difficulties experienced in obtaining visas to visit, and permits to collect specimens in Angola were a continuing challenge faced by foreign scientists. With the signing of an agreement between the South African National Biodiversity Institute (SANBI), the Angolan Ministry of Environment and the *Instituto Superior de Ciências da Educação* (ISCED), Lubango, in 2009, more ambitious cooperative biodiversity projects became possible. Initially designed as training exercises, the series of Rapid Biodiversity Assessments, in Huíla/Namibe (Huntley 2009), Lunda-Norte (Huntley 2011; Huntley and Francisco 2015) and across western Angola (Rejmánek et al. 2017), brought over 40 scientists from 14 countries to Angola to work with local students and researchers.

By the early 2010s, a wide diversity of major cooperative programmes had developed, including those of the Southern African Regional Environmental Programme (SAREP), the Southern African Science Service Centre for Climate Change and Adaptive Land Management (SASSCAL) (Revermann et al. 2018), the National Geographic Okavango Wilderness Project (NGOWP 2018), and conservation initiatives of NGOs such as *Elephants without Borders*, *Panthera*, Peace Parks Foundation, the Kavango-Zambezi Transfrontier Conservation Area (KAZA) project and several others. Collaboration between foreign museums and universities and Angolan counterparts stimulated additional specialist interests, collectively gaining momentum until the present. In October 2012, CIBIO (Research Centre in Biodiversity and Genetic Resources) at the University of Porto, Portugal, and ISCED-Huíla (Lubango) established a collaborative research, capacity building and advanced training project – the ISCED/CIBIO TwinLab initiative. The initiative was soon replicated in South Africa, Mozambique, Namibia and Zimbabwe, and the whole network of TwinLabs now forms a UNESCO Chair *Life on Land*, awarded at the end of 2017.

For much of the post-independence period, biodiversity research efforts had been un-coordinated and opportunistic. With the establishment of the *Instituto Nacional de Biodiversidade e Áreas de Conservação* (INBAC) in 2011, the opportunity arose for a greater level of coordination and priority setting. The *Plano Estratégico da Rede Nacional da Áreas de Conservação de Angola* (GoA 2011), provided a stimulus to studies in key biodiversity hotspots such as Mount Moco, Mount Namba, Serra da Neve, Serra Pingano, Cumbira, Lagoa Carumbo, and to the vast and very poorly researched catchments of the Cuando Cubango. While a greater level of inter-institutional collaboration is still possible, the momentum developed over the past decade has been unprecedented since 1975. The successes of the recent past are presented in this volume, often drawing on work that is still in progress, is unpublished, or is in press.

Chapter Outlines

Angola is a large country, and as emphasised throughout this volume, it has a rich diversity of landscapes, seascapes and associated biomes and ecoregions. The history of biodiversity research in Angola stretches over 200 years. The spatial, temporal and taxonomic scales embraced in this book results in it being structured in five parts. Part I, Chap. 1 (Huntley and Ferrand this chapter) provides an introduction to the book and its content. Chap. 2 (Huntley 2019) outlines the country's biogeography, drawing on the long history of geomorphological and landscape analysis in Angola, and describes the diversity of seven terrestrial biomes, 15 ecoregions and 32 vegetation types. In Chap. 3 Kirkman and Nsingi (2019) synthesise the findings of recent multi-national research activities on the Benguela Current Large Marine Ecosystem project and other studies on Angola's coastal and marine systems. The long history of the evolution of Angola's biota is introduced by Mateus et al. (2019) in Chap. 4, where the exciting recent discoveries in Angola's fossil record, most especially that of the Cretaceous, is described. A highlight was the discovery of the sauropod dinosaur *Angolatitan adamastor*, the first dinosaur to be found in Angola (Mateus et al. 2011). These authors emphasise the fact that for very long periods – hundreds of millions of years – the absence of fossiliferous rocks in Angola excludes the possibility of tracking animal and plant evolution in Angola over such long periods.

Part II presents an historical and contemporary analysis of our understanding of the country's flora and vegetation and on curious patterns and evolutionary processes in some typical Angolan plant communities. In Chap. 5 Goyder and Gonçalves (2019) note that the vascular flora now totals 6850 species, with 14.8% of these being endemic. The two early vegetation maps of Angola, prepared by the pioneers Gossweiler and Mendonça (1939) and Barbosa (1970), having served the country for many decades, now deserve renewed mapping efforts at a finer scale, using modern remote sensing and numerical analysis approaches, as recommended by Revermann and Finckh (2019) in Chap. 6. Of the many intriguing features of Angolan vegetation, the patterns of plant community/soil/animal associations, such as the 'fairy circles' of the Namib (Juergens 2013; Cramer and Barger 2013), 'fairy forests' of the miombo, and the influence of coastal fog on desert vegetation and fauna; are of special interest to ecologists. Few of these phenomena have been adequately interpreted, but Zigelski et al. (2019) in Chap. 7 presents recent studies on the 'underground forests' of the *chanas de ongote* of the Angolan plateau. The landscapes of Angola are not static, being subject to multiple processes of transformation. In Chap. 8, Mendelsohn (2019) uses results from satellite technologies and ground surveys to describe the dramatic impacts of deforestation, fires, mining and agricultural activities on vegetation, soils and water quality at landscape scales.

Part III details the results of surveys that have advanced rapidly over the past two decades on two invertebrate groups – dragonflies and butterflies. These colourful and taxonomically distinctive insects are known to be sensitive to subtle changes in environmental conditions, such as forest cover and water quality, and serve as effective indicators of change in environmental health. The chapters (9 and 10) on butterflies (Mendes et al. 2019) and on dragonflies (Kipping et al. 2019) have enriched Angola's knowledge of these important ecological groups. Prior to 2009, for example, only 158 species of dragonflies and damselflies were known from Angola. By 2018 this number had increased to 260. The butterfly checklist now stands at 792 species and subspecies, up by over 220 species and subspecies since the turn of the millennium. In contrast to the encouraging progress in these taxa, key environmental engineers – ants and termites – remain poorly documented and await study.

A major component of this volume has been devoted to the vertebrate taxa that have enjoyed the attention of scientists active in Angola since the mid-nineteenth century. Part IV presents detailed accounts of the pioneering work of such luminaries as Anchieta, Bocage, Boulenger, Machado, Rosa Pinto and Crawford-Cabral, but also of the very many other contributors to the inventory of Angola's vertebrate fauna. Skelton (2019), Chap. 11, provides a concise summary of what is known of Angola's ca. 358 species of freshwater fishes (of which 22% are endemic), and also presents a model of post-Cretaceous biogeography of Angola and the roles of regional tectonics and river capture on the speciation and distribution of the fish fauna. Baptista et al. (2019), Chap. 12, cover the amphibian fauna, noting that the group clearly deserves further survey given that thus far only 111 species have been recorded (compared to 128 species for similar-sized but much drier and cooler South Africa). In Chap. 13, Branch et al. (2019) offer a comprehensive account on the 278 species of Angolan reptiles and their patterns of diversity and endemism, documenting key reptile hotspots deserving further exploration. They predict that as many as 75 new species of lizards await discovery in Angola. Both Branch and Baptista demonstrate the value of molecular phylogenetics in clarifying taxonomic complexes in reptiles and frogs.

Angola, with ca. 940 bird species recorded, has in recent years become a favoured destination of ecotourists searching for the country's 29 endemic bird species, and Dean et al. (2019) provide in Chap. 14, a chronology of ornithological surveys, a listing of endemics and near-endemics, and sites of special interest to bird enthusiasts, both professional and amateur. They emphasise, as highlighted by Hall (Hall 1960), the faunistic importance of the Angolan Escarpment, and also of the relict Afromontane forests of the highlands (Vaz Silva 2015), as areas of critical importance to understanding the evolution of Africa's avifauna. These isolated, fragmented and rapidly declining forests merit the highest level of protection to secure their futures as evolutionary fingerprints of the past.

A team of ten mammal specialists, coordinated by Beja et al. (2019), present a major synthesis (Chap. 15) on Angola's 291 mammal species. This chapter fills a need felt since Hill and Carter's (1941) benchmark study, and the more recent coverage of ungulates by Crawford-Cabral and Veríssimo (2005). With 73 species of bats (one third of the bat species known for Africa) Angola has the highest number of species for southern Africa, despite the comparatively limited intensity of surveys undertaken to date. The most diverse mammal group, rodents, has 85 species listed for Angola, of which 13 are endemic or near-endemic. While the number of endemic mammal species is modest, the vulnerability of many species to extinction within Angola is high and deserves urgent conservation measures. Less well known to Angolans than the terrestrial mammals is the country's unusually rich marine mammal fauna. The 28 species of cetaceans (whales and dolphins) found off Angola's coast have been the subject of surveys and research undertaken by Weir (2019) since 2003. As noted in Chap. 16, the possible presence of a further seven species of cetaceans in Angolan waters makes the country globally important for marine mammal conservation.

The Angolan mammal that has enjoyed national and international attention is the Giant Sable Antelope, which has been the subject of an intense research and conservation project since 2002 (Chap. 17), led by Vaz Pinto (2019). The successful rescue and rehabilitation of this national icon, from the brink of extinction, is a conservation model for which Angola can justifiably be proud. The success of the Giant Sable Project needs replication for the many mammal species that are known to have been reduced to very low numbers, or have been hunted to extinction in Angola. These include most large carnivores – Cheetah, Lion, Wild Dog, plus many herbivores – Black-faced Impala, Red Hartebeest, Lichtenstein's Hartebeest, Tsessebe, Southern Lechwe, Puku, Forest Buffalo, Giraffe, Black Rhino, Western Gorilla, Chimpanzee, Forest Elephant and Manatee.

The final section of this volume (Part V) presents an overview of the country's conservation history and current opportunities for action, Chap. 18 (Huntley et al. 2019); and an introduction to the importance of natural history museums and herbaria in the country's biodiversity science and conservation agenda, Chap. 19 (Figueira and Lages 2019). What is abundantly clear, as expressed in the concluding chapter, (Russo et al. 2019), is that Angola is alive with research and conservation opportunities, stimulated by recent initiatives led by the Angolan government and supported by the international community.

This volume provides a first synthesis of what is known and published about Angola's diverse landscapes, biomes and ecosystems and the species that inhabit them. It is a humble attempt by its 46 contributors to place this knowledge before researchers and conservationists in Angola and beyond, especially those who might be stimulated to strengthen the imperfect understanding and vulnerable state of Angola's biodiversity. It is the fervent hope of this book's editors that this volume will provide an entry point for many young Angolan students to study the literature, be inspired by the dedication, tenacity and wisdom of the early pioneers and contemporary explorers, and enter careers in field-based biodiversity research and conservation in Angola.

References

Baptista N, Conradie W, Vaz Pinto P et al (2019) The amphibians of Angola: early studies and the current state of knowledge. In: Huntley BJ, Russo V, Lages F, Ferrand N (eds) Biodiversity of

Angola. Science & conservation: a modern synthesis. Springer Nature, Cham

Barbosa LAG (1970) *Carta Fitogeográfica da Angola*. Instituto de Investigação Científica de Angola, Luanda, 343 pp

Barnard P (1998) Biological diversity in Namibia: a country study. Namibian National Biodiversity Task Force, Windhoek, 325 pp

Beja P, Vaz Pinto P, Veríssimo L et al (2019) The mammals of Angola. In: Huntley BJ, Russo V, Lages F, Ferrand N (eds) Biodiversity of Angola. Science & conservation: a modern synthesis. Springer Nature, Cham

Branch WR, Vaz Pinto P, Baptista N et al (2019) The reptiles of Angola: history, diversity, endemism and hotspots. In: Huntley BJ, Russo V, Lages F, Ferrand N (eds) Biodiversity of Angola. Science & conservation: a modern synthesis. Springer Nature, Cham

Carr T, Carr P (1991) Surveys of the sea turtles of Angola. Biol Conserv 58(1):19–29

Cramer MD, Barger NN (2013) Are Namibian "fairy circles" the consequence of self-organizing spatial vegetation patterning? PLoS One 8(8):e70876

Crawford-Cabral J (1983) Esboço zoogeográfico de Angola. Unpublished manuscript. Instituto de Investigação Científica Tropical, Lisboa, 50 pp + 13 maps

Crawford-Cabral J, Veríssimo LN (2005) The ungulate fauna of Angola: systematic list, distribution maps, database report. Estudos, Ensaios e Documentos do Instituto de Investigação Científica Tropical 163:1–277

Davis DHS (ed) (1964) Ecological studies in Southern Africa. Junk, The Hague, 415 pp

de Andrade AAB (1985) O Naturalista José de Anchieta. Instituto de Investigação Científica Tropical, Lisboa, 187 pp

Dean WRJ, Melo M, Mills MSL (2019) The avifauna of Angola: richness, endemism and rarity. In: Huntley BJ, Russo V, Lages F, Ferrand N (eds) Biodiversity of Angola. Science & conservation: a modern synthesis. Springer Nature, Cham

Estes RD (1982) The giant sable and wildlife conservation in Angola. Report to IUCN Species Survival Commission. Gland, Switzerland

Figueira R, Lages F (2019) Museum and herbarium collections for biodiversity research in Angola. In: Huntley BJ, Russo V, Lages F, Ferrand N (eds) Biodiversity of Angola. Science & conservation: a modern synthesis. Springer Nature, Cham

Figueiredo E, Smith GF (eds) (2008) Plants of Angola / Plantas de Angola. Strelitzia 22:1–279

GoA (Government of Angola) (2011) Plano Estratégico da Rede Nacional de Áreas de Conservação de Angola (PLENARCA). Ministry of Environment, Luanda

Gossweiler J, Mendonça FA (1939) Carta Fitogeográfica de Angola. Ministério das Colónias, Lisboa, 242 pp

Goyder DJ, Gonçalves FMP (2019) The Flora of Angola: collectors, richness and endemism. In: Huntley BJ, Russo V, Lages F, Ferrand N (eds) Biodiversity of Angola. Science & conservation: a modern synthesis. Springer Nature, Cham

Gunn M, Codd LE (1981) Botanical exploration of Southern Africa. AA Balkema, Cape Town, 400 pp

Günther R, Feiler A (1986a) Zur phänologie, ökologie und morphologie angolanischer Vögel (Aves). Teil I: Non-Passeriformes Faunistische Abhandlungen aus dem Staatlichen Museum für Tierkunde in Dresden 13:189–227

Günther R, Feiler A (1986b) Zur phänologie, ökologie und morphologie angolanischer Vögel (Aves). Teil II: Passeriformes Faunistische Abhandlungen aus dem Staatlichen Museum für Tierkunde in Dresden 14:1–29

Hall BP (1960) The faunistic importance of the scarp of Angola. Ibis 102:420–442

Hawkins F (1993) An integrated biodiversity conservation project under development: the ICBP Angola Scarp Project. Proceedings of the VIII Pan-African Ornithological Congress: 279–284. Kigali, Rwanda, 1992. Koninklijk Museum voor Midden-Afrika, Tervuren

Hill JE, Carter TD (1941) The mammals of Angola, Africa. *Bulletin of the American Museum of Natural History* 78, 211 pp

Huntley BJ (ed) (1989) Biotic diversity in southern Africa: concepts and conservation. Oxford University Press, Oxford, 380 pp

Huntley BJ (2009) SANBI/ISCED/UAN Angolan biodiversity assessment capacity building project. Report on pilot project. Unpublished Report to Ministry of Environment, Luanda 97 pp,

27 figures

Huntley BJ (2011) Biodiversity rapid assessment of the Lagoa Carumbo area, Lunda-Norte, Angola. Expedition report. Ministry of Environment, Luanda

Huntley BJ (2019) Angola in outline: physiography, climate and patterns of biodiversity. In: Huntley BJ, Russo V, Lages F, Ferrand N (eds) Biodiversity of Angola. Science & conservation: a modern synthesis. Springer Nature, Cham

Huntley BJ, Francisco P (eds) (2015) Avaliação Rápida da Biodiversidade de Região da Lagoa Carumbo, Lunda-Norte – Angola / Rapid Biodiversity Assessment of the Carumbo Lagoon Area, Lunda-Norte – Angola. Ministério do Ambiente, Luanda, 219 pp

Huntley BJ, Matos L (1992) Biodiversity: Angolan environmental status quo assessment report. IUCN Regional Office for Southern Africa, Harare, 55 pp

Huntley BJ, Siebert SJ, Steenkamp Y, et al (2006) The achievements of the southern African botanical diversity network (SABONET) – a southern African botanical capacity building project. In: Ghazanfar SA, Beentje H (eds.) Taxonomy and ecology of African plants, their conservation and sustainable use: Proceedings of the 17th AETFAT Congress. Addis Ababa, Ethiopia, 2003. Royal Botanic Gardens, Kew, pp 531–543

Huntley BJ, Beja P, Vaz Pinto P et al (2019) Biodiversity conservation: history, protected areas and hotspots. In: Huntley BJ, Russo V, Lages F, Ferrand N (eds) Biodiversity of Angola. Science & conservation: a modern synthesis. Springer Nature, Cham

Jeffrey R (1996) A Phytosociological Survey of the Northern Sector of the Quicama National Park in Angola. B.Sc. (Hons.) Dissertation. Faculty of Biological and Agricultural Sciences, University of Pretoria, Pretoria

Juergens N (2013) The biological underpinnings of Namib Desert fairy circles. Science 339:1618–1621

Kipping J, Clausnitzer V, Fernandes Elizalde SRF et al (2019) The dragonflies and damselflies of Angola. In: Huntley BJ, Russo V, Lages F, Ferrand N (eds) Biodiversity of Angola. Science & conservation: a modern synthesis. Springer Nature, Cham

Kirkman SP, Nsingi KK (2019) Marine biodiversity of Angola: biogeography and conservation. In: Huntley BJ, Russo V, Lages F, Ferrand N (eds) Biodiversity of Angola. Science & conservation: a modern synthesis. Springer Nature, Cham

Maier K (2007) Angola: promises and lies. Serif, London, 224 pp

Mateus O, Jacobs LL, Schulp AS et al (2011) *Angolatitan adamastor*, a new sauropod dinosaur and the first record from Angola. An Acad Bras Cienc 83(1):221–233

Mateus O, Callapez P, Polcyn M et al (2019) Biodiversity in Angola through time: a paleontological perspective. In: Huntley BJ, Russo V, Lages F, Ferrand N (eds) Biodiversity of Angola. Science & conservation: a modern synthesis. Springer Nature, Cham

Mendelsohn JM (2019) Landscape changes in Angola. In: Huntley BJ, Russo V, Lages F, Ferrand N (eds) Biodiversity of Angola. Science & conservation: a modern synthesis. Springer Nature, Cham

Mendes L, Bivar-de-Sousa A, Williams M (2019) The butterflies and skippers of Angola. In: Huntley BJ, Russo V, Lages F, Ferrand N (eds) Biodiversity of Angola. Science & conservation: a modern synthesis. Springer Nature, Cham

Mills MSL (2010) Angola's central scarp forests: patterns of bird diversity and conservation threats. Biodivers Conserv 19:1883–1903

Mills MSL (2018) The special birds of Angola / as Aves Especiais de Angola. Go-away-birding/Kissama Foundation, Cape Town/Luanda

Monard A (1935) Contribution à la mammologie d'Angola et prodrome d'une faune d'Angola. Arquivos do Museu Bocage 6:1–314

Monteiro RFR (1970) Estudo da Flora e da Vegetação das Florestas Abertas do Planalto do Bié. Instituto de Investigação Cientifíca de Angola, Luanda, 352 pp

Morais M (2017) Projecto Kitabanga – Conservação de tartarugas marinhas. Relatório final da temporada 2016/2017. Universidade Agostinho Neto / Faculdade de Ciências, Luanda

NGOWP (2018) National Geographic Okavango Wilderness Project (2018) Initial findings from exploration of the upper catchments of the Cuito, Cuanavale and Cuando rivers in Central and South-Eastern Angola (May 2015 to December 2016). National Geographic Okavango Wilderness Project, 352 pp

OKACOM (2009) Final report: Okavango integrated river management project. The Permanent Okavango River Basin Water Commission, Maun

OKACOM (2011) Cubango-Okavango river basin transboundary diagnostic analysis. The Permanent Okavango River Basin Water Commission, Maun

Rejmánek M, Huntley BJ, le Roux JJ et al (2017) A rapid survey of the invasive plant species in western Angola. Afr J Ecol 55:56–69

Revermann R, Finckh M (2019) Vegetation survey, classification and mapping in Angola. In: Huntley BJ, Russo V, Lages F, Ferrand N (eds) Biodiversity of Angola. Science & conservation: a modern synthesis. Springer Nature, Cham

Revermann R, Krewenka KM, Schmeidel U et al (eds) (2018) Climate change and adaptive land management in Southern Africa – assessments, changes, challenges, and solutions. Biodivers Ecol 6:1–497

Russo V, Huntley BJ, Ferrand N (2019) Biodiversity research and conservation opportunities. In: Huntley BJ, Russo V, Lages F, Ferrand N (eds) Biodiversity of Angola. Science & conservation: a modern synthesis. Springer Nature, Cham

Sinclair ARE (2012) Serengeti story: life and science in the world's greatest wildlife region. Oxford University Press, Oxford, 270 pp

Skelton PH (2019) The freshwater fishes of Angola. In: Huntley BJ, Russo V, Lages F, Ferrand N (eds) Biodiversity of Angola. Science & conservation: a modern synthesis. Springer Nature, Cham

Swinscow TDV (1972) Friedrich Welwitsch, 1806–72: a centennial memoir. Biol J Linn Soc 4:269–289

Vaz da Silva B (2015) Evolutionary history of the birds of the Angolan highlands – the missing piece to understand the biogeography of the Afromontane forests. MSc Thesis. University of Porto, Porto

Vaz Pinto P (2018) Evolutionary history of the critically endangered giant Sable Antelope (Hippotragus niger variani): insights into its phylogeography, population genetics, demography and conservation. PhD Thesis. University of Porto, Porto

Vaz Pinto P (2019) The Giant sable Antelope: Angola's National Icon. In: Huntley BJ, Russo V, Lages F, Ferrand N (eds) Biodiversity of Angola. Science & conservation: a modern synthesis. Springer Nature, Cham

Walker JF (2004) A certain curve of horn. The hundred-year quest for the giant sable antelope of Angola. Grove/Atlantic Inc., New York, 514 pp

Weir CR (2019) The Cetaceans (Whales and Dolphins) of Angola. In: Huntley BJ, Russo V, Lages F, Ferrand N (eds) Biodiversity of Angola. Science & conservation: a modern synthesis. Springer Nature, Cham

Werger MJA, van Bruggen AC (eds) (1978) Biogeography and ecology of Southern Africa. The Junk, Hague, 1444 pp

Zigelski P, Gomes A, Finckh M (2019) Suffrutex dominated ecosystems in Angola. In: Huntley BJ, Russo V, Lages F, Ferrand N (eds) Biodiversity of Angola. Science & conservation: a modern synthesis. Springer Nature, Cham

2

Physical Geography and Biodiversity Characteristics of Angola

Brian J. Huntley

Abstract Angola is a large country of 1,246,700 km² on the southwest coast of Africa. The key features of the country's diverse geomorphological, geological, pedological, climatic and biotic characteristics are presented. These range from the ultra-desert of the Namib, through arid savannas of the coastal plains to a biologically diverse transition up the steep western Angolan Escarpment. Congolian rainforests are found in Cabinda and along the northern border with the Democratic Republic of Congo, with outliers penetrating southwards along the Angolan Escarpment, or up the tributaries of the Congo Basin. Above the escarpment, high mountains rise to 2620 m above sea level, with isolated remnants of Afromontane forests and grasslands. Extensive *Brachystegia/Julbernardia* miombo moist woodlands dominate the plateaus and peneplains of the Congo and Zambezi basins, and dry woodlands of *Colophospermum/Acacia* occur in the southeast towards the Cunene River, with *Baikiaea/Burkea/Guibourtia* woodlands dominating the Kalahari sands of the endorheic basins of the Cubango and Cuvelai rivers. Rainfall varies from lower than 20 mm per year in the southwest to over 1600 mm in the northwest and northeast. At a regional scale, Angola is notable for having representatives of seven of Africa's nine biomes, and 15 of the continent's ecoregions, placing Angola second only after South Africa for its diversity of African ecoregions.

Keywords Afromontane forest · Biogeography · Biomes · Climate change · Congolian forest · Ecoregions · Namib · Kalahari Basin · Zambezian savannas

Introduction

This chapter presents a general outline of the physical geography and biodiversity characteristics of Angola, as background to the chapters that follow. It draws on the work of the great Portuguese agro-ecologist Alberto Castanheira Diniz, who

B. J. Huntley (✉)
CIBIO-InBIO, Centro de Investigação em Biodiversidade e Recursos Genéticos, Universidade do Porto, Vairão, Portugal
e-mail: brianjhuntley@gmail.com

synthesised the diverse drivers of Angola's ecological systems and agricultural potential, based on his many decades of fieldwork in the country (Diniz and Aguiar 1966; Diniz 1973, 1991, 2006). Colonial records of climatic variables (Silveira 1967) are used in the absence of recent time series. The pioneer studies on Angolan vegetation by Gossweiler and Mendonça (1939) and Barbosa (1970) are fundamental to any account on Angola biodiversity. Surveys of Angola's protected areas and biodiversity 'hotspots' (Huntley 1974a, b, 2010, 2015, 2017) provide conservation context. This outline also draws on the recent regional geographies of Angola by Mendelsohn and co-workers (Mendelsohn et al. 2013; Mendelsohn and Weber 2013, 2015; Mendelsohn and Mendelsohn 2018). The chapter is also strengthened by material detailed in the specialist papers that form the core of this volume.

Location and Extent

As a large country of 1,246,700 km^2 on the southwest coast of Africa, Angola is roughly square in outline, lying between 4° 22′ and 18° 02′ south latitude, and 11° 41′ and 24° 05′ east longitude. It is bounded to the west by an arid 1600 km coastline along the Atlantic Ocean; to the north by the moist forest and savanna ecosystems of the Republic of Congo and the Democratic Republic of Congo (DRC); to the east by the moist savanna and woodland ecosystems of the DRC and Zambia; and by arid woodlands, savannas and desert along its 1200 km southern border with Namibia.

Geomorphology and Landscape Evolution

The general topography of Angola is illustrated in Fig. 2.1. In summary, coastal lowlands lying below 200 m altitude and of 10–150 km breadth occupy 5% of the country's land surface area, leading to a stepped and mountainous escarpment rising to 1000 m (23%), and an extensive interior plateau of 1000–1500 m (65%). Seven percent of the country lies above 1500 m, reaching its highest point at 2620 m on Mount Moco.

The ecological importance of the major physiographic divisions in Angola was recognised as early as the 1850s by the pioneer Austrian botanist Friedrich Welwitsch who categorised the 5000 plant species that he collected in Angola within three regions: *região litoral, região montanhosa*, and *região alto-plano* (Welwitsch 1859). Besides his remarkable contribution to the founding of Angolan botany, Welwitsch prepared detailed geological profiles across the landscapes inland of Luanda and Moçâmedes (Albuquerque and Figueirôa 2018), probably the first such analyses in western Africa (Fig. 2.2). His understanding of the patterns and relationships of geology, physiography and vegetation set a strong ecological tradition that has been followed by successive students of Angola's biodiversity.

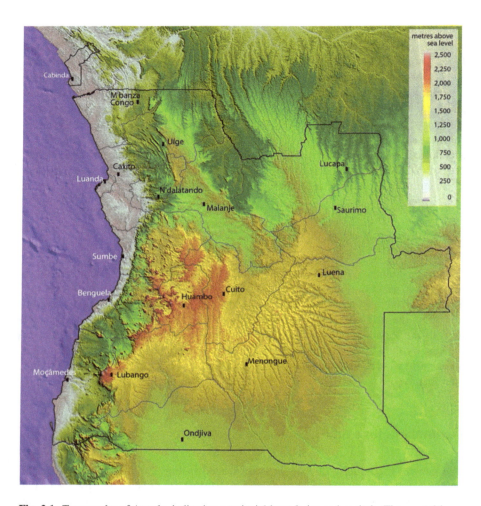

Fig. 2.1 Topography of Angola, indicating provincial boundaries and capitals. The coastal lowlands, western escarpments, central highlands and plateaus, and the major drainage basins of the Cuanza, Congo and Zambezi rivers are clearly revealed

A further detailed and indeed classic study of Angola's geomorphology and local ecology was that of the German geographer Otto Jessen (1936). Jessen undertook a series of 11 transects from the coast inland, traversing the escarpment to the interior plateau from Moçâmedes and thereafter at intervals northwards to Luanda. Describing, illustrating and mapping selected vegetation communities, geological exposures, landscapes, landuse and ethnological features of the country, Jessen's Angolan work remains unique in its diversity of interest and originality. He recognised five major erosional planation surfaces in western Angola at a time when geomorphology was evolving as a discipline, and he was recognised by King (1962) as one of the founders of peneplanation theory. Geomorphological studies in Angola

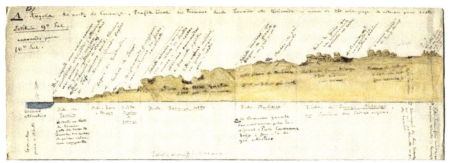
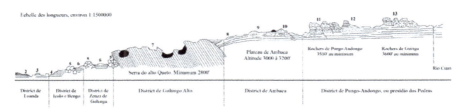

Fig. 2.2 Geological profile from Luanda to Quisonde, scanned from the original manuscript produced by Friedrich Welwitsch during his expeditions between 1853 and 1860. The lower profile is a redrafted version of the upper profile adapted from Choffat (1888), and reproduced with permission from Albuquerque and Figueirôa (2018) and of the Museums of the University of Lisbon Historical Archives

continued from the 1950s to 1970s by Portuguese researchers, including Marques (1963), Feio (1964) and Amaral (1969), whose work is summarised by Costa (2006).

More recent research, in particular that on the evolution and of the major tectonic and erosional patterns across southern Africa (Cotterill 2010, 2015; Cotterill and De Wit 2011) and on the biogeography of the freshwater fishes of Angola (Skelton 2019) provide a picture of a very dynamic landscape since the breakup of Gondwana in the late Cretaceous. These and other authors are providing an improved understanding of the processes of uplift, back-tilting, down-warping, deposition, erosion and river capture on the evolution of the Kalahari Basin. The impacts of sea-level fluctuations and of the flow of the Congo River on coastal waters and on the erosional forces of the Congo Basin as it impacts on the Zambezi Basin are guiding our interpretation of the dramatic events shaping the faunal and floral patterns of today. Cotterill (2015) presents a synthesis of hypotheses on the evolution of the Kalahari from the late Mesozoic into the early Cenozoic, events which were followed by the later overlying suite of younger Kalahari sediments – the world's largest sandsea. The interplay of geological and paleoclimatic drivers described by Cotterill (2010, 2015), through the pulsing of hot wet and cool dry episodes during the Plio-Pleistocene, was accompanied by the expansion and contraction of forest and savanna habitats responding to climatic and fire regimes.

The role of fire in shaping the landscapes of Angola – and particularly of the dominant miombo moist savanna biome – has become a topic of discussion in recent years (Zigelski et al. 2019). Maurin et al. (2014) provide evidence based on the dated phylogenies of 1400 woody species to support the proposal that the 'underground forests' (White 1976) that are so prominent across the moist miombo savannas and woodlands of the south-central African plateau, evolved in response to high fire frequency. They suggest that moist savannas pre-date the emergence of anthropogenic fire and deforestation, becoming a prominent component of tropical vegetation from the late Miocene (ca. 8 Ma). Maurin et al. (2014) conclude that the evolution of geoxyles ('underground trees') that characterise these moist savannas define the timing of the transition to fire-maintained savannas occurring in climates suitable for and previously occupied by forests. The further interpretation of these key drivers of evolution processes is fundamental to an improved understanding of the biogeography of Angola.

A major contribution towards an ecological understanding Angola's contemporary landscapes and natural regions, and their agro-forestry potential, was that of Castanheira Diniz. Diniz (Diniz and Aguiar 1966, Diniz 1973, 1991) provides a series of maps illustrating the key features of Angola's topography, geomorphology, geology, climate, soils, and phyto-geographic and bio-climatic zones. Diniz's 11 'mesological' units (Fig. 2.3) provide a useful framework for discussions on Angola's ecology and biodiversity. Indeed his mesological concept closely corresponds with current perceptions of ecoregions. He also delineated and described 32 agro-ecological zones (Diniz 2006). Although some of his 11 mesological units need more rigorous and objective definition and delineation, they have become widely adopted within Angola. Important aspects of these 11 broad units will be summarised here, integrating these with insights from other sources.

1. Coastal Belt (*Faixa litoranea* sensu *Diniz*). This is a mostly continuous platform at 10–200 m above sea level, broken occasionally by broad river valleys. In contrast to the situation on the east coast of Africa at similar latitudes, the Angolan coastline is notable for the absence of coral reefs and coastal dune forests. Long sandbars stretch northwards from rivers such as the Cunene and Cuanza. Mudflats and mangroves occur at most river mouths from Lobito northwards, increasing in dimension and diversity towards the Congo. Much of the coast is uplifted, resulting in sharp sea-cliffs of 10–100 m. In places as narrow as 10 km, the coastal belt is mostly of about 40 km width, broadening to 150 km northwards of Sumbe and up the lower Cuanza. The coastal plains are composed mostly of fossiliferous marine sediments of the Cabinda, Cuanza, Benguela and Namibe geological basins. The northern coastal platforms are covered by deep red Pleistocene sands (*terras de musseque*) of former beaches. Lying below the sands, and exposed over large areas, are Cretaceous to Miocene clays, gypsipherous marls, dolomitic limestones and sandstones. Important beds of Cretaceous fossils occur at Bentiaba and Iembe, the latter including the sauropod dinosaur *Angolatitan adamastor* (Mateus et al. 2011, 2019). The

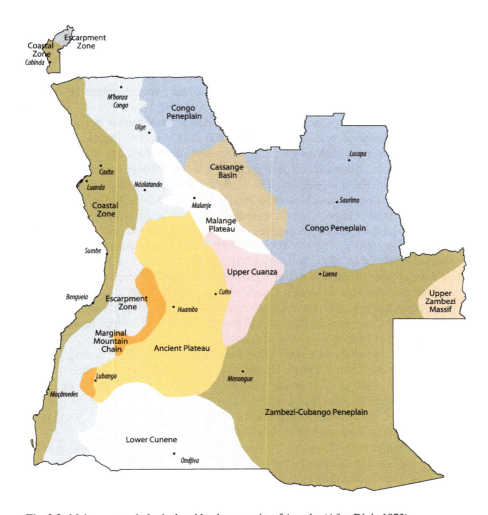

Fig. 2.3 Main geomorphological and landscape units of Angola. (After Diniz 1973)

southernmost segment of the Coastal Belt includes the mobile and mostly vegetation-less dunes of the Namib Desert.

2. Escarpment Zone (*Faixa subplanaltica* sensu *Diniz; região montanhosa* sensu *Welwitsch*). A broad transition belt lies between the coastal plains and the interior plateaus – variable in breadth and gradient. Over much of the zone, the transition advances up several steep steps of between 400 and 600 m. In the south, between Moçâmedes and Lubango, the escarpment of the Serra da Chela is very sharp, rising 1000 m at Tundavala and Bimbe. The geology of the Escarpment Zone is complex, comprising crystalline rocks of the Precambrian: granites, gneisses, schists, quartzites and amphibolites. The Escarpment Zone (also referred to as the Western Angolan Scarp) includes very hilly country,

with mountainous belts in the north, and some major inselbergs in the south, the most important of which is Serra de Neve, which rises to 2489 m from the surrounding plains and low hills. The Angolan Escarpment has long been recognised for its biogeographic importance (Humbert 1940, Hall 1960a, Huntley 1974a) and has been the centre of interest of many recent studies (Hawkins 1993, Dean 2000, Mills 2010, Cáceres et al. 2015, 2017).

3. Marginal Mountain Chain (*Cadeia Marginal de Montanhas* sensu *Diniz*). Residual mountain lands, mostly at 1800–2200 m, underlain mostly by Precambrian rocks such as gneiss, granites and migmatites, lie at the western margin of the extensive interior plateau, and are known as the Benguela, Huambo and Huíla Highlands. The highest peaks rise to 2420 m on Mount Namba, 2582 m on Serra Mepo and 2620 m at Mount Moco. The mountains are of biogeographic importance for their montane grasslands, with some elements of the Cape flora, and relict patches of Afromontane forests and endemic bird assemblages (Humbert 1940; Hall 1960b; Hall and Moreau 1962; Huntley and Matos 1994; Dean 2000; Mills et al. 2011, 2013; Vaz da Silva 2015).

4. Ancient Plateau (*Planalto Antigo*). This extensive plateau drops eastwards from below the Marginal Mountain Chain and encompasses the headwaters of the Cunene, Cubango, Queve and Cutato rivers, comprising rolling landscapes with wetlands and low ridges with scattered granitic inselbergs. It drops from 1800 m in the west to 1400 m in central Angola.

5. Lower Cunene (*Baixo Cunene*). This is a rather artificial unit, leading imperceptibly down from 1400 m on the 'Ancient Plateau' to the frontier with Namibia at 1000 m. The gentle gradient of the eastern half forms the very clearly defined Cuvelai Basin, which drains as an ephemeral catchment into the Etosha Pan. West of the Cunene the landscape is more broken, with pockets of Kalahari sands between low rocky hills.

6. Upper Cuanza (*Alto Cuanza*). The upper catchments of the Cuanza and its tributary the Luando, at altitudes between 1200 and 1500 m, form a distinct basin of slow drainage feeding extensive wetlands during the rain season.

7. Malange Plateau (*Planalto de Malange*). A gently undulating plateau at 1000–1250 m, dropping abruptly, on its northeastern margin, some several hundred metres to the *Baixa de Cassange* and the Cuango drainage. The escarpment ravines hold important moist forest outliers (such as at Tala Mungongo) that deserve investigation. To the west, the plateau is drained by the rivers flowing to the Atlantic, most spectacularly by a tributary of the Cuanza, the Lucala, that drops over 100 m at the famous Calandula Falls (formerly Duque da Bragança Falls).

8. Congo Peneplain (*Peneplanície do Zaire*). This is a vast sandy peneplain, drained by the northward flowing tributaries of the Cassai/Congo Basin, and stretching eastwards from the margins of the mountainous northern end of the Escarpment Zone in Uíge, to the extensive *Chanas da Borracha* of the Lundas. These gently dipping plains, mostly at 1100–800 m, are being aggressively dissected by the many northward flowing, parallel tributaries of the Congo Basin. The Cuango River, draining the *Baixa de Cassange*, drops to 500 m at the

frontier with the Democratic Republic of Congo. The southern boundary of this Congo Peneplain is defined imperceptibly by the watershed between the Zambezi and Congo basins, lying at ca. 1200 m.

9. Cassange Basin (*Baixa de Cassange*). A wide depression, several hundred metres below the surrounding plateaus, is demarcated by abrupt escarpments to the west and the densely dendritic catchment of the Cuango to the northeast. The geological substrate comprises Triassic Karoo Supergroup sediments of limestone, sandstone and conglomerates. Within the Basin, several large table-lands – remnants of the old planation surface – rise above the depression as extensive plateaus flanked by sheer, 300 m escarpments, exemplified by Serra Mbango, which awaits biological survey.

10. Zambezi-Cubango Peneplain (*Peneplanície do Zambeze-Cubango*). This is the vast peneplain draining deep Kalahari sands, with slow-flowing rivers that meander across the gently dipping plateau from 1200 m at the watershed with the Congo Basin to 1000 m at the frontier with Namibia. Within this extensive peneplain, the Bulozi Floodplain occupies an area in excess of 150,000 km² in Angola and Zambia.

11. Upper Zambezi Massif (*Maciço do Alto Zambeze*). The Calunda Mountains of eastern Moxico, composed of Precambrian schists and norites, dolorites, sandstones and limestones, rise to 1628 m above the Zambezi peneplain which lies at 1150 m. The mountains form a striking contrast to the almost featureless landscape that stretches some 800 km eastwards from Huambo to Calunda.

Rivers and Hydrology

Angolan river systems fall into two categories. First, coastal rivers drain the central and western highlands and flow rapidly westwards where they have penetrated the steep escarpment to the Atlantic Ocean. Most of these coastal rivers are relatively short, are highly erosive and carry high sediment loads. Backward erosion by some of these has produced minor basins, such as the amphitheatres of the upper Queve and Catumbela. The biogeographic importance of the river captures associated with these systems, especially the Congo, Cuanza, and Cunene, have been profound, as described by Skelton (2019). Most of the coastal rivers south of Benguela are ephemeral.

The second major category of river systems is that of the vast interior plateaus. Drained by nine large hydrographic basins, seven of which are transnational, Angola serves as the 'water tower' for much of southern and central Africa. Many of these rivers arise in close proximity on either side of the gently undulating watershed between the Cuanza, Cassai (Congo), Lungue-Bungo (Zambezi), Cunene, and endorheic Cubango (Okavango) basins. These rivers drain the vast and deep Kalahari sands, are slow moving and due to the filtering action of the sands, are crystal clear and nutrient poor. A separate ephemeral, endorheic system, the Cuvelai Basin, drains southwards into the Etosha Pan.

The conservation importance of the Angolan river systems is of great significance, feeding as they do two wetlands (Okavango and Etosha) of global importance, and the still under-researched Bulozi Floodplain of Moxico. This is possibly the largest ephemeral floodplain in Africa – 800 km from north to south and 200 km from east to west – straddling the Angola/Zambia frontier (Mendelsohn and Weber 2015).

Geology and Soils

The geological history and soil genesis of Angola is complex and interrelated, and influenced by rainfall, drainage, evaporation and wind. Mateus et al. (2019) provide a map and stratigraphic profile of the geology of Angola which summarises the major geological features of the country. The predominance of a broad belt of Precambrian systems along the western margin of the country, with Cenozoic systems occupying most of the eastern half, is striking. Over three-quarters of the country (Fig. 2.4) is covered by two main soil groups arenosols and ferralsols – an understanding of which provides an essential introduction to Angolan pedology. For simplicity, soils will be described with reference to their geological substrate.

First, Angola's main soil groups are the sandy arenosols (*solos psamíticos*) that cover more than 53% of the country. These sands are dominant features of three major landscapes: the dunes of the Namib Desert; the red *'terras de musseque'* of the coastal belt northwards from Sumbe; and the vast Kalahari Basin. The great majority of the arenosols lie to the east of approximately 18° longitude – the aeolian sands of the Kalahari Basin which cover nearly 50% of Angola and hides nearly all of the underlying geological formations. The Kalahari Basin, extending across 2500 km from the Cape in the south to the Congo Basin in the north, and up to 1500 km in breadth, is reputedly the largest body of sand in the world. The sands have been deposited by wind and water over the past 65 million years. Composed of quartz grains that hold no mineral nutrients, and with very little accumulated organic matter, they are thus of very low fertility and water-holding capacity. Waters passing through the vast catchments of the Congo, Cubango and Zambezi basins that drain the Kalahari are therefore extremely pure.

Second, the higher ground of the western half of Angola (the Ancient Massif) is dominated by ferralsols (*solos ferralíticos*) derived from underlying rocks (gneisses, granites, metamorphosed sediments of the Precambrian Basement Complex; and schists, limestones and quartzites of the West Congo System). Ferralsols cover approximately 23% of Angola. The soils are mostly of low water-holding capacity. Because they are heavily leached in higher rainfall areas, the loss of mineral nutrients and organic matter results in low fertility. They are characteristically reddish due to the oxidation of their high iron and aluminium content, which also accounts for the presence in many areas of ferricrete hardpan horizons a metre or two below the surface, impeding root and water penetration and resulting in the formation of extensive areas of laterite.

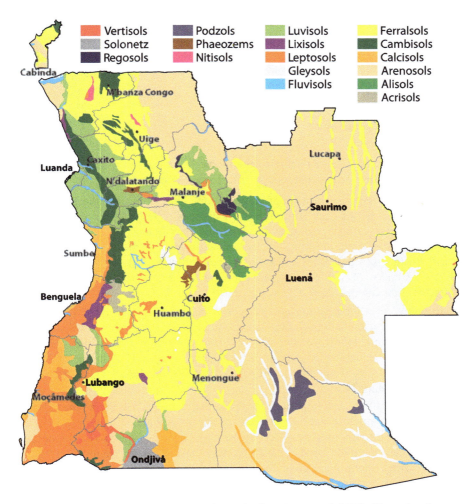

Fig. 2.4 Outline of the main soil types of Angola (from Jones et al. 2013), illustrating the predominance of arenosols in the eastern half of the country, and ferralsols across the western and central plateaus

These two low-fertility soil groups (arenosols and ferralsols) cover over 76% of the country, thus despite the adequacy of rainfall over most of Angola, agricultural production faces the challenges of low soil fertility (Neto et al. 2006; Ucuassapi and Dias 2006). The natural vegetation types that cover both arenosols and ferralsols – predominantly miombo woodlands – are well adapted to these soil conditions and the untransformed landscape gives the appearance of great vitality and luxuriance.

The next soil grouping in terms of landcover, occupying 6% of Angola, are the shallow regosols (*litosolos*) of rocky hills and gravel plains, most extensive in the arid southwest. Other important soil types include luvisols, calcisols and cambisols (*solos calcários, solos calcialíticos*), which provide fertile loam soils for crops (including the 'coffee forests' of the Escarpment Zone); alluvial fluvisols (*solos aluvionais*) in drainage lines with high organic content and high water retaining capacity, suitable for crops if not water-logged; gleysol clays (*solos hidromórficos*), typically acidic and waterlogged and occasionally very extensive – as on seasonally flooded plains such as Bulozi Floodplains.

Climate and Weather

The diverse climatic and weather conditions experienced across Angola result from many atmospheric, oceanic and topographic driving forces.

First, the geographic position of Angola, stretching from near the Equator to close to the Tropic of Capricorn, across 14 degrees of latitude, accounts for the overall decrease in solar radiation received and thus annual mean temperatures experienced from north to south. The latitudinal decrease in mean annual temperature is illustrated by data from stations in the hot northwest and northeast (Cabinda: 24.7 °C; Dundo: 24.6 °C), compared with stations in the milder southwest and southeast (Moçâmedes: 20.0 °C; Cuangar: 20.7 °C).

Secondly, both temperature and precipitation are influenced by altitude. The decrease in mean annual temperature can be illustrated from sites below the Chela escarpment to the highest weather station in the country: i.e. from Chingoroi: altitude 818 m, mean annual temperature 23.1 °C; Jau: altitude 1700 m, mean annual temperature 18.0 °C; and finally Humpata-Zootécnica: altitude 2300 m, mean annual temperature 14.6 °C.

Thirdly, and of greatest importance to the rainfall patterns that determine vegetation and habitat structure, are the influences of the atmospheric systems which dominate central and southern Africa. Circling the globe near the Equator is a belt of low pressure where the trade winds of both Northern and Southern Hemispheres converge, creating strong convective activity which generates the dramatic thunderstorms that characterise the inter-tropics. Known as the Inter-tropical Convergence Zone (ITCZ), the belt moves southwards over Angola during summer, and then returns northwards to the Equator as winter approaches. The rainfall season that is triggered by the ITCZ passes across northern Angola from early summer, reaching southern Angola in late summer. The climate is strongly seasonal, with hot wet summers (October to May) and mild to cool dry winters (June to September). Some stations in northern Angola receive two peaks of rainfall, early summer and late summer, often with a short drier period in mid-summer.

Moving in tandem with the ITCZ are two high-pressure systems – over the Atlantic and over southern Africa – the South Atlantic Anticyclone and the Botswana Anticyclone. In simple terms, these two anticyclones block the southward movement of moist air from the ITCZ during winter (preventing cloud formation) and as the high-pressure cells move southwards in summer, the conditions required for cloud formation return. This pulsing of rainfall systems is clearly illustrated in the series of rainfall maps prepared by Mendelsohn et al. (2013) from weather satellite imagery (Fig. 2.5).

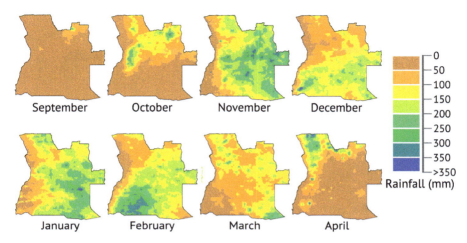

Fig. 2.5 The impact of the southwards and northwards pulsing of the Inter-tropical Convergence Zone on rainfall seasonality across Angola during 2009/2010. (From Mendelsohn et al. 2013)

During winter and early summer, the Botswana Anticyclone generates strong winds that blow across Angola from east to west, with impacts on micro-relief over much of the country. In the southwest, the winds pick up dust from the arid lands and create hot, choking dust storms that feed the sand dunes of the Namib. The winds are also notorious in the north, where they desiccate the grasslands of the Lundas. In the east, the winds and their sand deposits account for dune formation across the Bulozi Floodplain (Mendelsohn and Weber 2015).

Rainfall and temperature seasonality and other climatic parameters are illustrated by the climate diagrams in Fig. 2.6. The distribution of mean annual rainfall across Angola is summarised in Fig. 2.7.

Fourthly, as noted above, altitude and seasonality determine temperature conditions. However, an anomaly to this general rule occurs in the coastal belt of Angola, especially in the far south, through the influence of the temperature inversion created by the cold, upwelling Benguela Current. The Benguela Current has a stabilizing effect on the lower atmosphere and prevents the upward movement of moist, cloud-forming air off the ocean, accounting for the evolution of the Namib Desert. Its impact also extends as far north as Cabinda, where a narrow belt of arid savanna woodland and dry forest, of acacias, sterculias and baobabs, flanks the rainforests of the Maiombe.

Despite the aridity of the coastal zone, the cooling effect of the Benguela Current results in low stratus cloud and fog (*cacimbo*) through much of winter, with heavy dew condensing on vegetation along the coast even during the driest months of winter. The fog belt is most pronounced between Moçâmedes and Benguela, where epiphytic lichens reach great abundance in an otherwise desertic environment. The Benguela Current also results in a gradient of increasing precipitation from south to north and from west to east. The rainfall gradients are locally accentuated by the orographic influence of the escarpment and the highland mountain massifs. The

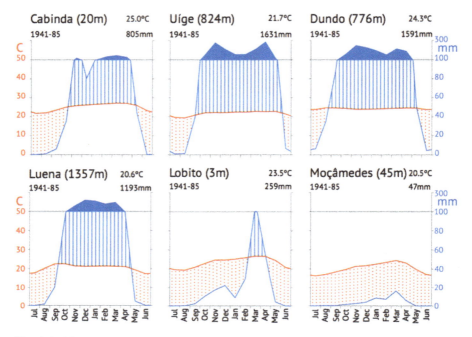

Fig. 2.6 Climate diagrams illustrating rainfall and temperature seasonality and other climatic parameters. Note weak bimodal rainfall maxima for stations in northern Angola and unimodal maxima in central and southern Angola

sharp relief of the escarpment creates conditions for orographic rainfall along most of this zone, supporting the 'coffee forests' of Seles, Gabela, Cuanza-Norte and Uíge.

Attempts to synthesise climatic characteristics into simple formulae or graphics have resulted in a wide range of classification systems. A synthesis of climatic data provided by the widely used Köppen and Thornthwaite classification systems was undertaken by Azevedo (1972) to map and quantify, at a national scale, the climatic regions of Angola, based on the substantial data set available at that time. Interestingly, despite some of its shortcomings, the Azevedo map provides a closer fit with general features of Angola's bio-climatic patterns than a much more recent map (Peel et al. 2007). The latter map is based on a global synthesis and review of the Köppen system, and draws on a very limited data set for Angola (5 stations for temperature; 16 stations for precipitation). The northern region of Angola is typical of Köppen's Tropical Wet Savanna (Aw) group, the plateau of the Temperate Mesothermal (Cw) group, and the southwest and coastal plain the Dry Desert and Semi-desert (Bsh, Bwh) group.

Mean annual rainfall and mean monthly temperatures for hottest and coldest months illustrate a few climatic characteristics of the Köppen regions (Table 2.1). The absence of data on extreme minimum temperatures and of frost occurrence is regrettable, as these factors, in tandem with fires and herbivory, play significant roles in the floristic composition and physiognomic structure of Angolan vegetation (Zigelski et al. 2019).

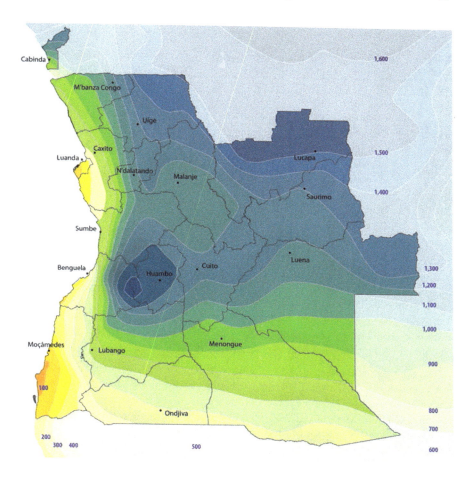

Fig. 2.7 Mean annual rainfall in Angola

Climate Change

Studies on the climate of Angola have been frustrated in recent years by the collapse of the extensive network of weather stations maintained during the colonial era by the then Meteorological Services of Angola. The publication by Silveira (1967) of recordings from 184 stations across all 18 provinces provides an invaluable record of the country's climate. According to the Ministry of Environment's Initial National Report to the UNFCCC (GoA 2013) the weather recording network collapsed from 225 'climatological posts' in 1974 to zero posts in 2010, while synoptic stations decreased from 29 in 1974 to 23 operational stations in 2010, 12 being automatic and 11 conventional. The network has since been strengthened by 22 automatic

Table 2.1 Representative climatic data following the Köppen climate classification system

Köppen symbol	Station	Altitude (m)	Precipitation (mm)	Mean of hottest month °C	Mean of coldest month °C
Aw	Belize	245	1612	26.7	22.2
Aw	Saurimo	1081	1355	23.8	20.3
Bsh	Ondgiva	1150	577	26.4	16.7
Bsh	Cuangar	1050	596	24.6	15.0
Bsh'	Chitado	1000	405	27.4	19.2
Bsh'	Luanda	44	405	27.0	20.1
Bwh	Moçâmedes	44	37	24.2	15.5
Bwh	Tômbwa	4	12	24.2	14.5
Bwh'	Benguela	7	184	26.3	18.0
Bwh'	Caraculo	440	123	26.4	17.2
Cwa	Menongue	1348	965	23.4	14.5
Cwa	Luena	1328	1182	22.7	17.0
Cwb	Huambo	1700	1210	20.6	15.7
Cwb	Lubango	1760	802	20.7	15.3

Data from Silveira (1967)

stations established by the Southern African Science Service Centre for Climate Change and Adaptive Land Management (SASSCAL).

The poor national coverage and reliability of climatic data collected over the past four decades is a challenge for climate change research. However, a recent study (Carvalho et al. 2017) provides the first analysis and comparison of a set of four Regional Climate Models (RCMs) with data from 12 meteorological monitoring stations in Angola. Scenarios of future temperature and precipitation anomaly trends and the frequency and intensity of droughts are presented for the twenty-first century. While there is a difference in the performance of the four RCMs, in particular for precipitation, consistent results were found for temperature projections, with an increase of up to 4.9 °C by 2100. The temperature increases are lowest for the northern coastal areas and highest for the southeast. In contrast to temperature rises, precipitation was projected to fall over the century, with an average of −2% across the country. Again, the strongest change was projected for the southeast, with decreases of up to −4%. Due principally to the projected increase in Sea Surface Temperatures by approximately 3 °C over the Atlantic during the twenty-first century, the central coastal region is expected to have a slight increase in precipitation.

Carvalho et al. (2017) highlight the extreme climate vulnerability of Angola, as previously noted by other studies (Brooks et al. 2005; Cain and Cain 2015). They conclude that climate change in Angola will bring stronger and more frequent droughts through the century, with impacts on water resources, agricultural productivity and wildfire potential. These factors will no doubt play out in negative ways on the current trends of land transformation and degradation as described by Mendelsohn (2019).

Biogeography, Biomes and Ecoregions

Overview

Angola's geographic location, geological history, climate and physiography account for its rich biological diversity. The comparative paucity of research focused on or within Angola explains the dependence of descriptions of the country's biogeography on broader regional reviews. While a full synthesis and interpretation of the evolution of the country's fauna and flora awaits development (Cotterill 2010, 2015), recent workers have advanced towards consensus on the main patterns, as discussed in general terms for terrestrial biota in this chapter, for marine systems by Kirkman and Nsingi (2019) and Weir (2019), and for freshwater fishes by Skelton (2019) in other chapters of this volume.

In brief, three marine ecoregions (Spalding et al. 2007) are within or overlap with Angola's marine environment, namely the Guinea South, Angolan, and Namib Ecoregions, the first two of which belong to the tropical Gulf of Guinea biogeographical province whereas the latter is part of the Benguela biogeographical province (Kirkman and Nsingi 2019). Most of Angola's EEZ falls within the Benguela Current Large Marine Ecosystem, with only Cabinda in the far north being included in the Guinea Current Large Marine Ecosystem.

The freshwater ecoregions of Africa have been classified and mapped by Thieme et al. (2005) and the eight ecoregions found within Angola are described in this volume (Skelton 2019). Skelton (2019) provides an elegant biogeographic model to explain the patterns and dynamics of freshwater fish faunas of Angola. Neither the floral nor vegetation patterns reflect the complexities and subtleties embedded in the ichthyological zoogeography of Angola, given the mobility of terrestrial plant dissemination. Here I will confine discussion to the terrestrial biota and ecosystems.

Angola lies between and within two major terrestrial biogeographic regions: the moist forests and savannas of the Congolian region; and the woodlands, savannas and floodplains of the Zambezian region. These two major divisions occupy over 97% of Angola. Gallery and escarpment forests of Congolian affinity penetrate southwards into the Zambezian savannas and woodlands of the Angolan *planalto* along deeply incised tributaries of the Congo Basin, and form a broken chain of forests southwards along the western escarpment. In the south, the extensive *Brachystegia/Julbernardia* miombo woodlands that occupy most of central Angola transition to *Baikiaea/Guibourtia/Burkea* savannas and woodlands. In the southwest, the arid *Acacia/Commiphora/Colophospermum* savannas, dwarf shrublands and desert of the Karoo-Namib region are found, penetrating northwards as a narrowing wedge along the coastal lowlands to Cabinda. The smallest of Africa's centres of botanical endemism – the *Podocarpus* Afromontane forests and montane grasslands – are represented by extremely restricted, relict patches in the mountains of the Benguela, Huambo and Huíla highlands.

Early Studies

Beyond general agreement on the above brief outline, botanists and zoologists have described and debated as many systems of biogeographic classification and of terminologies for Angola and for Africa as there are authors of the papers on the topic (Werger 1978). The pioneering works of Welwitsch (1859); Gossweiler and Mendonça (1939) and Barbosa (1970) provided the basis for several subsequent attempts to integrate the vegetation of Angola within a regional framework (Monteiro 1970; White 1971, 1983; Werger 1978). Zoogeographic classifications (Chapin 1932; Frade 1963; Monard 1937; Hellmich 1957; Crawford-Cabral 1983) are, with some minor exceptions, compatible with the overall systems of botanists (Werger 1978; Linder et al. 2012), (but see Branch et al. 2019, for comments on lizards). The Africa-wide synthesis of White (1983) is particularly useful in considering Angola's floristic (and in general terms, zoological) patterns and affinities. In broad terms, and following White's terminology, Angola includes representation of four 'regional centres of endemism'. They comprise the following centres with estimates of the percentage of their total area in Angola from Huntley (1974a, 2010):

- *Guineo-congolian regional centre of endemism* - mosaics of forests, thickets, tall grass savannas – 25.7% (This is Linder et al.'s Congolian Region and includes their Shaba sub-region);
- *Zambezian regional centre of endemism* – moist woodlands, savannas, grasslands and thickets – 71.6% (The Zambezian Region *sensu* Linder et al.);
- *Karoo-Namib regional centre of endemism* – desert, shrublands, arid savannas, woodlands and thickets – 2.6% (Most of this is placed in Linder et al.'s Southern African Region as their Southwest Angola sub-region); and
- *Afromontane archipelago-like regional centre of endemism* – forests, savannas and grasslands – 0.1%. (This is related to Linder et al.'s Ethiopian Region).

Statistical Regionalisation

Attempts have recently been made to use the massive databases of species distribution records held by museums and herbaria to bring objectivity and consistency to the classification of Africa's floral and faunal regions. A major step towards such regionalisation is provided by the statistical definition of biogeographical regions of sub-Saharan Africa by Linder et al. (2012). Using data for 1877 grid cells of one-degree resolution, the study included data for over a million records of 1103 species of mammals, 1790 species of birds, 769 species of amphibians, 480 species of reptiles and 5881 species of vascular plants. The databases were analysed using cluster analysis techniques to define biogeographical units that "comprise grid cells that are more similar in species composition to each other than to any other grid cells" (Linder et al. 2012). They proposed seven biogeographical regions for sub-Saharan Africa: Congolian, Zambezian, Southern African, Ethiopian, Somalian, Sudanian and Saharan. Their analyses demonstrated that patterns of richness and endemism are positively and significantly correlated among plants, mammals, amphibians, birds and reptiles and with the overall biogeographical regions revealed by the sum of the data sets.

The use of modern cluster analysis techniques was taken further, at an Angolan level, by Rodrigues et al. (2015). Based on a cluster analysis of data for 9880 records of 140 species of ungulates, rodents and carnivores at a quarter degree resolution, the study found general congruence with that of Linder et al. (2012) and the earlier divisions of Angola's biogeography (Beja et al. 2019). Rodrigues et al. (2015) identify 18 indicator species for their four main divisions, which agree with the groupings based on field surveys undertaken in the 1970s (Huntley 1973) that also included the enclave of Cabinda, which was not included in the Rodrigues et al. (2015) analyses.

Both of the above very detailed and objective cluster analyses confirmed the general patterns of biogeographical regionalisation used for many decades across Africa, as described at the head of this section, even though terminology and detail of boundaries and transitions between regions differ from one author to the next. While objective, it is possible that the cluster analysis approach lacks the subtlety and flexibility of scale that classical expert systems permit. A particular challenge is the paucity of geo-referenced data for Angolan taxa, as experienced in a recent botanical analysis at inter-tropical scale (Droissart et al. 2018). Both cluster analyses and expert systems remain works in progress.

Biomes and Ecoregions

The chorological studies of White (1983) and statistical analyses of Linder et al. (2012) capture some of the evolutionary history and relationships of Africa's flora and fauna, but they do not fully reflect the continent's diversity of biomes, habitats and ecosystems – which are based on structural and functional rather than evolutionary relationships. The most comprehensive recent synthesis on African habitats (Burgess et al. 2004) has been widely adopted as a basis for conservation planning and is of use for any study of African biomes, ecoregions and habitats (MacKinnon et al. 2016). At the first level, a global classification and map of the world's ecoregions (Olson et al. 2001) was used to identify the nine biomes of Africa's three main biogeographic divisions (Palearctic, Afrotropical and Cape). The biome concept used was defined as "vegetation types with similar characteristics grouped together as habitats, and the broadest global habitat categories are called biomes" (Olson et al. 2001). Of the nine biomes recognised, seven are represented in Angola – the largest range of biomes represented in any African country. These are:

- Tropical and subtropical moist forests;
- Montane grasslands and shrublands;
- Tropical and subtropical grasslands, savannas, shrublands, and woodlands;
- Tropical and subtropical dry and broadleaf forests;
- Deserts and xeric shrublands;
- Mangroves; and
- Flooded grasslands and savannas.

Within the biomes, Burgess et al. (2004) defined a total of 119 terrestrial ecoregions for Africa and its islands. Ecoregions are defined as "large units of land or water that contain a distinct assemblage of species, habitats and processes, and whose boundaries attempt to depict the original extent of natural communities before major land-use change" (Dinerstein et al. 1995). It is impressive to note that based on the Burgess et al. (2004) assessment, Angola has not only the largest diversity of biomes, but also the second largest representation of ecoregional diversity in Africa (Table 2.2, Fig. 2.8).

Figure 2.8 (Burgess et al. 2004) provides a useful framework for the understanding of Angola's biodiversity patterns. Despite its coarse grain, it allows a general synthesis to be refined as new information becomes available. The relationship between biomes and ecoregions (*sensu* Burgess et al. 2004) and the vegetation types of Barbosa (1970) is summarised in Table 2.3. The very brief notes on key genera found within the Barbosa vegetation units provide an idea of the floristic composition that characterises the ecoregion. The photos presented in Fig. 2.9 provide examples of the main vegetation types and habitats

The Biological Importance of the Angolan Escarpment

While the classification of White (1983) and Linder et al. (2012) are useful at a continental scale, a more detailed and subtle analysis of the major biomes and habitat groupings is needed at a national scale for both research and conservation

Table 2.2 Representation of biomes and ecoregions in southern African countries

Country	Biomes	Ecoregions number and total (T)	T
Angola	7	8, 32, 42, 43, 49, 50, 51,55, 56, 63, 81, 82, 106, 109, 116	15
Botswana	3	54, 57, 58, 63, 68, 105	6
Congo Republic	3	8, 12, 13, 43,116	5
D.R. Congo	5	8, 13, 14, 15, 16, 17, 42, 43, 49, 50, 73, 116	12
Mozambique	3	21, 22, 52, 53, 54, 64, 76, 117	8
Namibia	3	51, 55, 58, 67, 105, 106, 107, 108, 109, 110	10
South Africa	5	22, 23, 24, 54, 57, 58, 77, 78, 79, 80, 89, 90, 91, 105, 108, 110, 117	17
Zambia	4	32, 50, 53, 54, 56, 63, 74	7
Zimbabwe	2	51, 53, 54, 57, 58, 76	6

From Burgess et al. (2004)

Physical Geography and Biodiversity Characteristics of Angola

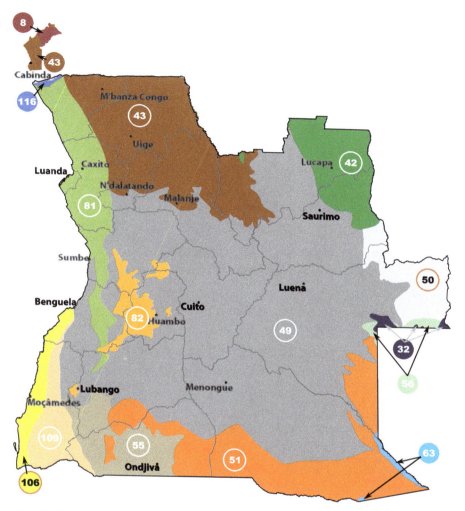

Fig. 2.8 Ecoregions of Angola
8 Atlantic Equatorial Coastal Forest • 32 Zambezian *Cryptosepalum* Dry Forest • 42 Southern Congolian Forest-Savanna Mosaic • 43 Western Congolian Forest-Savanna Mosaic • 49 Angolan Miombo Woodland • 50 Central Zambezian Miombo Woodland • 51 Zambezian *Baikiaea* Woodland • 55 Angola Mopane Woodland • 56 Western Zambezian Grassland • 63 Zambezian Flooded Grasslands • 81 Angolan Scarp Savanna and Woodland • 82 Angolan Montane Forest-Grassland Mosaic • 106 Kaokoveld Desert • 109 Namib Escarpment Woodlands • 116 Central African Mangroves. (After Burgess et al. 2004, map used with permission)

Biodiversity Conservation (Volume I)

Table 2.3 African biomes and ecoregions (as defined by Burgess et al. 2004) and Angolan vegetation types (Barbosa 1970) with indicative genera

Ecoregion n°	Biome	Ecoregion	Barbosa n°, name and key genera
8	Tropical and Subtropical Forest	Atlantic Equatorial Coastal Forest	1,2. Closed Forest
			Gilbertiodendron, Librevillea, Tetraberlinia
32	Tropical and Subtropical	Zambezian *Cryptosepalum*	4. Closed Forest
	Dry Broadleaf Forest	Dry Forest	*Cryptosepalum, Brachystegia, Erythrophleum*
42	Tropical and Subtropical Grasslands, Savannas, Shrublands and Woodlands	Southern Congolian Forest-Savanna Mosaic	8. Forest-Savanna Mosaic
			Marquesia, Berlinia, Daniella, Hymenocardia
43	Tropical and Subtropical Grasslands, Savannas, Shrublands and Woodlands	Western Congolian	3. Closed Forest
			Celtis, Albizia, Celtis
		Forest-Savanna Mosaic	13. Thicket-Forest Mosaic
			Annona, Piliostigma, Andropogon, Hyparrhenia
49	Tropical and Subtropical Grasslands, Savannas, Shrublands and Woodlands	Angolan Miombo Woodland	16, 17, 18. Woodland
			Brachystegia, Julbernardia, Guibourtia, Burkea, Pterocarpus
50	Tropical and Subtropical Grasslands, Savannas, Shrublands and Woodlands	Central Zambezian Miombo Woodland	17, 19. Woodland
			Brachystegia, Julbernardia, Cryptosepalum
51	Tropical and Subtropical Grasslands, Savannas, Shrublands and Woodlands	Zambezian Baikiaea Woodland	25. Tree and Shrub Savanna
			Baikiaea, Guibourtia, Pterocarpus, Combretum
55	Tropical and Subtropical Grasslands, Savannas, Shrublands and Woodlands	Angola Mopane Woodland	20. Woodland
			Colophospermum, Croton, Combretum, Sclerocarya, Acacia
56	Tropical and Subtropical Grasslands, Savannas, Shrublands and Woodlands	Western Zambezian Grassland	31. Grasslands
			Loudetia, Monocymbium, Tristachya, Parinari, Syzygium
63	Flooded Grasslands and Savannas	Zambezian Flooded Grasslands	31. Grasslands
			Loudetia, Echinochloa, Oryza
81	Montane Grasslands and Shrublands	Angolan Scarp Savanna and Woodland	10, 11, 22, 23. Forest-Savanna-Woodland-Thicket Mosaic
			Adansonia, Acacia, Albizia, Celtis, Piliostigma
82	Montane Grasslands and Shrublands	Angolan Montane Forest-Grassland Mosaic	6, 32. Relict Forest, Grasslands
			Podocarpus, Apodytes, Pittosporum, Protea, Erica

(continued)

Table 2.3 (continued)

Ecoregion n°	Biome	Ecoregion	Barbosa n°, name and key genera
106	Deserts and Xeric Shrublands	Kaokoveld Desert	28, 29. Desert, Steppes
			Welwitschia, Zygophyllum, Stipagrostis, Odyssea
109	Deserts and Xeric Shrublands	Namib Escarpment Woodlands	27. Steppes
			Acacia Commiphora, Colophospermum, Sesamothamnus, Rhigozum
116	Mangroves	Central African Mangroves	14 A. Mangroves
			Rhizophora, Avicennia, Raphia, Elaeis

planning in Angola (Revermann and Finckh 2019). What is equally important in biogeographic analysis is the detection of patterns of endemism and diversity at dispersed scales – such as the Angolan Escarpment Zone – described by Hall (1960a) and subsequently recognised by many workers as of great biodiversity and evolutionary importance (Huntley 1973, 1974a, 2017; Hawkins 1993; Mills 2010; Clark et al. 2011). Indeed, each taxon-based account in this volume, on plants (Goyder and Gonçalves 2019), odonata (Kipping et al. 2019), lepidoptera (Mendes et al. 2019), fishes (Skelton 2019), birds (Dean et al. 2019), amphibians (Baptista et al. 2019), reptiles (Branch et al. 2019) and mammals (Beja et al. 2019) draws attention to the importance of the Angolan Escarpment as a centre of endemism and speciation. Hall (1960a) explained her recognition of the importance of the Angolan Escarpment as the major speciation hotspot for birds in Angola by it: (i) creating a barrier between arid-adapted species of the coastal plains and of the miombo woodlands of the plateau, (ii) creating a steep ecological gradient, and (iii) functioning as a refuge for moist forest specialists that were isolated here during the dry periods of the glacial cycles. Dean et al. (2019) note that 75% of Angola's endemic birds are found in this zone.

The Angolan Escarpment and the remote, isolated and fragmentary remnants of Afro-montane forests of the Angolan Highlands offer ideal testing grounds for biogeographic models, as recently explored by Vaz da Silva (2015). The Angolan Escarpment biogeographic unit awaits clear definition, description and demarcation, but its scientific importance is matched only by the vulnerability of its threatened forest habitats (Cáceres et al. 2015). Linder et al. (2012) similarly recognise the importance of the Angolan Escarpment, and that of the transition from Congolian to Zambezian regions along the northern border of Angola (which they place in their Shaba sub-region). High species replacement values are found across these biologically rich areas, emphasising the urgency for their protection.

Fig. 2.9 Examples of some of the Ecoregions of Angola with numbering as per map in Fig. 2.8 and summary in Table 2.3. 8. Maiombo Forest, Cabinda; 42. Congolian gallery forest and moist miombo woodlands and savanna grasslands, Lunda-Norte; 49. *Brachystegia/Julbernardia* woodland Luando Strict Nature Reserve, Malange; 51. *Baikiaea/Guibourtia* Woodland Mucusso, Cuando Cubango; 56. Wetlands of the Bulozi Floodplain, Moxico; 81. Angolan Escarpment at Serra da Chela, Tundavala, Huíla; 82. Remnant patches of Afromontane forest in ravines on Mount Moco, Huambo; 106 Grasslands of the intermontane plains of central Iona National Park, Namibe. (Photos: Bulozi – JM Mendelsohn, others by BJ Huntley)

Conclusions

This brief outline of the biogeography of Angola demonstrates the country's unusual diversity of landscapes, climates and ecoregions, with Angola embracing the highest number of biomes represented within any African state.

The many classifications and terminologies applied to Angola's biogeographic units over the past century have not yet resulted in a nationally adopted nomenclature for its biomes and habitats. This situation prevails despite the existence of strong traditions in Angola's ethnic groups of indigenous taxonomies for habitats, such as those of the Chokwe of the Lundas, that are as perfect and detailed as modern systems (Redinha 1961; Huntley 2015). Furthermore, while many vernacular terms (*mato de panda, anharas do alto, floresta cafeeira, muxitos, mulolas, chanas da borracha*, etc.) enjoy wide use, they are imprecise and inadequate for Angola's great diversity of biomes and habitats.

The absence of a uniform system of nomenclature limits the use of information attached to biological collections, which in most cases provide only site locality data, and more recently, geo-referencing. Several southern African countries have nationally accepted biome and vegetation maps (e.g. South Africa, Lesotho and Swaziland – Mucina and Rutherford 2006) with clear descriptors for each biome and vegetation unit, facilitating communication between researchers and conservation planners. As Angola re-assesses its biodiversity wealth, and the need to protect and sustainably utilise these resources, the development of a new map of its vegetation, ecosystems and biomes becomes a high priority. Equally urgent but similarly daunting is the study of the evolutionary processes and relationships of the biota of the Angolan Escarpment and Afromontane forests, and the effective protection of these fingerprints of the past.

References

Albuquerque S, Figueirôa S (2018) Depicting the invisible: Welwitsch's map of travellers in Africa. Earth Sci Hist 37:109–129

Amaral I (1969) 'Inselberge' (ou montes-ilhas) e superfícies de aplanação na bacia do Cubal da Hanha em Angola. Garcia da Orta 17:474–526

Azevedo AL (1972) Caracterizacão Sumária das Condições Ambientais de Angola. Universidade de Luanda, Luanda, 106 pp

Baptista N, Conradie W, Vaz Pinto P et al (2019) The amphibians of Angola: early studies and the current state of knowledge. In: Huntley BJ, Russo V, Lages F, Ferrand N (eds) Biodiversity of Angola. Science & conservation: a modern synthesis. Springer Nature, Cham

Barbosa LAG (1970) Carta Fitogeográfica da Angola. Instituto de Investigação Científica de Angola, Luanda, 343 pp

Beja P, Vaz Pinto P, Veríssimo L et al (2019) The mammals of Angola. In: Huntley BJ, Russo V, Lages F, Ferrand N (eds) Biodiversity of Angola. Science & conservation: a modern synthesis. Springer Nature, Cham

Branch WR, Vaz Pinto P, Baptista N et al (2019) The reptiles of Angola: history, diversity, endemism and hotspots. In: Huntley BJ, Russo V, Lages F, Ferrand N (eds) Biodiversity of Angola. Science & conservation: a modern synthesis. Springer Nature, Cham

Brooks N, Adger WN, Kelly PM (2005) The determinants of vulnerability and adaptive capacity at the national level and the implications for adaptation. Glob Environ Chang 15:151–163

Burgess N, Hales JD, Underwood E et al (2004) Terrestrial ecoregions of Africa and Madagascar – a conservation assessment. Island Press, Washington DC, 499 pp

Cáceres A, Melo M, Barlow J et al (2015) Threatened birds of the Angolan Central Escarpment: distribution and response to habitat change at Kumbira Forest. Oryx 49:727–734

Cáceres A, Melo M, Barlow J et al (2017) Drivers of bird diversity in an understudied African centre of endemism: the Angolan Escarpment Forest. Bird Conserv Int 27:256–268

Cain A, Cain A (2015) Climate change and land markets in coastal cities of Angola. In 2015 World Bank conference on land and poverty. The World Bank, Washington, DC

Carvalho SCP, Santos FD, Pulquério M (2017) Climate change scenarios for Angola: an analysis of precipitation and temperature projections using four RCMs. Int J Climatol 37:3398–3412

Chapin JP (1932) The birds of the Belgian Congo. Bull Am Mus Nat Hist 65:1–756

Choffat P (1888) Dr. Welwitsch: Quelques notes sur la géologie d'Angola coordonnées et annotées par Paul Choffat. Separata das Comunicações dos Serviços Geológicos de Portugal 19:1–24

Clark VR, Barker NP, Mucina L (2011) The Great Escarpment of southern Africa: a new frontier for biodiversity exploration. Biodivers Conserv 20:2543–2561

Costa FL (2006) O conhecimento geomorfológico de Angola. In: Moreira I (ed) Angola: Agricultura, Recursos Naturais e Desenvolvimento. ISA Press, Lisboa, pp 477–495

Cotterill FPD (2010) The evolutionary history and taxonomy of the Kobus leche species complex of south-central Africa in the context of Palaeo-drainage dynamics. Unpublished PhD thesis, University of Stellenbosch

Cotterill FPD (2015) Biogeographical overview of the Lunda region, northeast Angola. In: Huntley BJ, Francisco P (eds) Avaliação Rápida da Biodiversidade de Região da Lagoa Carumbo, Lunda-Norte – Angola/Rapid biodiversity assessment of the Carumbo Lagoon Area, Lunda-Norte – Angola. Ministério do Ambiente, Luanda, pp 77–99

Cotterill F, De Wit M (2011) Geoecodynamics and the Kalahari Epeirogeny: linking its genomic record, tree of life and palimpsest into a unified narrative of landscape evolution. S Afr J Geol 114:489–514

Crawford-Cabral J (1983) Esboço zoogeográfico de Angola. Unpublished manuscript, Lisbon, 50 pp + 13 maps

Dean WRJ (2000) The birds of Angola: an annotated checklist. BOU Checklist No. 18. British Ornithologists' Union. Tring, UK

Dean WRJ, Melo M, Mills MSL (2019) The Avifauna of Angola: richness, endemism and rarity. In: Huntley BJ, Russo V, Lages F, Ferrand N (eds) Biodiversity of Angola. Science & conservation: a modern synthesis. Springer Nature, Cham

Dinerstein E, Olson D, Graham A et al (1995) A conservation assessment of the ecoregions of Latin America and the Caribbean. World Bank, Washington, DC

Diniz AC, Aguiar FB (1966) Geomorfologia, solos e ruralismo de região central angolana. Agronomia Angolana 23:11–17

Diniz AC (1973) Características mesológicas de Angola. Missão de Inquéritos Agrícolas de Angola, Nova Lisboa

Diniz AC (1991) Angola, o meio físico e potencialidades agrárias. Instituto para a Cooperação Económica, Lisboa, 189 pp

Diniz AC (2006) Características mesológicas de Angola. Instituto Português de Apoio ao Desenvolvimento, Lisbon, 546 pp

Droissart V, Dauby G, Hardy OJ et al (2018) Beyond trees: biogeographical regionalization of tropical Africa. J Biogeogr 45:1153–1167

Frade F (1963) Linhas gerais da distribuição geográfica dos vertebrados em Angola. Memórias da Junta de Investigações do Ultramar 43:241–257

Feio M (1964) A evolução da escadaria de aplanações do sudoeste de Angola. Garcia da Orta 12:323–354

GoA (Government of Angola) (2013) Angola initial national communication to the United Nations framework convention on climate change. Ministry of the Environment, Luanda, 194 pp

Gossweiler J, Mendonça FA (1939) Carta Fitogeográfica de Angola. Ministério das Colónias, Lisboa, 242 pp

Goyder DJ, Gonçalves FMP (2019) The flora of Angola: collectors, richness and endemism. In: Huntley BJ, Russo V, Lages F, Ferrand N (eds) Biodiversity of Angola. Science & conservation: a modern synthesis. Springer Nature, Cham

Hall BP (1960a) The faunistic importance of the scarp of Angola. Ibis 102:420–442

Hall BP (1960b) The ecology and taxonomy of some Angolan birds. Bull Br Mus (Nat Hist) Zool 6:367–463

Hall BP, Moreau RE (1962) The rare birds of Africa. Bull Br Mus (Nat Hist) Zool 8:315–381

Hawkins F (1993) An integrated biodiversity conservation project under development: the ICBP Angola Scarp Project. In: Proceedings of the VIII Pan-African Ornithological Congress, pp 279–284. Kigali, Rwanda, 1992. Koninklijk Museum voor Midden-Afrika, Tervuren

Hellmich W (1957) Herpetologische Ergebnisse einer Forschungsreise in Angola. Veröffentlichungen der Zoologischen Staatssammlung München 5:1–92

Humbert H (1940) Zones et Étages de Végétation dans le Sud-Ouest de l'Angola. Compte-rendu Sommaire des Séances de la Societé de Biogéographie 17:47–57

Huntley BJ (1973) Distribution and Status of the Larger Mammals of Angola, with particular reference to Rare and Endangered species: First Progress Report. December 1973. Repartição Técnica da Fauna, Serviços de Veterinária, Luanda, Mimeograph report, 14 pp

Huntley BJ (1974a) Vegetation and Flora Conservation in Angola. Ecosystem Conservation Priorities in Angola. Ecologist's Report 22. Repartição Técnica da Fauna, Serviços de Veterinária, Luanda, Mimeograph report, 13pp

Huntley BJ (1974b) Ecosystem Conservation Priorities in Angola. Ecologist's Report 26. Repartição Técnica da Fauna, Serviços de Veterinária, Luanda, Mimeograph report

Huntley BJ (2010) Estratégia de Expansão da Rede de Áreas Protegidas da Angola/Proposals for an Angolan Protected Area Expansion Strategy (APAES). Unpublished report to the Ministry of Environment, Luanda, 28 pp + map

Huntley BJ (2015) Biophysical profile of Lunda-Norte. In: Huntley BJ, Francisco P (eds) Avaliação Rápida da Biodiversidade de Região da Lagoa Carumbo, Lunda-Norte – Angola / Rapid Biodiversity Assessment of the Carumbo Lagoon Area, Lunda-Norte – Angola. Ministério do Ambiente, Luanda, pp 31–75

Huntley BJ (2017) Wildlife at war in Angola: the rise and fall of an African Eden. Protea Book House, Pretoria, 432 pp

Huntley BJ, Matos EM (1994) Botanical diversity and its conservation in Angola. Strelitzia 1:53–74

Jessen O (1936) Reisen und Forschungen in Angola. Dietrich Reimer Verlag, Berlin

Jones A, Breuning-Madsen H, Brossard M et al (2013) Soil atlas of Africa. Publications Office of the European Union, Brussels

King LC (1962) Morphology of the Earth. Oliver & Boyd, London, 699 pp

Kipping J, Clausnitzer V, Fernandes Elizalde SRF et al (2019) The dragonflies and damselflies of Angola. In: Huntley BJ, Russo V, Lages F, Ferrand N (eds) Biodiversity of Angola. Science & conservation: a modern synthesis. Springer Nature, Cham

Kirkman SP, Nsingi KK (2019) Marine biodiversity of Angola: biogeography and conservation. In: Huntley BJ, Russo V, Lages F, Ferrand N (eds) Biodiversity of Angola. Science & conservation: a modern synthesis. Springer Nature, Cham

Linder HP, de Klerk HM, Born J et al (2012) The partitioning of Africa: statistically defined biogeographical regions in sub-Saharan Africa. J Biogeogr 39:1189–1925

MacKinnon J, Aveling C, Olivier R et al (2016) Inputs for an EU strategic approach to wildlife conservation in Africa – regional analysis. European Commission, Directorate-General for International Cooperation and Development, Brussels

Marques MM (1963) Notas sobre a geomorfologia da Angola 1. Significado morfológico de algumas 'anharas do alto'. Garcia da Orta 11:541–560

Maurin O, Davies TJ, Burrows JE et al (2014) Savanna fire and the origins of the 'underground forests' of Africa. New Phytol 204(1):201–214

Mateus O, Callapez P, Polcyn M et al (2019) Biodiversity in Angola through time: a paleontological perspective. In: Huntley BJ, Russo V, Lages F, Ferrand N (eds) Biodiversity of Angola. Science & conservation: a modern synthesis. Springer Nature, Cham

Mateus O, Jacobs LL, Schulp AS et al (2011) *Angolatitan adamastor*, a new sauropod dinosaur and the first record from Angola. Ann Braz Acad Sci 83(1):221–233

Mendelsohn JM (2019) Landscape changes in Angola. In: Huntley BJ, Russo V, Lages F, Ferrand N (eds) Biodiversity of Angola. Science & conservation: a modern synthesis. Springer Nature, Cham

Mendelsohn JM, Mendelsohn S (2018) Sudoeste de Angola: um Retrato da Terra e da Vida. South West Angola: a portrait of land and life. Raison, Windhoek

Mendelsohn J, Weber B (2013) An atlas and profile of Huambo: its environment and people. Development Workshop, Luanda, 80 pp

Mendelsohn J, Weber B (2015) Moxico: an atlas and profile of Moxico, Angola. Raison, Windhoek, 44 pp

Mendelsohn J, Jarvis A, Robertson T (2013) A profile and atlas of the Cuvelai-Etosha Basin. Raison & Gondwana Collection, Windhoek, 170 pp

Mendes L, Bivar-de-Sousa A, Williams M (2019) The butterflies and skippers of Angola. In: Huntley BJ, Russo V, Lages F, Ferrand N (eds) Biodiversity of Angola. Science & conservation: A modern synthesis. Springer Nature, Cham

Mills MSL (2010) Angola's central scarp forests: patterns of bird diversity and conservation threats. Biodivers Conserv 19:1883–1903

Mills MSL, Melo M, Vaz A (2013) The Namba mountains: new hope for Afromontane forest birds in Angola. Bird Conserv Int 23:159–167

Mills MSL, Olmos F, Melo M et al (2011) Mount Moco: its importance to the conservation of Swierstra's Francolin *Pternistis swierstrai* and the Afromontane avifauna of Angola. Bird Conserv Int 21:119–133

Monard A (1937) Contribution à l'herpétologie d'Angola. Arquivos do Museu Bocage 8:19–154

Monteiro RFR (1970) Estudo da Flora e da Vegetacão das Florestas Abertas do Planalto do Bié. Instituto de Investigacão Científica de Angola, Luanda, 352 pp

Mucina L, Rutherford MC (2006) The vegetation of South Africa, Lesotho and Swaziland. Strelitzia 19:1–807

Neto AG, Ricardo RP, Madeira M (2006) O alumínio nos solos de Angola. In: Moreira I (ed) Angola: Agricultura, Recursos Naturais e Desenvolvimento. ISA Press, Lisboa, pp 121–143

Olson DM, Dinerstein E, Wikramanayake ED et al (2001) Terrestrial ecoregions of the World: a new map of life on Earth. Bioscience 51:933–938

Peel MC, Finlayson BL, McMahon TA (2007) Updated world map of the Köppen-Geiger climate classification. Hydrol Earth Syst Sci 11:1633–1644

Redinha J (1961) Nomenclaturas nativas para as formações botânicas do nordeste de Angola. Agronomia Angolana 13:55–78

Revermann R, Finckh M (2019) Vegetation survey, classification and mapping in Angola. In: Huntley BJ, Russo V, Lages F, Ferrand N (eds) Biodiversity of Angola. Science & conservation: a modern synthesis. Springer Nature, Cham

Rodrigues P, Figueira R, Vaz Pinto P et al (2015) A biogeographical regionalization of Angolan mammals. Mammal Rev 45:103–116

Silveira MM (1967) Climas de Angola. Serviço Meteorólogico de Angola, Luanda, 44 pp

Skelton PH (2019) The Freshwater Fishes of Angola. In: Huntley BJ, Russo V, Lages F, Ferrand N (eds) Biodiversity of Angola. Science & conservation: a modern synthesis. Springer Nature, Cham

Spalding MD, Fox HE, Allen GR et al (2007) Marine ecoregions of the world: a bioregionalization of coastal and shelf areas. Bioscience 57:573–583

Thieme ML, Abell R, Stiassny ML et al (eds) (2005) Freshwater ecoregions of Africa and Madagascar: a conservation assessment. Island Press, Washington DC

Ucuassapi AP, Dias JCS (2006) Acerca da fertilidade dos solos de Angola. In: Moreira I (ed) Angola: Agricultura, Recursos Naturais e Desenvolvimento. ISA Press, Lisboa, pp 477–495

Vaz da Silva B (2015) Evolutionary history of the birds of the Angolan highlands – the missing piece to understand the biogeography of the Afromontane forests. MSc Thesis, University of Porto, Porto

Weir CR (2019) The whales and dolphins of Angola. In: Huntley BJ, Russo V, Lages F, Ferrand N (eds) Biodiversity of Angola. Science & conservation: a modern synthesis. Springer Nature, Cham

Welwitsch F (1859) Apontamentos phyto-geographicos sobre a Flora da Provincia de Angola na Africa Equinocial servindo de relatório preliminar acerca da exploração botanica da mesma provincia. Annaes do Conselho Ultramarino (Ser. 1):527–593

Werger MJA (1978) Biogeographical division of southern Africa. In: Werger MJA, van Bruggen AC (eds) Biogeography and ecology of Southern Africa. Junk, The Hague, pp 145–170

White F (1971) The taxonomic and ecological basis of chorology. Mitteilungen Botanischen Staatsammlung München 10:91–112

White F (1976) The underground forests of Africa: a preliminary review. The Gardens' Bull Singapore 24:57–71

White F (1983) The vegetation of Africa – a descriptive memoir to accompany the UNESCO/AETFAT/UNSO vegetation map of Africa. UNESCO, Paris, 356 pp

Zigelski P, Gomes A, Finckh M (2019) Suffrutex dominated ecosystems in Angola. In: Huntley BJ, Russo V, Lages F, Ferrand N (eds) Biodiversity of Angola. Science & conservation: a modern synthesis. Springer Nature, Cham

3

Biogeography and Conservation of Angolan Coastal and Marine Systems

Stephen P. Kirkman and Kumbi Kilongo Nsingi

Abstract Some major physical and oceanographic features of the Angolan marine system include a narrow continental shelf, the warm, southward flowing Angola Current, the plume of the Congo River in the north and the Angola-Benguela Front in the south. Depth, substrate types and latitude have been shown to account for species differences in demersal faunal assemblages including fish, crustaceans, and cephalopods. The extremely narrow shelf between Tômbwa (15°48′S) and Benguela (12°33′S) may serve as a barrier for the spreading of shelf-occurring species between the far south, which is influenced by the Angola-Benguela Front, and the equatorial waters of the central and northern areas. A similar pattern is evident for coastal and shallow-water species, including fishes, intertidal invertebrates and seaweeds, with species that have temperate affinities found in the far south and tropical species further to the north. In general the fauna and flora of the littoral zone appears to be consistent with a pattern of relatively low diversity of the shore and near-shore areas, that is characteristic of West Africa, but paucity of data for Angola may make such comparisons of diversity with other areas inappropriate at this stage. The Congo River delta and many features that are interspersed along the coast such as estuaries and associated floodplains, wetlands, lagoons, salt marshes and mangroves, support a rich suite of species, many of which are rare, endemic, migratory, and/or threatened, and provide important ecosystem services. While the ecological value of many areas or features is recognised, lack of any legal protection in the form of marine protected areas (MPAs) has been identified as one of the main challenges facing conservation and sustainable use of Angola's marine and coastal biodiversity and habitats, in the face of multiple threats. A current process to identify and describe ecologically or biologically significant marine areas (EBSAs) could provide a foundation for designating some MPAs in future.

S. P. Kirkman (✉)
Department of Environmental Affairs, Oceans and Coastal Research,
Cape Town, South Africa
e-mail: skirkman@environment.gov.za

K. K. Nsingi
Benguela Current Convention, Swakopmund, Namibia
e-mail: kkilongo@gmail.com

Keywords Benguela current · Ecologically or biologically significant marine areas · Important bird areas · Fish · Marine protected areas · Marine spatial planning · Seaweed · Systematic conservation planning · West Africa

Physical and Oceanographic Context

The coastline of Angola, which is approximately 1650 km long, consists of sandy and rocky stretches of coastline, punctuated by numerous coastal features such as estuaries, mangroves, coastal lakes, wetlands and tidal flats (Harris et al. 2013). Between Rio Bero (north of Moçâmedes) in Namibe Province and to the north of Rio Coporolo in Benguela Province are rocky shores; the rest of the coastline is predominantly sandy although there are some scattered rocky shores further north of Lobito (Harris et al. 2013). The continental shelf, which extends to about 200 m depth, is relatively narrow especially near the south where it is as little as 6 km wide and very steep in parts of Namibe and Benguela, but it widens further north to 33 km width near the mouth of the Congo River, and in the south it widens a little between Tômbwa and Cunene (Figure 1, Bianchi 1992). The neritic zone (i.e. waters overlying the continental shelf) covers about one third of Angola's Exclusive Economic Zone (EEZ), which also includes extensive bathyal and abyssal zones, with depths up to 4000 m in the latter (Nsiangango et al. 2007).

Hutchings et al. (2009) describe the marine system of Angola's continental shelf area as a subtropical transition zone between the Equatorial Atlantic to the north and the Benguela's wind-driven upwelling system to the south. The conspicuous, dynamic but relatively shallow Angola-Benguela Front at 17°S in the south of Angola forms the boundary with the upwelling system, and the boundary to the north is near the plume of the Congo River. Seasons are well-defined and there is intermediate productivity; moderate to weak upwelling occurs year-round in the south and all along the coast in winter with strengthening of southeast trade winds. The major oceanographic feature of the system is the warm (>24 °C) Angola Current, which flows southward along the shelf and slope as an extension of the South Equatorial Counter-current, extending down to 200 m depth and with a mean flow at 50 m depth of 5–8 cm s^{-1} (Kopte et al. 2017). During winter and spring the Angola Current tends to retreat to the northwest and is replaced by slightly cooler waters; this is linked to the intensity of wind-driven upwelling off the Namibian coast (Meeuwis and Lutjeharms 1990; O'Toole 1980).

Other important drivers of the system are Kelvin waves propagating from the Equatorial Atlantic and the South Equatorial Counter Current (Florenchie et al. 2004; Shillington et al. 2006), as well as southward flow of brackish water with high nutrient loads from the Congo River outflow and solar heating (Veitch et al. 2010), Both result in stratification of the water column (Kirkman et al. 2016), with thermocline depth ranging from 10 m in the north down to about 50 m off central Angola (Bianchi 1992). Another feature of the Angolan system is the cold water Angola Dome, found offshore of the Angola Current. This is a cyclonic eddy that causes doming of the thermocline, centred near 10°S and 9°E (Lass et al. 2000). The Angola Dome has lower salinity and concentrations of oxygen than surrounding waters, but it does not exist in winter and its width and extension depend on the intensity and horizontal shear of southeasterly trade winds (Signorini et al. 1999). Phytoplankton production associated with the Angola Dome strongly influences the shelf ecosystem throughout northern Angola (Monteiro et al. 2008).

Biodiversity and Biogeography

There is a high diversity of demersal species in Angola relative to the temperate Benguela Ecosystem to the south, with species richness greatest at about 100 m depth according to research surveys (Kirkman et al. 2013). Demersal fish stocks are exploited by a multispecies bottom trawl fishery extending from southern to northern Angola, that exploits over 30 species of fish belonging to the families Sparidae (seabreams), Scianidae (croakers), Serranidae (groupers), Haemulidae (grunts) and Merlucidae (hakes). Some of the most commercially important species include Benguela Hake *Merluccius polli* and demersal sparid fish such as *Dentex* spp. (Kirkman et al. 2016); there is also bottom fishing for crustaceans, most importantly deep-sea crab *Chaceon maritae* and shrimps *Aristeus varidens* and *Parapenaeus longirostris* (Japp et al. 2011; Kirkman et al. 2016). The most important fish targeted by small pelagic fisheries include Kunene Horse Mackerel *Trachurus trecae* and *Sardinella* species, with most large pelagic fishing (tuna spp.) taking place in the south (Japp et al. 2011, Kirkman et al. 2016). Several of the above stocks are targeted both by industrial and artisanal fisheries (Duarte et al. 2005; Japp et al. 2011). The Angolan fisheries are described by Hutchings et al. (2009) as being of moderate intensity with stocks generally declining. There is also a rapidly growing local and foreign recreational shore-fishery sector in southern Angola targeting mainly Leerfish *Lichia amia*, West Coast Dusky Kob *Argyrosomus coronus* and Shad *Pomatomus saltatrix* (Potts et al. 2009).

Bianchi (1992) and later Nsiangango et al. (2007) studied the structure of demersal assemblages of the continental shelf and upper Angolan slope, including fish, crustaceans and cephalopods, based on trawl surveys. It was shown that thermal, depth-dependent stratification explained the main faunal groupings, with certain species generally restricted to waters shallower than where the lower limit of the thermocline meets the shelf, and others usually occurring in deeper waters than this. Species such as Shallow-water Croaker *Pteroscion peli*, Red Pandora *Pagellus bellotti*, Lesser African Threadfin *Galeiodes decadactylus* and Grunt *Pomadasys incisus* dominated in the shallower demersal water (coast to 100 m), with some sea breams in low densities, whereas deeper waters of the shelf and upper slope were dominated by such species as Splitfin *Synagrops microlepis*, Atlantic Green-eye *Chlorophthalmus atlanticus,* Angolan Dentex *D. angolensis* and *M. polli*. Within the different depth strata, substrate type and latitudinal gradients were the main factors affecting the composition of species assemblages, and a major latitudinal shift both in shallow- and deep-water assemblages was shown to occur in southern Angola between Tômbwa and Cunene where the shelf widens and where Large-eyed Dentex *D. macrophthalmus* dominated the catches. Bianchi (1992) related the shift to the southern limit of warmer equatorial waters, the presence of the Angola-Benguela Front where there is year-round upwelling and cooler waters, and the extremely narrow shelf to the north of Tômbwa (up to Benguela), which may serve as a barrier to the spreading of northern species to the south and vice versa.

While deep-water coral reefs have been documented for Angola's continental slope (Le Guilloux et al. 2009), shallow-water coral reefs are absent and in general the fauna and flora of littoral zone seems to be consistent with the pattern of relatively low diversity of the shore and near-shore areas of West Africa (John and Lawson 1991). Factors that could account for this include the lack of hard substrata (most of the shoreline being sandy), upwelling of cooler water in areas, high turbidity and sediment input from a massive river such as the Congo, or loss of species associated with reductions in sea-temperatures that considerably reduced the tropical zone during Pleistocene glaciations (van den Hoek 1975; John and Lawson 1991). However, while recent studies (e.g. Hutchings et al. 2007; Anderson et al. 2012) have added to the existing species lists (e.g. Lawson et al. 1975; Penrith 1978 and others) for coastal fishes, sandy beach macrofauna, rocky shore invertebrates and seaweeds, at this stage the paucity of information in Angola may not make comparison with other areas appropriate. In general the data that exist both for coastal and estuarine fishes (Whitfield 2005; Hutchings et al. 2007) but also offshore fish species (Kirkman et al. 2013; Yemane et al. 2015) of Angola show decreasing species richness from north to south, seemingly supporting the established trend of decreasing diversity with latitude as one moves polewards from tropical regions (e.g. Rex et al. 2000; Willig et al. 2003).

Based on the latitudinal distributions of intertidal fauna of rocky shores (Kensley and Penrith 1980), the southern limit of tropical biota has previously been reported to be around the border of Angola and Namibia. Lawson (1978) on the other hand, based on analyses of seaweed flora, considered Angola to be intermediate in nature between tropical and temperate. Results of surveys of intertidal invertebrates and seaweeds by Hutchings et al. (2007) however, showed that although there was a marked discontinuity between the biota of Angola and that of northern Namibia, which supports a cool-temperate intertidal flora up to nearby the Cunene River (Rull Lluch 2002), a number of taxa found in the south of Angola had temperate affinities. This led the authors to suggest that the inshore biota of the south of Angola may be intermediate in nature, and that of the north truly tropical. This is confirmed by Anderson et al. (2012) who conclude that the overall affinities of the Angolan seaweed flora is Tropical West African, but with a well-developed temperate element in southern Angola (from about 13°S) comprising mainly cooler-water species. Broadly, this supports the division of the Angolan coast into at least two sub-areas, with the more temperate south influenced by the cooler waters of the Angola-Benguela Front. This is similar to the division between demersal assemblages of north and south (Bianchi 1992) and also congruent with a break in the pelagic ecosystem of the inshore as determined from classification of key oceanographic variables and depth (Lagabrielle 2011). It also ties in with global mapping classifi-

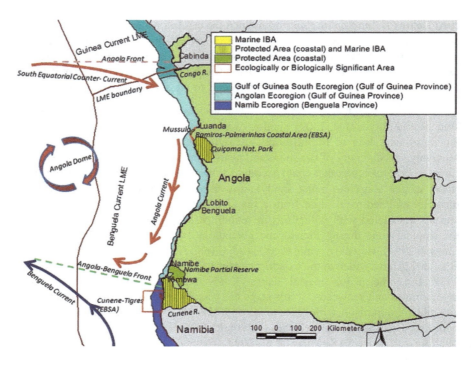

Fig. 3.1 Delineations of marine ecoregions (Spalding et al. 2007) and large marine ecosystems (Sherman 2014) that coincide with Angola. The ecoregions extend from the coast to the shelf edge. Also shown are recognised marine and coastal biodiversity areas and the approximate situations of important oceanographic processes

cation of coastal and shelf areas based on species distributions and levels of endemism of benthic and pelagic biota (Spalding et al. 2007; Briggs and Bowen 2012), that puts the divide between the temperate Benguela Province and the tropical Gulf of Guinea Province, near Moçâmedes (Fig. 3.1). Spalding et al. (2007) situate the majority of the Angolan EEZ in the Angolan ecoregion of the Gulf of Guinea Province, but include the area north of 6°30′S in the more tropical Gulf of Guinea south ecoregion. This is slightly incongruent with the mapping of large marine ecosystems (LMEs) of the world (which is expert- rather than data-derived), whereby most of Angola is included in the Benguela Current LME bounded to the north by the Angola Front (ca. 5°S), and only Cabinda in the far north included in the Guinea Current LME (Sherman 2014).

Marine and Coastal Biodiversity Hotspots, Threats and the Need for Protection

Whereas the Angolan coastal and shallow habitats are considered to be relatively low in biodiversity, coastal features such as the Congo River Delta, estuaries such as the Cuanza, Catumbela, Longa and Cunene, and associated floodplains, wetlands, lagoons, salt marshes and (north of Lobito) mangroves support a rich suite of species, often in high abundance (Hughes and Hughes 1992; van Niekerk et al. 2008; Harris et al. 2013). This includes several rare, endemic, migratory, and/or threatened fauna such as the African Manatee *Trichechus senegalensis*, turtle species and diverse waterbird species. Recognised ecosystem services of such features include (amongst others) providing habitat for important food-fish and crustacean species and their critical life stages (e.g. performing important nursery functions for many marine fishes) or providing plant species that are useful for medicinal, subsistence or construction purposes (Hughes and Hughes 1992; van Niekerk et al. 2008). While Angola is not currently a contracting party of the Ramsar Convention, some coastal wetland sites have been identified as being potential Ramsar sites, including Quiçama National Park between the Cuanza and Longa Rivers (Fig. 3.1), which is also a confirmed Important Bird and Biodiversity Area (IBA). Other confirmed coastal IBAs in Angola include Mussulo just south of Luanda and the Iona National Park in the south between the Cunene and Curoca Rivers. These IBAs are important for numerous waterbirds and are frequented by wintering seabird species that breed further south on the sub-continent such as Cape gannet *Morus capensis* (IUCN Red List – Endangered) and Damara tern *Sternula balaenarum* (Vulnerable) (Birdlife International 2002), the latter of which is known to also breed in the Iona National Park (Simmons 2010).

While the ecological value of these and other areas is recognised, the lack of formal protection of key biodiversity areas or features in Angola's marine and coastal environments has been noted as a concern (e.g. Tarr et al. 2007). As part of a regional systematic conservation planning (SCP) project involving all three member states of the Benguela Current Convention (BCC; a legally constituted collaborative mechanism representing Angola, Namibia and South Africa), Holness et al. (2014) showed that Angola is particularly poorly off in terms of spatial protection of its marine systems, with 102 out of 133 identified ecosystem types having no protection at all. Whereas some legislative protection of coastal ecosystem types in the Cuanza, Cunene and Tômbwa areas may be afforded by terrestrial national parks (Quiçama and Iona) or reserves (Namibe Partial Reserve), this may have value for conservation of the adjacent marine areas if effective management of these areas is achieved through the provision of increased human and financial resources. Holness et al. (2014) therefore opined that a programme of rapid expansion of protected areas for the Angolan marine systems is urgently required, and the ultimate product of their study was the prioritisation of sites for protection (ideally within a MPA network).

The current lack of marine protected areas (MPAs) was described by Tarr et al. (2007) as amongst the main challenges facing conservation and sustainable use of Angola's marine and coastal biodiversity and habitats, in light of multiple threats to the ecosystem that are likely to worsen over time. These threats include (but are not limited to) rapid, unplanned coastal urbanisation causing habitat destruction and a severe problem with waste management along the coast, particularly in the area of Luanda; escalation in over-exploitation of living marine resources related to rapid urbanisation and human migration to the coastal nodes, especially since the end of the civil war; industrial pollution caused e.g. by deposition of industrial wastes in catchment areas or cleaning of ships; offshore oil exploitation in the north, with potential for oil spills; loss of mangroves, which includes threats from pollution and wood collection for firewood and construction; rapid growth of the tourism industry; and impacts of climate change (Tarr et al. 2007; Heileman and O'Toole 2009).

With such threats in mind, Angola, like the other two member states of the BCC, has committed to implementing ecosystem-based management (EBM) of the marine environment to address responsible use of its ocean and its resources and put in practice the principles of sustainable development (BCC 2014). EBM is an integrative approach to management that takes into account all interactions in the ecosystem (including those involving human activities) and their cumulative impacts in space and time (Long et al. 2015). To be able to assist EBM with regard to the allocation and siting of ocean uses or protection measures, there is an initiative to implement marine spatial planning (MSP) in Angola and the other countries of the region (Kirkman et al. 2016). A pilot area for an experimental MSP project, covering an area of approximately 107,000 km^2 between south of Palmerinhas and the Tapado River mouth (GNC-OEM 2018), has recently been identified. A key element of the process is to identify and describe a network of Ecologically or Biologically Significant Marine Areas (EBSAs) - geographically or oceanographically discrete areas that have been identified as important for the services that they provide and for the healthy functioning of oceans (Dunstan et al. 2016) and to include these in marine spatial plans.

Currently, only two Angolan EBSAs have been described and subsequently endorsed by CBD (CBD 2014), namely the Ramiros-Palmerinhas Coastal Area which partly adjoins the Mussulo Peninsula south of Luanda, and the Cunene-Tigres EBSA which overlaps with northern Namibia and is adjacent to the Iona National Park on the Angola side (Fig. 3.1). The former includes estuaries with mangroves and salt marshes and has special importance for bird aggregations and breeding turtles. The latter includes the Cunene estuary and associated wetland as well as the Baía dos Tigres Island-Bay complex to the north of it, and has special importance for migratory birds and in terms of its nursery function for many marine species. Both of these areas have undergone thorough assessment processes with a view to expanding their areas in order to include other relevant features such as estuaries, sensitive coastline, canyons and seamounts.

Currently Angola is in the process of describing new potential EBSAs, in coastal and offshore areas, as part of a collaborative regional project with Namibia and South Africa, coordinated by the BCC (http://www.benguelacc.org). Currently, five new areas have been proposed as EBSAs which include coastal and offshore areas in the provinces of Cabinda, Zaire, Luanda, Cuanza-Sul and Namibe. Although EBSA status itself does not carry any conservation or protection interventions, legal protection is among the management measures that can be applied to secure the persistence of these special features and their ecosystem services. Therefore the process of expanding the EBSA network could provide a foundation for initiating a network of MPAs in Angola. In this regard, there is a recent project proposal for the establishment of the first MPA in Angola in the offshore area adjacent to the Iona National Park.

References

Anderson RJ, Bolton JJ, Smit AJ et al (2012) The seaweeds of Angola: the transition between tropical and temperate marine floras on the west coast of southern Africa. Afr J Mar Sci 34:1–13

BCC (Benguela Current Commission) (2014) Strategic action programme 2015–19. Swakopmund, Namibia, 36 pp. http://benguelacc.org/index.php/en/publications

Bianchi G (1992) Demersal assemblages of the continental shelf and upper slope of Angola. Mar Ecol Prog Ser 81:101–120

BirdLife International (2002) Important bird areas and potential Ramsar sites in Africa. BirdLife International, Cambridge, MA, 136 pp + appendices

Briggs JC, Bowen BW (2012) A realignment of marine biogeographic provinces with particular reference to fish distributions. J Biogeogr 39:12–30

CBD (Convention on Biological Diversity) (2014) Decision adopted by the Conference of the Parties to the Convention on Biological Diversity. XII/22. Marine and Coastal Biodiversity: Ecologically or Biologically Significant Marine Areas (EBSAs). Twelfth meeting of the Conference for the Parties, 6–17 October 2014, Pyeongchang, Republic of Korea. UNEP/CBD/COP/DEC/XII/22

Duarte A, Fielding P, Sowman M, et al (2005) Overview and analysis of socio-economic and fisheries information to promote the management of artisanal fisheries in the Benguela Current Large Marine Ecosystem (BCLME) region (Angola). Unpublished Final Report. Rep. No. LMRAFSE0301B. Cape Town Environmental Evaluation Unit, University of Cape Town, Cape Town

Dunstan PK, Bax NJ, Dambacher JM et al (2016) Using ecologically or biologically significant marine areas (EBSAs) to implement marine spatial planning. Ocean Coast Manag 121:116–127

Florenchie P, Reason CJC, Lutjeharms JRE et al (2004) Evolution of interannual warm and cold events in the Southeast Atlantic Ocean. J Clim 17:2318–2334

Grupo Nacional de Coordenação para o Ordenamento do Espaço Marinho (GNC-OEM) (2018). Relatório Preliminar sobre o Ordenamento do Espaço Marinho em Angola: Área Experimental Palmeirinhas – Tapado. Unpublished report

Harris L, Holness S, Nel R et al (2013) Intertidal habitat composition and regional-scale shoreline morphology along the Benguela coast. J Coast Conserv 17:143–154

Heileman S, O'Toole MJ (2009) I West and Central Africa: I-1 Benguela current LME. In: Sherman K, Hempel G (eds) The UNEP large marine ecosystems report: a perspective on changing conditions in LMEs of the World's regional seas. UNEP Regional Seas Report and Studies No. 182. United Nations Environment Programme, Nairobi, pp 103–115

Holness S, Kirkman S, Samaai T, et al (2014) Spatial biodiversity assessment and spatial management, including marine protected areas. Final report for the Benguela Current Commission project BEH 09-01, 105 pp + annexes

Hughes RH, Hughes JS (1992) A directory of African wetlands. IUCN/UNEP/WCMC, Gland/Cambridge/Nairobi/Cambridge, xxiv + 820 pp, 48 maps

Hutchings K, Clark B, Steffani, Anderson R (2007) Identification of communities, biotopes and species in the offshore areas and along the shoreline and in the shallow subtidal areas in the BCLME region. Section B. Angola field trip report. Final report for Benguela Current Large Marine Ecosystem Programme project BEHP/BAC/03/03

Hutchings L, van der Lingen CD, Shannon LJ et al (2009) The Benguela current: an ecosystem of four components. Prog Oceanogr 83:15–32

Japp DW, Purves MG, Wilkinson S (eds) (2011) State of stocks review. Report No. 2 (Updated by C Kirchner). Benguela Current Large Marine Ecosystem State of Stocks Report 2011, 105 pp

John DM, Lawson GW (1991) Littoral ecosystems of tropical western Africa. In: Mathieson AC, Nienhuis PH (eds) Ecosystems of the world, vol 24. London. New York, Tokyo, pp 297–321

Kensley B, Penrith ML (1980) The constitution of the fauna of rocky intertidal shores of south West Africa. Part III. The north coast from False Cape Frio to the Kunene River. Cimbebasia (Series A) 5:201–214

Kirkman SP, Blamey L, Lamont T et al (2016) Spatial characterisation of the Benguela ecosystem for ecosystem based management. Afr J Mar Sci 38:7–22

Kirkman SP, Yemane D, Kathena J et al (2013) Identifying and characterizing of demersal biodiversity hotspots in the BCLME: relevance in the light of global changes. ICES J Mar Sci 70:943–954

Kopte R, Brandt P, Dengler M et al (2017) The Angola current: flow and hydrographic characteristics as observed at 11°S. J Geophys Res Oceans 122:1177–1189

Lagabrielle E (2011) A pelagic bioregionalisation of the Benguela current system. Appendix 4 In: Holness S, Kirkman S, Samaai T, et al (2014) Spatial Biodiversity Assessment and Spatial Management, including Marine Protected Areas. Final report for the Benguela Current Commission project BEH 09-01, 105 pp + annexes

Lass HU, Schmidt M, Mohrholz V et al (2000) Hydrographic and current measurements in the area of the Angola-Benguela front. J Phys Oceanogr 30:2589–2609

Lawson GW (1978) The distribution of seaweed floras in the tropical and subtropical Atlantic Ocean: a quantitative approach. Bot J Linn Soc 76(3):177–193

Lawson GW, John DM, Price JH (1975) The marine algal flora of Angola: its distribution and affinities. Bot J Linn Soc 70(4):307–324

Le Guilloux E, Olu K, Bourillet JF et al (2009) First observations of deep-sea coral reefs along the Angola margin. Deep Sea Research Part II: Tropical Studies in Oceanography 56:2394–2403

Long RD, Charles A, Stephenson RL (2015) Key principles of marine ecosystem-based management. Mar Policy 57:53–60

Meeuwis JM, Lutjeharms JRE (1990) Surface thermal characteristics of the Angola-Benguela front. S Afr J Mar Sci 9:261–279

Monteiro PMS, van der Plas AK, Mélice J-L et al (2008) Interannual hypoxia variability in a coastal upwelling system: ocean-shelf exchange, climate and ecosystem-state implications. Deep-Sea Res I 55:435–450

Nsiangango S, Shine K, Clark B (2007) Identification of communities, biotopes and species in the offshore areas and along the shoreline and in the shallow subtidal areas in the BCLME region. Section C3. Biogeographic patterns and assemblages of demersal fishes on the coast of Angola. Final report for Benguela Current Large Marine Ecosystem Programme project BEHP/BAC/03/03

O'Toole MJ (1980) Seasonal distribution of temperature and salinity in the surface waters off south West Africa, 1972–1974. Investig Rep S Afr Sea Fish Inst 121:1–25

Penrith MJ (1978) An annotated check-list of the inshore fishes of southern Angola. Cimbebasia (Series A) 4:179–190

Potts WM, Childs AR, Sauer WHH et al (2009) Characteristics and economic contribution of a developing recreational fishery in southern Angola. Fish Manag Ecol 16:14–20

Rex MA, Stuart CT, Coyne G (2000) Latitudinal gradients of species richness in the deep-sea benthos of the North Atlantic. Proc Natl Acad Sci USA 97:4082–4085

Rull Lluch JR (2002) Marine benthic algae of Namibia. Sci Mar 66. (suppl. 3:5–256

Sherman K (2014) Toward ecosystem-based management (EBM) of the world's large marine ecosystems during climate change. Environ Dev 11:43–66

Shillington FA, Reason CJC, Duncombe Rae CM et al (2006) Large scale physical variability of the Benguela Current Large Marine Ecosystem (BCLME). In: Shannon V, Hempel G, Malanotte-Rizzoli P et al (eds) Benguela: predicting a large marine ecosystem, vol 14. Elsevier, Amsterdam, pp 49–70

Signorini SR, Murtuguddo RG, McClain CR et al (1999) Biological and physical signatures in the tropical and subtropical Atlantic. J Geophys Res 104:18367–18382

Simmons RE (2010) First breeding records for Damara Terns and density of other shorebirds along Angola's Namib Desert coast. Ostrich 81:19–23

Spalding MD, Fox HE, Allen GR et al (2007) Marine ecoregions of the world: a bioregionalization of coastal and shelf areas. Bioscience 57:573–583

Tarr P, Krugmann H, Russo V, Tarr J, et al (2007) Analysis of threats and challenges to marine biodiversity and marine habitats in Namibia and Angola. Final report for Benguela Current Large

Marine Ecosystem Programme project BEHP/BTA/04/01. 132 pp + annexes

Van den Hoek C (1975) Phytogeographic provinces along the coast of the northern Atlantic Ocean. Phycologia 14:317–330

Van Niekerk L, Neto DS, Boyd AJ, et al (2008) Baseline surveying of species and biodiversity in estuarine habitats. BCLME project BEHP/BAC/03/04. 118 pp + appendices

Veitch JA, Penven P, Shillington F (2010) Modeling equilibrium dynamics of the Benguela Current System. J Phys Oceanogr 40:1942–1964

Whitfield AK (2005) Preliminary documentation and assessment of fish diversity in sub-Saharan African estuaries. Afr J Mar Sci 27(1):307–324

Willig MR, Kaufman DM, Stevens RD (2003) Latitudinal gradients of biodiversity: pattern, process, scale, and synthesis. Annu Rev Ecol Evol Syst 34:273–309

Yemane D, Mafwila SK, Kathena J et al (2015) Spatio-temporal trends in diversity of demersal fish species in the Benguela Current Large Marine Ecosystem (BCLME) region. Fish Oceanogr 24. (Suppl. 1:102–121

4

Angola and its Alpha Paleobiodiversity

Octávio Mateus, Pedro M. Callapez, Michael J. Polcyn, Anne S. Schulp, António Olímpio Gonçalves and Louis L. Jacobs

Abstract This chapter provides an overview of the alpha paleobiodiversity of Angola based on the available fossil record that is limited to the sedimentary rocks, ranging in age from Precambrian to the present. The geological period with the highest paleobiodiversity in the Angolan fossil record is the Cretaceous, with more than 80% of the total known fossil taxa, especially marine molluscs, including ammonites as a majority among them. The vertebrates represent about 15% of the known fauna and about one tenth of them are species firstly described based on specimens from Angola.

O. Mateus (✉)
GeoBioTec, Faculdade de Ciências e Tecnologia, Universidade Nova de Lisboa, Lisbon, Portugal

Museu da Lourinhã, Rua João Luis de Moura, Lourinhã, Portugal
e-mail: omateus@fct.unl.pt

P. M. Callapez
CITEUC; Departamento de Ciências da Terra, Faculdade de Ciências e Tecnologia, Universidade de Coimbra, Coimbra, Portugal
e-mail: callapez@dct.uc.pt; jacobs@smu.edu

M. J. Polcyn · L. L. Jacobs
Roy M. Huffington Department of Earth Sciences, Southern Methodist University, Dallas, TX, USA
e-mail: mpolcyn@smu.edu

A. S. Schulp
Naturalis Biodiversity Center, Leiden, The Netherlands

Faculty of Earth and Life Sciences, VU University Amsterdam, Amsterdam, The Netherlands
e-mail: anne.schulp@naturalis.nl

A. O. Gonçalves
Departamento de Geologia, Faculdade de Ciências, Universidade Agostinho Neto, Luanda, Angola
e-mail: antonio.goncalves@geologia-uan.com

Keywords Ammonites · Benguela Basin · Cenozoic · Cretaceous · Cuanza Basin · Dinosaur · Invertebrates · Mammals · Mollusca · Mosasaur · Namibe Basin · Paleobiodiversity · Pleistocene · Vertebrate · Plesiosaur · Turtle

Studies of Paleobiodiversity

The study of paleobiodiversity, i.e., the development of biodiversity through geological time, is challenging at multiple levels. In addition to the issues and biases affecting the study of the diversity of modern life, understanding paleobiodiversity faces extra challenges, mostly because of the dependency on the fossil record. Glimpses of entire ecosystems and clades may never reach the paleontologist's eyes if appropriate rocks of that exact time and space were not formed, or if formed, did not preserve fossils, or are eroded away or otherwise inaccessible (see Jackson and Johnson 2001; Crampton et al. 2003).

The study of life's diversity in the past is filtered by the remains that can leave traces and fossilize, remains that actually fossilized, fossils existing today, fossils accessible today, fossils collected (number of fossils accessible to scientists), and species recognised (Fig. 4.1). Moreover, the definition and discrimination of species in the fossil record can be problematic.

Angola has no known fossiliferous rocks from the Paleozoic (541–251 Ma – millions of years ago) nor the Jurassic (199–145 Ma) leaving only windows to life in the territory during the Triassic (251–199 Ma), Cretaceous (145–66 Ma), and Cenozoic (66 Ma -Present) (Fig. 4.2). The chances of finding Paleozoic or Jurassic fossils from Angola are essentially zero. Thus, within the last 550 Ma, the known

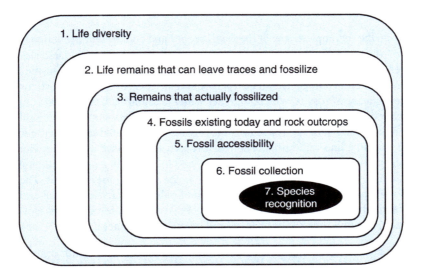

Fig. 4.1 The seven layers of filters of uneven preservation in the fossil record that obscure accurate reconstruction of paleobiodiversity

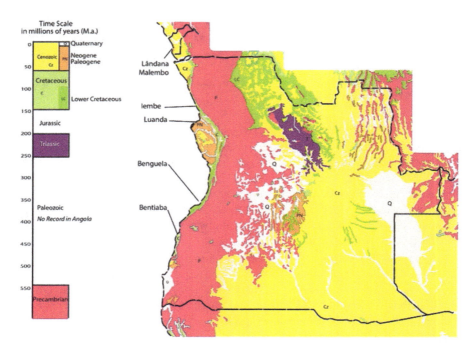

Fig. 4.2 The geological record of Angola, in a stratigraphic log (left) and geological map (right), leaves more than 350 Ma of blank geological record. Map extracted after Africa Geological Map 1:30.000.000 by U.S. Geological Survey, 2002 available at www.uni-koeln.de/sfb389/e/e1

rocks of Angola represent less than 196 million years of geologic time, leaving more than 354 million years (64% of the time) with no known fossil record.

Despite the incompleteness of the fossil record and consequent limitations to the study of the paleobiodiversity, cooperative research and modern databases can, however, improve approximation of the estimated number of species in the fossil record. The PaleoBiology Database (paleobiodb.org) is, by far, the most comprehensive database of fossils, which, together with the scientific literature and our own research, contributed to the Supplementary Material and its summary in Table 4.1 that compiles the list of the fossil taxa in Angola, with updated taxonomy, geological age, locality, and references. The fossil record can be used as a lower limit for the alpha paleobiodiversity for specific times and locations in Angola, although it is likely an underestimate of true paleodiversity in the vast majority of cases. The total of all fossil species is a gross underestimation of paleodiversity for the full extent of the time involved, exacerbated by missing intervals of fossiliferous rocks. However, the pattern through time can inform an understanding of general trends.

Fossil collecting in Angola has been conducted since the nineteenth century with Friederich Welwitsch, José de Anchieta (1885), Freire de Andrade, Augusto Eduardo Neuparth and others (Brandão 2008, 2010; Silva and Geirinhas 2010; Callapez et al. 2011; Masse and Laurent 2016). Numerous paleontologists have contributed to the

Table 4.1 Summary of the known fossil record of Angola

COUNT (Genus/sp, Family Indet)		Period							
		Triassic	Cretaceous	Paleogene	Neogene	Quaternary	Triassic	Grand Total	
Bacteria	Cyanobacteria		1					1	0%
Bacterial total			1					1	0 %
Plantae	Chlorophyta		1					1	0%
	Cycadophyta		1					1	0%
	Ginkgophyta		1					1	0%
	Pinophyta	1	1				1	2	0%
	Pteridosperma-tophyta	4					4	4	0%
	Rhodophyceae		2					2	0%
Plantae total		5	6				5	9	1%
Protista and Invertebrates	Foraminifera		207					207	16%
	Anthozoa		1	1	2	2		6	0%
	Brachiopoda		1					1	0%
	Mollusca		677	27	22	72		798	61%
	Echinodermata		61	1	4	1		67	5%
	Arthropoda	13	3	1	1	2	13	20	2%
Protista and invertebrates total		13	950	30	29	77	13	1102	84%
Vertebrata	Chondrichthyes	2	63	1	18		2	84	6%
	Actinopterygii	6	10		5		6	21	2%
	Sarcopterygii	3					3	3	0%
	Reptilia		21	5				26	2%
	Mammalia			12	1	54		67	5%
Vertebrata total		11	94	18	24	54	11	201	15%
Grand Total		**29**	**1052**	**48**	**53**	**131**	**29**	**1313**	**100%**
		2%	80%	4%	4%	10%	2%	100%	

understanding the paleobiodiversity of Angola, since the first explorations and studies: Fernando Mouta, Paul Choffat, Carlos Teixeira, Carlos Freire de Andrade, Edgard Casier, A Jamotte, M Leriche, Heitor de Carvalho, E Dartevelle, Miguel Telles Antunes, Alexandre Borges, Philippe Brebion, Gaspar Soares de Carvalho, Louis Dollo, Henri Douvillé, Otto Haas, Manuel Mascarenhas Neto, Arménio Tavares da Rocha, Gumerzindo Henriques da Silva, LF Spath, António Ferreira Soares, Maurice Collignon, among many others (see bibliography compiled by Nunes 1991). In vertebrate paleontology the work of Miguel Telles Antunes and co-authors is noteworthy. Today, various researchers work on the paleontology of Angola, among them is the team of the Projecto PaleoAngola (paleolabs.org/paleoangola), with regular yearly scientific expeditions since 2005 (Jacobs et al. 2006, 2016).

A Brief Geological History and Context of Angola

The most significant geological event governing the paleogeography of Angola is the opening of the South Atlantic Ocean, in which Africa and South America rifted apart beginning in the Early Cretaceous Epoch about 134 million years ago and the subsequent drifting apart of these continents as the South Atlantic grew (Guiraud and Maurin 1991; Buta-Neto et al. 2006; Quirk et al. 2013; Pérez-Díaz and Eagles 2017). After about 120 million years ago, marine deposition along the coast began to preserve fossils. Africa's place in Gondwana prior to this time resulted in the lack of a marine record for the entire Paleozoic Era and the consequent lack of a fossil record for that time.

As the South Atlantic opened it was colonised by species moving in from the southern ocean and, as a connection between the North Atlantic and South Atlantic oceans developed, from the north. Sea turtles (*Angolachelys mbaxi*) and mosasaurs (*Angolasaurus bocagei* and *Tylosaurus iembeensis*), with relatives to the north, first occur at about 88 million years ago. Along with them were plesiosaurs, probably with southern affinities. Washed into the sea and found with marine creatures, is the sauropod dinosaur *Angolatitan adamastor*, which is probably a remnant of a more broadly distributed Gondwanan dinosaur assemblage. At that time, coastal Angola lay approximately 10–12 degrees further south than today.

Geologic uplift along the coast resulted in the erosion of rock and loss of the fossil record from effected strata. Permian and Jurassic uplift eliminated those intervals from the terrestrial fossil record. Early and Late Cretaceous uplift also occurred, prior to burial of the remaining Cretaceous with up to 1.5–2 km of sediments, since removed by erosion accompanying mid-Cenozoic uplift beginning around 30 Ma and 20 Ma (Green and Machado 2015). Uplift from as young as 45,000 years ago also has been recorded along the Angolan coast (Walker et al. 2016), resulting in a large diversity of fossiliferous raised-beach deposits, frequently associated with pre-historic shell-middens with Paleolithic industries.

The Precambrian – The First Fossils

Precambrian Era (geologic time since the formation of Earth and prior to 541 Ma) rocks worldwide are mostly devoid of fossils as life was unicellular for most time and only macroscopic in the last stages. The shale-limestone series of the Bembe System in Angola includes dolomitic limestones, mostly devoid of fossils, but containing levels with concentrations of coalescent stromatolites (structures due to cyanophilic activity) attributable to the genera *Collenia* and *Conophyton* in Mavoio, Alto Zambeze, and Humpata (Vasconcelos 1951; Antunes 1970; Duarte et al. 2014).

In Angola, no fossils are known from the Paleozoic (Cambrian through Permian periods), which represents a time gap of more than 290 million years (Fig. 4.2).

The Triassic – Inner Basins

The Triassic is a geological period stretching from about 250 to 200 Ma. The begin_ning and end of this interval are both marked by mass extinction events, the older Permian-Triassic extinction event, concomitant with the Siberian Magmatic event, marking the initiation of the Triassic Period. The vast supercontinent Pangea existed until the Triassic, after which it gradually began to break-up apart, separating the two masses of land, Laurasia to the north and Gondwana to the south. The global climate during the Triassic Period was warm and dry, with deserts covering much of the interior of Pangea. The end of the Triassic was marked by another major mass extinction related with the Central Atlantic Magmatic Province and the early open_ing of the North Atlantic. Therapsids (a large group containing mammals and their extinct relatives) and archosaurs (dinosaurs, birds, crocodiles and their relatives) were the major terrestrial vertebrates during this time. The dinosaurs first appeared in the Triassic. The first true mammals, which are derived therapsids, evolved dur_ing this period, as well as the first flying vertebrates, the pterosaurs.

Triassic outcrops of Angola are restricted to the Baixa de Cassange (Cassange Depression) in Malanje and Lunda-Norte in rocks referred to the Karoo Supergroup geological unit.

Plants and Invertebrates

The Triassic paleoflora of Angola includes the extinct genera *Glossopteris*, *Sphenopteris*, and *Noeggerathiopsis* (Teixeira 1948a, 1961). A few fossils indicate a freshwater environment and a Triassic age. These include the coleopteran Coptoclavidae insect *Coptoclavia africana* and 12 conchostracan crustaceans: *Estheriella moutai* Leriche 1932, *E. cassambensis* Teixeira 1958, *Estheria anchietai* Teixeira 1947, *E. (Echinestheria) marimbensis* Marliére 1950, *E. (Euestheria) man-galiensis* Jones 1862, *Palaeolimnadiopsis reali* Teixeira 1958, *Palaeolimnadia*

(Palaeolimnadia) wianamattensis (Mitchell), *P. (Grandilimnadia) oesterleni* Tasch 1987, *P. (G.) africania* Tasch 1987, *Gabonestheria gabonensis* (Marliére 1950), *Cornia angolata* Tasch 1987, and *Estheriina (Nudusia)* cf. *rewanensis* Tasch 1979.

Vertebrates

The only Triassic vertebrates known from Angola are fishes, including the Elasmobranchii *Lissodus cassangensis,* the paleoniscoids *Perleidus lutoensis* Teixeira 1947, Palaeonisciformes canobiid *Marquesia moutai,* Halecostomi *Angolaichthys lerichei* Teixeira, the ray-finned *Teffichthys lehmani* and *T. lutoensis,* and the Sarcopterygii lungfish *Microceratodus angolensis.*

This faunal assemblage indicates a freshwater environment with insects and a nearly exclusively endemic fauna. Based on the fishes, a Lower Triassic age (252–247 Ma) is indicated for Lunda and Baixa de Cassange rocks (Murray 2000; Antunes et al. 1990). No tetrapods have been collected so far.

The Early Cretaceous – The Opening of South Atlantic

Most of the fossils and outcrops of Cretaceous age in Angola are in Mesozoic-Cenozoic basins of coastal Angola: the Cabinda, Zaire, Cuanza, Benguela, and Namibe sedimentary basins (Antunes 1964; Séranne and Anka 2005; Guiraud et al. 2010), bordered by basement rocks. Almost all formations are mostly marine except for the ichnofauna from the Catoca mine (Marzola et al. 2014; Mateus et al. 2017), in Lunda-Sul. The oldest fossiliferous Cretaceous formations seen in outcrop seem to be Barremian to Aptian lacustrine deposits that contain unidentified gastropods (Ceraldi and Green 2016).

Plants, Protists and Invertebrates

Cretaceous plant remains in Angola seem to be rare (see Supplementary Material), but several examples of unstudied field contexts with transitional facies, namely above the Cuvo units, might come to reveal new fossil sites. The same situation is likely for palynomorphs and dinoflagellates of Early Cretaceous and more recent ages. Plants are known from pollen taxa such as *Classopolis* sp. and *Eucommiidites* sp. *Pachypteris montenegroi* Teixeira 1948 is a Ginkgophyta Umkomasiaceae from Early Albian lagoonal sediments of the Cuanza Basin (Teixeira 1948b; Antunes 1964; Neto 1970; Nunes 1991) and first found in Angola. One species of Chlorophyta, one Cycadophyta, one Rhodophyceae, one Ginkgophyta, and two Pinophyta have been reported from Cabinda, Benguela and Cuanza Basin (Antunes 1964; Neto 1970; Nunes 1991; Araújo and Guimarães 1992; Tavares 2006).

More than 200 foraminifera taxa were identified in the Cretaceous of Angola, and many indicate an Albian age, such as *Globotruncana ventricosa* (Rocha 1984; Antunes and Cappetta 2002; Antunes 1964; Jacobs et al. 2006).

Crustaceans are known from the Decapoda *Parapirimela angolensis* Van Straelen 1937 from the Albian of Iela beach, Benguela Basin (Ferreira 1957; Van Straelen 1937; Antunes 1964) and the ostracods *Chloridella angolia* and *Petrobrasia tenuistriata longinsuela* from Quiçama, Cuanza Basin, (Berry 1939; Antunes 1964) and Cabinda (Araújo and Guimarães 1992), respectively.

The Early Cretaceous of Angola yielded fossils of seven species of Scleractinean corals, one of brachiopod and eight crustaceans.

Of the more than known 600 species of molluscs from the Cretaceous of Angola, the vast majority are ammonites, many are unique to Angola and received related specific epithets such as the *Anisoceras teixeirai, Durnovarites autunesi* Collignon 1978, *Durnovarites netoi, Elobiceras lobitoense* Spath 1922, *Hamitoides angolanus* (Tavares 2006), followed by bivalves such as *Neithea angoliensis* Newton 1917, and gastropods such as *"Cerithium" monteroi* Choffat. By far, the ammonites are the most relevant portion of the Cretaceous biodiversity of Angola (Tavares et al. 2007; Haas 1942, 1943) and also age-indicators for geologists and paleontologists. Haas (1942) described and reported many ammonites from the Albian, some as new species including *Hysteroceras falcicostatum* Haas 1942 and *H. intermedium* Haas 1942. See the Supplementary Material for the complete list.

Echinoderms, mostly echinoids, are remarkably common in the Early Cretaceous of Angola, with about 50 taxa known, but that number depends on the validity and synonymisation of the taxa addressed (see the Supplementary Material for the complete list). A few echinoderms received names after Angolan toponyms and researchers such as the *Douvillaster benguellensis* Loriol 1888 and *D. carvalhoi* Loriol 1888 and *Epiaster catumbelensis* Loriol 1888, and *Holaster domboensis* Loriol 1888 from the Lower Cretaceous of Dombe Grande, Catumbela and Praia da Hanha (Loriol 1888; Ferré and Granier 2001; Tavares 2006; Tavares et al. 2007).

Vertebrates

In the Catoca Diamond Mine, in Lunda-Sul Province, mammaliamorph, crocodylomorph, and sauropod tracks were discovered in Early Cretaceous crater lake sediments. One sauropod track has skin impressions preserved. These are the only fossil vertebrate tracks known in Angola. The most surprising feature is the unexpectedly large size of the mammaliamorph footprints considering the age (Mateus et al. 2017). The Catoca Diamond Mine has an eruption age of around 118 Ma (Aptian; Robles-Cruz et al. 2012). A fragment of a caudal vertebra of a sauropod dinosaur was recovered from Tzimbio, in northern Namibe, from strata that are likely Albian in age.

The Late Cretaceous – Marine Reptiles Flourish

The Late Cretaceous is the time of the Cretaceous Period between 100.5 Ma and 66 million years ago. It is subdivided into the Cenomanian, Turonian, Coniacian, Santonian, Campanian and Maastrichtian ages, from the oldest to the most recent. The climate was warmer than present, although with a cooling trend throughout the period. In the oceans, where the sea-level was much higher than today, mosasaurs (a group of marine lizards) suddenly appeared and underwent a spectacular evolutionary radiation (Polcyn et al. 2014). Modern grade sharks also appeared and plesiosaurs diversified. These predators fed on the numerous teleost fishes, which in turn evolved into new advanced and modern forms (Neoteleostei). Ichthyosaurs and pliosaurs (a group of short-necked plesiosaurs), on the other hand, became extinct during the Cenomanian-Turonian anoxic event (Schlanger et al. 1987) and are not known from Angola. The end of Cretaceous is marked by the mass extinction of some three-quarters of plant and animal species on Earth, known as the Cretaceous-Paleogene (K/Pg) event (Archibald et al. 2010).

Protists and Invertebrates

The Late Cretaceous Foraminifera of Angola were studied by various researchers including Ferreira and Rocha (1957), Lapão and Simões (1972), Rocha (1984) and Blake et al. (1996) who list more than 180 taxa. Foraminifera are known from 15 taxa of Granuloreticulosea, such as the Gavelinellidae *Anomalia berthelini* from the Aptian to Cenomanian of Cabinda, Cuanza and Benguela Basins (Araújo and Guimarães 1992; Rocha 1984).

Molluscs of the Late Cretaceous count more than 240 known taxa, mostly ammonites (see Supplementary Material for the full list). Some were named after Angolan toponyms or after paleontologists that worked in Angola (Borges 1946; Carvalho 1961; Haas 1943; Howarth 1965; Cooper 1972, 1982; Cooper and Kennedy 1979): *Acera choffati* Rennie 1945, *Axonoceras angolanum* Haas 1943, *Didymoceras* cf. *angolaense* Haughton 1924 (Howarth 1965), *Eutrephoceras egitoense* Miller and Carpenter 1956, *Kitchinites angolaensis* Howarth 1965, *Libycoceras dandense* Howarth 1965, *Lucina egitoensis* Rennie 1945, *L. angolensis* Rennie 1929, *Mammites mocamedensis* Howarth 1966, *Nostoceras mariatheresianum* Haas 1943, *Oiophyllites angolaensis* Spath 1953, *Prionocyclus carvalhoi* Howarth 1966, *Protacanthoceras angolaense* Spath 1931, *Prohysteroceras hanhaense* (Haas 1942), *P. angolaense* (Boule et al. 1907), *Protocardia moutai* Rennie 1945, *Pseudocalycoceras angolaense* (Spath 1931), *Pseudomelania salenasensis* Rennie, *Pterotrigonia borgesi* Rennie 1945, *Mortoniceras (Angolaites) stolikzcai* (Spath 1922), *Mortoniceras (?) rochai* Collignon 1978, *M. (Deiradoceras) reali* Collignon 1978, *Collignoniceras (Selwynoceras) reali* Collignon 1978, and *Solenoceras bembense* Haas 1943. The ammonites are found in almost all Upper Cretaceous sections, including those of the Quissonde Formation from the beaches of Quimbala, Chamure, Cabeça da Baleia, Egito in Benguela Basin, Teba, Bembe in Cuanza Basin, Bentiaba and Salinas in Namibe, and Iembe in Bengo Province (Segundo et al. 2014).

Rennie (1945) describes ten species of gastropods and bivalves from Cabeça da Baleia, Egito-Praia: *Trigonia (Scabrotrigonia) borgesi, Lucina egitoensis, Protocardia moutai, Pseudomelania egitoensis, Confusiscala angolensis, Acirsa (Plesioacirsa?) egitoensis, Dicroloma (Perissoptera) o'donnelli, Paleopsephaea o'donnelli, Acera choffati,* and *Ringicula moutai* (Lapão and Pereira 1971).

More than ten echinoderm taxa are known from the Late Cretaceous of Angola, mostly echinoids, some received Angola-related names, such as the Toxasteridae *Epiaster angolensis* Haughton 1924, collected 150 m below the Itombe Formation at Zenza do Itombe, and *E. carvalhoi* Dartevelle 1953 (Haughton 1924; Kier and Lawson 1978; Néraudeau and Mathey 2000), *Leiostomaster angolanus* Greyling and Cooper 1995, Palaeostomatidae and *Tholaster carvalhoi* Greyling and Cooper 1995, and Holasteridae from the Middle Campanian of Egito Praia, near Quimbala.

Vertebrates

In Angola, a peak of vertebrate paleobiodiversity is observed in the Late Cretaceous, with more than 100 taxa recognised (Antunes 1964; Antunes and Cappetta 2002; Jacobs et al. 2006, 2009, 2016). This peak is in comparison with other time intervals and is likely an artifact of the inadequacies of the fossil record rather than a statement of biological reality. Most known Late Cretaceous sedimentary rocks from Angola are marine, thus the fossil assemblage also reflects this ecosystem. Non-marine organisms are occasionally drifted into a marine setting, including the sauropod dinosaur *Angolatitan adamastor* Mateus et al. (2011) from Iembe, in Bengo Province, originally considered Turonian in age but following the discovery of the ammonite *Protexanites* sp. at Iembe, it is now considered Coniacian. Pterosaurs are known from the Maastrichtian marine sediments of Bentiaba, in Namibe Province. *Angolatitan adamastor* is the first dinosaur found in Angola. Its generic name means 'Titan of Angola' and the specific name refers to Adamastor made famous by Luís de Camões in the *Lusíadas*. It was 13 m long and lived in an arid environment (Mateus et al. 2011). A number of bones of large flying reptiles, pterosaurs, collected in the marine sediments of the Maastrichtian of Bentiaba are the only know pterosaur remains in sub-saharian Africa for this stage.

The fossil record of chondrichthyans shows a peak in Late Cretaceous deposits with 64 species known, including taxa only known in Angola or with Angolan toponyms such as *Angolabatis angolensis, A. benguelaensis, Chlamydoselachus gracilis, Cretascymnus quimbalaensis,* and *Echinorhinus lapaoi* named by Antunes and Cappetta (2002). The diversity of chondrichthyans includes representatives of most of the Cretaceous lineages: Hexanchiformes, Squaliformes, Rajiformes, Orectolobiformes, Odontaspidida, Lamniformes, Carcharhiniformes, and Squaliformes (Antunes and Cappetta 2002).

In contrast to the diversity of chondrichthyans, the Late Cretaceous bony ray-finned fish from Angola are represented by ten taxa only, most likely due to the difficulty of identification of isolated remains, in comparison with the abundance of shark and ray teeth. The teleost genus *Enchodus* is known from various species (*Enchodus bursauxi, E. crenulatus, E. elegans, E. faujasi* and *E. libycus*). Other teleosts include *Eodiaphyodus lerichei, Pseudoegertonia bebianoi*, and *Stephanodus libycus* from the Late Campanian and Maastrichtian in Benguela and Namibe provinces (Antunes and Cappetta 2002) (Fig. 4.3).

Two Upper Cretaceous fossil localities from Angola are remarkably rich in large reptiles: Iembe in Bengo Province (Fig. 4.4) of Coniacian age (~88 Ma), and Bentiaba in Namibe (Maastrichtian). The Coniacian provided the bizarre durophagous cryptodiran turtle *Angolachelys mbaxi* Mateus et al. (2009) that justified the erection of its own clade Angolachelonia, and the sauropod dinosaur *Angolatitan adamastor*. Indeterminate plesiosaur remains have also been recovered from this region. Mosasaurs are represented by *Angolasaurus bocagei* (Fig. 4.3), *Tylosaurus iembeensis* Antunes (1964) and an indeterminate halisaurine.

The richest vertebrate locality in Angola is near Bentiaba in Namibe province (Strganac et al. 2014, 2015a, b). The reptile list includes the cheloniid turtles *Euclastes* sp., *Protostega* sp. and *Toxochelys* sp. Terrestrial reptiles include isolated remains that allow undetermined large pterosaurs, an undetermined sauropod based on a metapodial, and a possible hadrosaur based on an isolated phalanx (Mateus et al. 2012). The squamate mosasaurs are, by far, the most abundant and speciose group of tetrapods, and include *Globidens phosphaticus, Prognathodon kianda, Halisaurus* sp., *Mosasaurus* sp., *Phosphorosaurus* sp., and '*Platecarpus*' *ptychodon* (Polcyn et al. 2007, 2010, 2014; Schulp et al. 2006, 2008, 2013). The plesiosaur

Fig. 4.3 The mosasaur *Angolasaurus bocagei* skull and anterior postcrania (Museum of Geology, Universidade Agostinho Neto). (Photo: Hillsman Jackson, Southern Methodist University)

Fig. 4.4 Reconstruction of the fauna during the Late Cretaceous, based on the Coniacian fauna of Iembe. Illustration using acrylic painting with brush by Fabio Pastori

Elasmosauridae are also abundant at Bentiaba where two taxa are known: Aristonectinae indet. and *Cardiocorax mukulu* (Araújo et al. 2015a, b).

The Paleogene – Mammals Take Over

The Paleogene Period begins at the end of the Cretaceous (66 Ma) and lasted until the Neogene Period 23.03 Ma. The Paleogene is the interval of Earth's history in which mammals diversified and flourished after the K/Pg mass extinction, when most large reptiles, and belemnites and ammonites went extinct. There is a dearth of terrestrial vertebrates, especially mammals, from the Paleogene of Angola. During the global Paleocene-Eocene Thermal Maximum (PETM), 55.8 million years ago, there occurred a sudden change of the climate that marked the end of the Paleocene and the beginning of the Eocene, one of the most significant periods of climate change in the Cenozoic era (Zachos et al. 2005). The best known Paleogene geologic section in Angola is that of Lândana in Cabinda Province. A recent study of the Paleocene and Eocene biota and strata of Lândana indicated that the PETM is missing from that section, either because the event was too short to be recorded at the sampling interval used or that it falls within one of the several stratigraphic gaps documented in the Lândana section (Solé et al. 2018).

Protista and Invertebrates

Among the protists, foraminifera stand out as the most important and well-known taxonomic group in the Angolan marine series of Paleogene age, due to their importance for biostratigraphic correlations in offshore oil drilling and their equivalence with landward outcrops. Important works include those of Rocha (1973) and Kender et al. (2008), which include many Eocene and Oligocene characteristic taxa such as: *Cyclammina* cf. *compressa*, *Nonion centrosulcatum*, *Cassidulina subglobosa*, *Globigerina ampliapertura*, *Bolivina* cf. *pygmaea*, *Bulimina alsatica*, *B. kacksonensis* and *B. nkomi*.

In the Paleogene of Angola, molluscs remain the most speciose clade of invertebrates, comprising at least three nautiloids, 14 bivalves, and 21 gastropods (see Supplementary Material). The bivalves are mainly known from the Eocene (Lutetian) Quimbriz Formation along the Luculo River (Tavares et al. 2007) and include *Leda africae*, *Noetia veatchi*, '*Cardium' luculensis*, '*C.' sandigii*, *Crassatella schoonoverae*, *Lucina* cf. *landanensis*, *Macrocallista palmerae*, *Metis olssoni*, *Pitar quimbrizensis*, *P. quipayensis*, *Protonoetia nigeriensis*, *Raetomya schweinfurthi*, *Venericardia angolae*, and *V. heroyi* (Tavares et al. 2007). The three nautiloids are *Cimomia landanensis*, *Deltoidonautilus caheni*, *Hercoglossa diderrichi* from the Danian of Cabinda Basin, Landana (Soares 1965), and the gastropods are also mainly known from the Eocene, Lutetian of Luculo River (Tavares et al. 2007): *Ficula roscheni*, *Fulguroficus harrisi*, *Pleurotoma angolae*, *P. rebeccae*, *Polinices (Neverita) angolae*, *Ringicula hughesae*, *Sinum dusenberryi*, *Surcula* cf. *ingens*, and *Turricula (Knefastia) angolensis*.

Other groups such as Anthozoa, Arthropoda and Echinodermata, exist but are low in numbers (Dartevelle 1953).

Vertebrates

Paleogene marine sediments have been studied mostly in the Benguela Basin and in Cabinda. Adnet et al. (2009) recognized a new species of Eocene lamniform shark, *Xiphodolamia serrata*, from Benguela. Sharks and bony fishes from Cabinda have been listed by Solé et al. (2018), and Taverne (2016) has provided new information about osteoglossid fishes from Lândana. The tetrapods from the Paleogene of Angola are mostly from the fossil sites of Lândana (Solé et al. 2018) and Malembo in Cabinda. The section begins with Lower Paleocene strata at Lândana. Reptiles comprise the dyrosaurid crocodylomorph *Congosaurus bequaerti*, and indeterminate crocodilians (Jouve and Schwarz 2004; Schwarz 2003; Schwarz et al. 2006), the turtles *Taphrosphys congolensis*, a toxochelyid, and *Cabindachelys landanensis* (Myers et al. 2017). Along the Chiloango River, Cabinda, a vertebra of the snake *Palaeophis* was reported by Antunes (1964).

The youngest portion of the Cabinda section is at Malembo Point, south of Lândana, which was originally considered Miocene. The Malembo mammal fauna is comparable to the Early Oligocene fauna of the Fayum, Egypt because of the presence of the embrithopod *Arsinoitherium*, hyracoids such as *Geniohyus* aff. *Mirus* and *Bunohyrax* aff. *Fajumensis*, the proboscidean cf. *Phiomia* or *Hemimastodon,* the sirenian *Halitherium,* and a reported anthropoid canine (Hooijer 1963; Pickford 1986; Jacobs et al. 2016). Recent findings by *Projecto PaleoAngola* in Malembo include a ptolemaiidan molar more similar to *Kelba* from the Miocene of Songhor, Kenya (19.5 Ma) than to Fayum *Ptolemaia*, and an isolated premolar tooth of a large primate comparable in size to that of a female gorilla, likely an undescribed taxon (Jacobs et al. 2016), and not representing any of the numerous Fayum primate taxa. In addition, *Arsinoitherium* is now known from the latest Oligocene of Kenya. The presence of a Kelba-like ptolemaiadan, a unique primate, and Arsinoitherium may indicate a latest Oligocene or even earliest Miocene age for Malembo Point. The assemblage certainly has differences from the Fatum fauna and may indicate the presence of a lowland West Africa faunal province near the Paleogene-Neogene boundary in age and distinct from other regions such as the East African Rift Valley or the Fayum.

The Neogene – The Founding of Modern Biodiversity

The Neogene began about 23 Ma and extends until the Pleistocene (1.8 Ma). It is divided into Miocene (23 Ma to 5.3 Ma) and Pliocene (5.3 Ma to 2.6 Ma), from the oldest to the more recent. This period saw the expansion of the large mammals, and the appearance of hominids. In the Miocene the climate warms again and grasslands and savannas spread. In the Pliocene, the Earth had become similar to the one we know today.

Protists and Invertebrates

The Neogene foraminifera of the Angolan coastal basins, including those from the Miocene Quifandongo series of the Cuanza Basin, are known from a diversity of planktonic and benthic taxa widely used in oil industry offshore correlations or as palaeoenvironmental indicators. Rocha (1957), Graham et al. (1965), Mcmillan and Fourie (1999) and Kender et al. (2009), among others describe the essentials of these West African foraminiferal assemblages, which include planktonic taxa such as *Globigerina praebulloides*, *Globigerinella obesa*, *Globigerinoides bisphericus*, *G. immaturus, G. trilobus, Globorotalia peripheroronda* and *Orbulina bilobata*.

Invertebrates are surprisingly poorly known and comprise, at least, the nautiloid mollusc *Aturia luculoensis,* the decapod crustacean *Callianassa floridana* from the Miocene Burdigalian of Cabinda Basin (Newton 1917), several taxa of Miocene echinoid echinoderms such as *Clypeaster borgesi, Echinolampas antunesi, Rotula deciesdigitata, Rotuloidea vieirai, Amphiope neuparthi,* and *Plagiobrissus* sp. (Loriol 1905; Dartevelle 1953; Gonçalves 1971; Kroh 2010; Silva and Pereira 2014; Pereira and Stara 2018), and two species of anthozoan corals *Flabellum extensum* and *Stylophora raristella* (Chevalier 1970). Nevertheless, bivalve and gastropod molluscs are undoubtedly the most diverse and abundant invertebrate taxa of the marine Neogene of Luanda, Benguela and Namibe, with several rich fossil sites, some of them presently in course of study. The molluscan faunas of these coastal basin areas, including new species such as *Pereiraea africana, Clavatula loandensis* or *Chlamys silvai* Antunes (1964), were the focus of Douvillé (1933), Keller (1934), Dartevelle (1952, 1953), Dartevelle and Roger (1954), Soares (1961, 1962), Silva (1962), Silva and Soares (1962), Antunes (1964), and more recently Antunes (1984), Lozouet and Gourgues (1995), among others.

Vertebrates

Mammals are known from Benguela and the Cuanza provinces where Projecto PaleoAngola collected skulls of fossil baleen whales. An odontocete has also been found from Barra da Cuanza.

In Angola the most abundant group of Neogene vertebrates are the Elasmobranchii chondrichthyans (18 taxa; see Supplementary Material). The following taxa are from the Pliocene of Farol das Lagostas (Cuanza Basin): *Aetobatus, Carcharhinus egertoni, Carcharhinus priscus, Carcharias taurus, Carcharocles megalodon, Carcharodon carcharias, Galeocerdo cuvier, Hemipristis serra, Isurus benedeni, Isurus oxyrinchus, Mitsukurina, Myliobatis, Negaprion brevirostris, Paragaleus, Pristis, Pteromylaeus bovina, Rhinoptera brasiliensis,* and *Sphyrna zygaena* (Antunes 1964). Five bony fishes are known, the actinopterygians *Cybium, Sparus, Sphyraena barracuda, Tachysurus,* and *Tetrodon.*

The Quaternary – The Dominance of Humans

The Quaternary (2.6 Ma to present, including the Pleistocene and Holocene) is the third geological period of the Cenozoic era and the most recent in the geological time scale. This period is characterised by the return of glaciations at higher elevations and latitudes, the dominant role of the genus *Homo* in all terrestrial habitats and the extinction of much of the megafauna.

Invertebrates

In Angola, the Quaternary biodiversity is again marked by the high number of mollusc taxa (73 or more), of which 29 are bivalves such as *Arcopsis afra, Barbatia complanata, Cardium indicum, Chama crenulata, Glycymeris concentrica, Lutraria senegalensis, Noetiella congoensis, Ungulina cuneata* from the Middle Pleistocene of Pipas (Namibe Basin) and 44 gastropods mainly known from Namibe Basin, such as *Cantharus viverratus, Columbella adansoni, Conus babaensis, Siphonaria capensis*, and *Terebra senegalensis* (Miller and Carpenter 1956; Sessa et al. 2013). Other invertebrates such as corals, arthropods and echinoderms are known but reduced to a handful of known taxa, such as *Cladangia carvalhoi* from the Pleistocene of Salinas de Bero, Saco, Namibe (Wood 1973). In most situations they occur in a variety of raised-beach and lagoonal deposits related to coastal uplift and major sea-level changes (Carvalho 1961). The post-glacial Holocene is marked by the accretion of sand-spits and deltaic facies with rich coquinas, including the bivalve *Senilia senilis* as a typical species (Dinis et al. 2016). See Supplementary Material for the updated list.

Vertebrates

A remarkable fossilized jaw (dentary bone) of Blue Whale *Balaenoptera musculus* present in the *Museu Nacional de História Natural* in Luanda measuring nearly seven meters in length is not only the largest known fossilized bone but also one of the largest whales, thus animals, ever recorded. Large land mammals, including *Bubalus, Syncerus* cf. *nanus* Boddaert; *Phacochoerus* sp., *Equus, Hippotigris* cf. *zebra* have been reported from the site called *Cemitério dos Ossos*, north of Luanda (Antunes 1961).

The caves of Humpata, in Huíla Province, in southern Angola, are formed in Chela Dolomite that hosts fossiliferous caves and fissures (Amaral 1973; Antunes 1965; Arambourg and Mouta 1952; França 1964; Mouta 1950). Pickford et al. (1990, 1992, 1994) listed taxa of mammals from the Humpata Caves. These include the insectivore *Crocidura*, a Macroscelididae, the chiropteran *Rhinolophus, Miniopterus, Nycteris*, 19 genera of rodents (*Uranomys, Acomys, Dasymys, Aethomys, Thallomys, Zelotomys, Mus, Pelomys, Malacomys, Praomys, Grammomys, Dendromus, Steatomys, Petromyscus, Tatera, Otomys, Cryptomys, Graphiurus*, and *Hystrix*), the lagomorph *Serengetilagus*, Mustelidae, Viverridae, Canidae, and the Hyaenidae cf. *Chasmoporthetes*, the Hyracoidea *Gigantohyrax* and *Procavia*, Rhinocerotidae, Equidae, Suidae *Metridiochoerus andrewsi* and the Bovidae Hippotragini and *Connochaetes*.

The most thoroughly studied of the Humpata fossils are those of the extinct baboon. Cercopithecid primates from Humpata caves include *Soromandrillus quadratirostris*, cf. *Theropithecus* sp., and *Cercopithecoides* sp. (dated as ca. 2.0–3.0 Ma) (Minkoff 1972; Jablonski 1994; Jablonski and Frost 2010; Gilbert 2013).

Pleistocene deposits in Namibe provided remains of fossilized ostrich *Struthio* eggshells, artiodactyl bones and numerous human artifacts - including Acheulean hand axes (amygdaloid bifaces) suggesting the presence of early humans, such as *Homo ergaster* or *H. erectus*, whereas bones of early humans are not known in Angola.

Final Remarks on the Fossil Record and Paleobiodiversity

Measuring paleobiodiversity is challenging due to the sparse and limited availability of data, compared with modern extant faunas. The paleobiodiversity of Angola is mostly known from Cretaceous and Cenozoic fossils that comprise 90% or more of all known fossil records of Angolan taxa (see Table 4.1 and Supplementary Material). The vast preponderance of the fossil taxa is marine which is consistent with the geological settings and paleogeography, related to the opening of the South Atlantic and repeated inundation of the Angolan continental margin.

For this study, we compiled a list of taxa using species, genus or the lowest known taxonomical clade reported in the scientific literature for Angola (see Supplementary Material). Of the resulting list of more than 1300 fossil taxa, many may require systematic revision and the final number will depend on the validity of the taxonomy.

By far the most speciose group are the molluscs (about 61% of taxa, more than half being Cretaceous ammonites) and Cretaceous foraminiferans (16%), followed by vertebrates with about 15% of taxa. Chondrichthyes and mammals represent 6% and 5% of taxa, respectively.

About 10% of vertebrate taxa listed are unique or were first recognised in Angola, most of them receiving species names after localities in Angola or of geologists that worked in the country. According to current knowledge, at least 67 taxa (6.1%) of invertebrates are endemic or were first mentioned from Angola.

References

Adnet S, Hosseinzadeh R, Antunes MT et al (2009) Review of the enigmatic Eocene shark genus *Xiphodolamia* (Chondrichthyes, Lamniformes) and description of a new species recovered from Angola, Iran and Jordan. J Afr Earth Sci 55(3–4):197–204

Amaral L (1973) Nota sobre o "karst" ou carso do Planalto do Humpata (Huila), no Sudoeste de Angola. Garcia de Orta 1:29–36

Anchieta J (1885) Traços geológicos da África Occidental Portuguesa. Tipografia Progresso, Benguela, 15 pp

Antunes MT (1961) A jazida de vertebrados fósseis do Farol das Lagostas: II Paleontologia. Boletim dos Serviços de Geologia e Minas de Angola 3:1–18

Antunes MT (1964) O Neocretácico e o Cenozóico do litoral de Angola. Junta de Investigações Ultramar, Lisboa, 255 pp

Antunes MT (1965) Sur la faune de vertébrés du Pléistocène de Leba, Humpata (Angola). Actes du Ve Congrès Panafricain de Préhistoire et de l'Étude du Quaternaire. Tenerife, 127–128

Antunes MT (1970) Paleontologia de Angola. In: Curso de Geologia do Ultramar. Junta de Investigações do Ultramar, Lisboa, pp 126–143

Antunes MT (1984) Étude d'une faune gastéropodes miocène récoltés par M. M Feio dans le Sud de l'Angola Comunicações dos Serviços Geológicos de Portugal 70(1):126–128

Antunes MT, Cappetta H (2002) Sélaciens du Crétacé (Albien-Maastrichtien) d'Angola. *Palaeontographica*, Abteilung A 264 (5–6):85–146

Antunes MT, Maisey JG, Marques MM, et al (1990) Triassic fishes from the Cassange depression (R.P. de Angola). Ciências da Terra (UNL), special number 1:1–64

Arambourg C, Mouta F (1952) Les grottes et fentes à ossements du sud de l'Angola. Actes du IIème Congrès Panafricain de Préhistoire d'Alger 12:301–301

Araújo AG, Guimarães F (1992) Geologia de Angola, Notícia explicativa da Carta Geológica à escala 1: 1 000 000. Serviço Geológico de Angola, Luanda, 140 pp

Araújo R, Polcyn MJ, Lindgren J et al (2015a) New aristonectine elasmosaurid plesiosaur specimens from the Early Maastrichtian of Angola and comments on paedomorphism in plesiosaurs. Neth J Geosci 94(1):93–108

Araújo R, Polcyn MJ, Schulp AS et al (2015b) A new elasmosaurid from the early Maastrichtian of Angola and the implications of girdle morphology on swimming style in plesiosaurs. Neth J Geosci 94(1):109–120

Archibald JD, Clemens WA, Padian K et al (2010) Cretaceous extinctions: multiple causes. Science 328(5981):973–973

Berry CT (1939) A summary of the fossil Crustacea of the Order Stomatopoda, and a description of a new species from Angola. Am Midl Nat 21(2):461–471

Blake DB, Breton G, Gofas S (1996) A new genus and species of Asteriidae (Asteroidea; Echinodermata) from the Upper Cretaceous (Coniacian) of Angola, Africa. Paläontol Z 70(1–2):181–187

Borges A (1946) A costa de Angola da Baía da Lucira à Foz do Bentiaba (entre Benguela e Mossâmedes). Boletim da Sociedade Geológica de Portugal 5(3):141–150

Boule M, Lemoine P, Thevenin A (1907) Paléontologie de Madagascar. III Céphalopodes crétacés des environs de Diègo-Suarez. Annales de Paléontologie 2:1–56

Brandão JM (2008) "Missão Geológica de Angola": contextos e emergência. Memórias e Notícias, new series 3:285–292

Brandão JM (2010) O "Museu de Geologia Colonial" das Comissões Geológicas de Portugal: contexto e memória. Revista Brasileira de História da Ciência 3(2):184–199

Buta-Neto A, Tavares TS, Quesne D et al (2006) Synthèse préliminaire des travaux menés sur le bassin de Benguela (Sud Angola): implications sédimentologiques et structurales. Áfr Geosci Rev 13(3):239–250

Callapez PM, Gomes CR, Serrano Pinto M et al (2011) O contributo do Museu e Laboratório Mineralógico e Geológico da Universidade de Coimbra para os estudos de Paleontologia Africana. In: Neves LF, Pereira AC, Gomes CR, Pereira LCG, Tavares AO (eds) Modelação de Sistemas Geológicos. Homenagem ao Professor Doutor Manuel Maria Godinho. Laboratório de Radioactividade Natural da Universidade de Coimbra, Coimbra, pp 159–174

Carvalho GS (1961) Geologia do deserto de Moçâmedes, (Angola): Uma contribuição para o conhecimento dos problemas da orla sedimentar de Moçâmedes. Memórias da Junta de Investigações do Ultramar 26:1–227

Ceraldi TS, Green D (2016) Evolution of the South Atlantic lacustrine deposits in response to Early Cretaceous rifting, subsidence and lake hydrology. In: Ceraldi S, Hodgkinson RA, Backe G (eds) Petroleum geoscience of the West Africa Margin. Geological Society, London, Special Publications, 438, doi: https://doi.org/10.1144/SP438.10

Chevalier JP (1970) Les Madreporaires du Neogene et du Quaternaire de l'Angola [Neogene and Quaternary corals from Angola]. Annalen Koninklijk Museum voor Midden-Afrika 8: Geologische Wetenschappen 68:13–33

Collignon M (1978) Ammonites du Crétacé Moyen-Supérieur de l'Angola. 2° Centenário Academia das Ciências. Estudos de Geologia e Paleontologia e de Micologia, Academia das Ciências, Lisboa, pp 1–75

Cooper MR (1972) The Cretaceous stratigraphy of San Nicolau and Salinas, Angola. Ann S Af Mus 6(8):245–251

Cooper MR (1982) Lower Cretaceous (Middle Albian) ammonites from Dombe Grande, Angola. Ann S Af Mus 89:265–314

Cooper MR, Kennedy WJ (1979) Upper most Albian (Stoliczkaia dispar zone) ammonites from the Angolan littoral. Ann S Af Mus 77:175–308

Crampton JS, Beu AG, Cooper RA et al (2003) Estimating the rock volume bias in paleobiodiversity studies. Science 301(5631):358–360

Dartevelle E (1952) Echinides fossiles du Congo et de l'Angola. Partie 1: Introduction historique et stratigraphique. Annales du Musée Royal du Congo Belge, série 8, Sciences Géologiques 12:1–70

Dartevelle E (1953) Echinides fossiles du Congo et de l'Angola. Partie 2: description systématique des échinides fossiles du Congo et de l'Angola. Annales du Muséum Royal du Congo Belge, série 8, Sciences Géologiques 13:1–240

Dartevelle E, Roger J (1954) Contribution à la connaissance de la faune du Miocène de l'Angola. Comunicações dos Serviços Gelógicos de Portugal 35:227–312

Dinis P, Huvi J, Cascalho J, Garzanti E, Vermeesch P, Callapez P (2016) Sand-spits systems from Benguela region (SW Angola). An analysis of sediment sources and dispersal from textural and compositional data. J Afr Stud 117:181–192

Douvillé H (1933) Le Tertiaire de Loanda. Boletim do Museo e Laboratorio Mineralógico e Geológico da Universidade de Lisboa, 1:63–118

Duarte LV, Callapez PM, Kalukembe A, et al (2014) Do Proterozoico da Serra da Leba (Planalto da Humpata) ao Cretácico da Bacia de Benguela (Angola). A geologia de lugares com elevado valor paisagístico. Comunicações Geológicas 101(Especial III):1255–1259

Ferré B, Granier B (2001) Albian roveacrinids from the southern Congo Basin off Angola. J S Am Earth Sci 14:219–235

Ferreira JM, Rocha AT (1957) Foraminíferos do Senoniano de Catumbela (Angola). Garcia de Orta 5(3):517–545

Ferreira OV (1957) Acerca de Parapirimela angolensis Van Straelen nas Camadas de Iela, Angola. Comunicações dos Serviços Geológicos de Portugal 38:465–468

França JC (1964) Nota preliminar sobre uma gruta pré-histórica do Planalto da Humpata (Angola). Junta de Investigações do Ultramar 2(50):59–67

Gilbert CC (2013) Cladistic analysis of extant and fossil African papionins using craniodental data. J Hum Evol 64(5):399–433

Gonçalves F (1971) Echinolampas antunesi, nov. sp. Cassidulidae, échinide nouveau du Miocène de la région de Luanda, Angola. Revista da Faculdade de Ciências, C – Ciências Naturais 16(2):307–310

Graham JJ, Klasz I, Rerat D (1965) Quelques importants foraminifères du Tertiaire du Gabon (Afrique Equatoriale). Revue de Micropalèontologie 8:71–84

Green PF, Machado V (2015) Pre-rift and synrift exhumation, post-rift subsidence and exhumation of the onshore Namible margin of Angola revealed from apatite fission track analysis. In: Sabato Ceraldi T, Hodgkinson RA, Backe G (eds) Petroleum geoscience of the West Africa Margin. Geological Society, London, Special Publications 438, pp 99–118

Greyling MR, Cooper MR (1995) Two new irregular echinoids from the Upper Cretaceous (mid-Campanian) of Angola. Durban Museum Novitates 20(1):63–71

Guiraud R, Maurin JC (1991) Le rifting en Afrique au Crétacé Inférieur: Synthèse structural, mise en évidence de deux phases dans la genèse des bassins, relations avec les ouvertures océaniques péri-africaines. Bulletin de la Société Géologique de France 165(5):811–823

Guiraud M, Buta-Neto A, Quesne D (2010) Segmentation and differential post-rift uplift at the Angola margin as recorded by the transform – rifted Benguela and oblique-to-orthogonal-rifted Kwanza basins. Mar Pet Geol 27:1040–1068

Haas O (1942) The vernay collection of cretaceous (Albian) ammonites from Angola. Bull Am Mus Nat Hist 81(1):1–224

Haas O (1943) Some abnormally coiled ammonites from the Upper Cretaceous of Angola. Am Mus Novit 1222:1–17

Haughton SH (1924) Notes sur quelques fossiles crétacés de l'Angola (Céphalopodes et Échinides). Comunicações dos Serviços Geológicos de Portugal 15:79–106

Hooijer DA (1963) Miocene mammals of Congo. Annales du Museum Royal d'Afrique Centrale, Series 8 Sciences Géologiques 46:1–77

Howarth MK (1965) Cretaceous ammonites and nautiloids from Angola. Bull Br Mus Nat Hist Geol 10:335–412

Howarth MK (1966) A mid-Turonian ammonite fauna from the Moçâmedes desert, Angola. Garcia de Orta 14(2):217–228

Jablonski NG (1994) New fossil cercopithecid remains from the Humpata Plateau, southern Angola. Am J Phys Anthropol 94(4):435–464

Jablonski NG, Frost S (2010) Cercopithecoidea. In: Werdelin L, Sanders WJ (eds) Cenozoic Mammals of Africa. University of California Press, Berkeley, pp 393–428

Jackson JB, Johnson KG (2001) Measuring past biodiversity. Science 293(5539):2401–2404

Jacobs LL, Mateus O, Polcyn MJ et al (2006) The occurrence and geological setting of Cretaceous dinosaurs, mosasaurs, plesiosaurs, and turtles from Angola. J Paleontol Soc Korea 22:91–110

Jacobs LL, Mateus O, Polcyn MJ et al (2009) Cretaceous paleogeography, paleoclimatology, and amniote biogeography of the low and mid-latitude South Atlantic Ocean. Bull Geol Soc Fr 180(4):333–341

Jacobs LL, Polcyn MJ, Mateus O et al (2016) Post-Gondwana Africa and the vertebrate history of the Angolan Atlantic Coast. Mem Mus Vic 74:343–362

Jones TR (1862) A Monograph of the fossil Estheriae. Palaeontol Soc 14:1–134

Jouve S, Schwarz D (2004) *Congosaurus bequaerti*, a Paleocene dyrosaurid (Crocodyliformes; Mesoeucrocodylia) from Landana (Angola). Bulletin de l'Institut Royal des Sciences Naturelles de Belgique, Sciences de la Terre 74:129–146

Keller A (1934) Contribution a la géologie de l'Angola. Le Tertiaire dc Luanda. Description des especes, Mollusque. Lamellibranches. Boletim do Museo e Lab. Mineral. e Geol. do Universidade de Lisboa, la Ser, 3, pp 219–250

Kender S., Kaminski MA, Jones, BW (2008) Oligocene deep-water agglutinated foraminifera from the Congo Fan, offshore Angola: Palaeoenvironments and assemblage distributions. In: Kaminski MA, Coccioni R (eds) Proceedings of the seventh international workshop on agglutinated foraminifera. Grzybowski Foundation Special Publication 13, London, pp 107–156

Kender S, Kaminski MA, Jones BW (2009) Early to middle Miocene foraminifera from the deep-sea Congo Fan, offshore Angola. Micropaleontology 54(6):477–568

Kier PM, Lawson M (1978) Index of living and fossil echinoids, 1924-1970. Smithson Contrib Paleobiol 34:1–182

Kroh A (2010) Index of Living and Fossil Echinoids 1971-2008. Annalen des Naturhistorischen Museums in Wien 112:195–470

Lapão LGP, Pereira ES (1971) Notícia explicativa Carta Geológica de Angola, escala 1:100 000, folha n° 206, Egito Praia. Direcção Provincial dos Serviços de Geologia e Minas, Luanda, 42 pp

Lapão LGP, Simões MC (1972) Notícia Explicativa da Carta Geológica de Angola, escala 1:100 000, Folha n°184, Novo Redondo. Direcção Provincial dos Serviços de Geologia e Minas, Luanda, 54 pp

Leriche M (1932) Sur les premiers fossiles découverts au nord de l'Angola, dans le prolongement des couches du Lubilash et des couches du Lualaba. Association Française pour l'Avancement des Sciences, Compte Rendu 56éme session: 1–6

Loriol P (1888) Matériaux pour l'étude stratigraphique et paléontologique de la province d'Angola. Description des Echinides. Mémoires de la Société de Physique et d'histoire naturelle de Genève 30(2):97–114

Loriol P (1905) Notes pour servir à l'étude des échinodermes. Fasc. 2 (3). Georg Editeur, Bâle/Genève, pp 119–146

Lozouet P, Gourgues D (1995) Senilia (Bivalvia: Arcidae) et Anazola (Gastropoda: Olividae) dans le Miocène d'Angola et de France, témoins d'une paléo-province Ouest-Africaine. Haliotis, 24, pp 101–108

Marliére R (1950) Ostracodes and Phyllopodes au Système du Karroo au Congo Belge et les régions avoisinantes. Annales du Muséum Royal du Congo Belge, Sciences Géologiques, Série 8(6):1–43

Marzola M, Mateus O, Schulp A, et al (2014) Early Cretaceous tracks of a large mammaliamorph, a crocodylomorph, and dinosaurs from an Angolan diamond mine. J Vertebr Paleontol, Program and Abstracts, 181

Masse P, Laurent O (2016) Geological exploration of Angola from Sumbe to Namibe: a review at the frontier between Geology, natural resources and the history of Geology. C R Geosci 348(1):80–88

Mateus O, Jacobs LL, Polcyn MJ et al (2009) The oldest African eucryptodiran turtle from the Cretaceous of Angola. Acta Paleontol Pol 54:581–588

Mateus O, Jacobs LL, Schulp AS et al (2011) *Angolatitan adamastor*, a new sauropod dinosaur and the first record from Angola. Anais da Academia Brasileira de Ciências 83(1):221–233

Mateus O, Polcyn MJ, Jacobs LL, et al (2012) Cretaceous amniotes from Angola: dinosaurs, pterosaurs, mosasaurs, plesiosaurs, and turtles. V Jornadas Internacionales sobre Paleontología de Dinosaurios y su Entorno, Salas de los Infantes, Burgos, pp 75–105

Mateus O, Marzola M, Schulp AS et al (2017) Angolan ichnosite in a diamond mine shows the presence of a large terrestrial mammaliamorph, a crocodylomorph, and sauropod dinosaurs in the Early Cretaceous of Africa. Palaeogeogr Palaeoclimatol Palaeoecol 471:220–232

Mcmillan IK, Fourie A (1999) Kwanza Basin coastal stratigraphy with atlas of Albian to Holocene Foraminifera species. De Beers Marine (Pty) Limited, Luanda, 167 pp

Miller AK, Carpenter LB (1956) Cretaceous and Tertiary Nautiloids from Angola. Estudos, Ensaios e Documentos da Junta de Investigações do Ultramar 21:1–48

Minkoff EC (1972) A fossil baboon from Angola, with a note on *Australopithecus*. J Paleontol 46(6):836–844

Mouta F (1950) Sur la présence du Quaternaire ancien dans les hauts plateaux du Sud de l'Angola (Humpata, Leba). Compte Rendu sommaire des Scéances de la Société géologique de France 14:261–262

Murray AM (2000) The palaeozoic, mesozoic and early cenozoic fishes of Africa. Fish Fish 1(2):111–145

Myers TS, Polcyn MJ, Mateus O et al (2017) A new durophagous stem cheloniid turtle from the lower Paleocene of Cabinda, Angola. Pap Palaeontol 2017:1–16

Néraudeau D, Mathey B (2000) Biogeography and diversity of South Atlantic Cretaceous echinoids: implications for circulation patterns. Palaeogeogr Palaeoclimatol Palaeoecol 156(1–2):71–88

Neto MGM (1970) O sedimentar costeiro de Angola. Algumas notas sobre o estado actual do seu conhecimento. In: Curso de Geologia do Ultramar, vol 2. Publicações da Junta de Investigações do Ultramar, Lisboa, pp 193–232

Newton RB (1917) On some Cretaceous Brachiopoda and Mollusca from Angola, Portuguese West Africa. Earth and environmental science. Trans R Soc Edinb 51(3):561–580

Nunes AF (1991) A Investigação Geológico-Mineira em Angola. Ministérios dos Negócios Estrangeiros, Ministério das Finanças, Instituto para a Cooperação Económica, Lisboa, 387 pp

Pereira P, Stara P (2018) Redefinition of *Amphiope neuparthi* de Loriol, 1905 (Echinoidea, Astriclypeidae) from the early-middle Miocene of Angola. Comunicações Geológicas, 104(1): in press

Pérez-Díaz L, Eagles G (2017) South Atlantic paleobathymetry since early Cretaceous. Sci Rep 7:11819. https://doi.org/10.1038/s41598-017-11959-7

Pickford M (1986) Première découverte d'une faune mammalienne terrestre paléogène d'Afrique sub-saharienne. Comptes rendus de l'Académie des Sciences Paris Série II 19:1205–1210

Pickford M, Fernandez T, Aço S (1990) Nouvelles découvertes de remplissages de fissures à primates dans le 'Planalto da Humpata', Huilà, Sud de l'Angola. Comptes Rendus de l'Académie des Sciences Paris, Série II 310:843–848

Pickford M, Mein P, Senut B (1992) Primate-bearing Plio-Pleistocene cave deposits of Humpata, southern Angola. Hum Evol 7:17–33

Pickford M, Mein P, Senut B (1994) Fossiliferous Neogene karst fillings in Angola, Botswana and Namibia. S Afr J Sci 90:227–230

Polcyn M, Jacobs L, Schulp A, et al (2007) *Halisaurus* (Squamata: Mosasauridae) from the Maastrichtian of Angola. J Vertebr Paleontol 27(Suppl. to 3):130A

Polcyn MJ, Jacobs LL, Schulp AS et al (2010) The North African Mosasaur *Globidens phosphaticus* from the Maastrichtian of Angola. Hist Biol 22(1–3):175–185

Polcyn MJ, Jacobs LL, Araújo R et al (2014) Physical drivers of mosasaur evolution. Palaeogeogr Palaeoclimatol Palaeoecol 400:17–27

Quirk DG, Hertle M, Jeppesen JW, et al (2013) Rifting, subsidence and continental break-up above a mantle plume in the central South Atlantic. In: Mohriak WU, Danforth A, Post PJ, et al (eds) Conjugate Divergent Margins. Geological Society, London, Special Publication, 369: 185–214

Rennie JVL (1929) Cretaceous fossils from Angola (Lamellibranchia and Gastropoda). Ann S Afr

Mus 28:1–54

Rennie JVL (1945) Lamelibrânquios e gastrópodes do Cretácico Superior de Angola (vol. 1). Junta das Missões Geográficas e de Investigações Coloniais 1:1–141

Robles-Cruz SE, Escayola M, Jackson S et al (2012) U-Pb SHRIMP geochronology of zircon from the Catoca kimberlite, Angola: Implications for diamond exploration. Chem Geol 310-311:137–147. https://doi.org/10.1016/j.chemgeo.2012.04.001

Rocha AT (1957) Contribuição para o estudo dos foraminíferos do Terciário de Luanda. Garcia de Orta 5(2):297–312

Rocha AT (1973) Contribution à l'étude des foraminifères paléogènes du bassin du Cuanza (Angola). Memórias e Trabalhos do Instituto de Investigação Científica de Angola 12:1–309

Rocha AT (1984) Notas micropaleontológicas sobre as formações sedimentares da orla meso-cenozóica de Angola - V. O Maestrichtiano inferior da mancha de Cabeça da Baleia (a norte de Egito-Praia). Garcia de Orta, Série Geológica 7(1–2):97–108

Schlanger SO, Arthur MA, Jenkyns HC et al (1987) The Cenomanian-Turonian Oceanic Anoxic Event, I. Stratigraphy and distribution of organic carbon-rich beds and the marine δ13C excursion. Geol Soc Lond, Spec Publ 26(1):371–399

Schulp AS, Polcyn MJ, Mateus O et al (2006) New mosasaur material from the Maastrichtian of Angola, with notes on the phylogeny, distribution and palaeoecology of the genus *Prognathodon*. On Maastricht Mosasaurs. Publicaties van het Natuurhistorisch Genootschap in Limburg 45(1):57–67

Schulp AS, Polcyn MJ, Mateus O et al (2008) A new species of *Prognathodon* (Squamata, Mosasauridae) from the Maastrichtian of Angola, and the affinities of the mosasaur genus Liodon. In: Proceedings of the Second Mosasaur Meeting, vol 3. Fort Hays State University Hays, Kansas, pp 1–12

Schulp AS, Polcyn MJ, Mateus O et al (2013) Two rare mosasaurs from the Maastrichtian of Angola and the Netherlands. Neth J Geosci 92(1):3–10

Schwarz D (2003) A procoelous crocodilian vertebra from the lower Tertiary of Central Africa (Cabinda enclave, Angola). Neues Jahrbuch für Geologie und Paläontologie, Monatshefte 2003:376–384

Schwarz D, Frey E, Martin T (2006) The postcranial skeleton of the Hyposaurinae (Dyrosauridae; Crocodyliformes). Palaeontology 49(4):695–718

Segundo J, Duarte LV, Callapez PM (2014) Lithostratigraphy of the Quissonde Formation marl-limestone succession (Albian) of the Ponta do Jomba-Praia do Binge sector (Benguela Basin, Angola). Comunicações Geológicas 101(Especial III):567–571

Séranne M, Anka Z (2005) South Atlantic continental margins of Africa: a comparison of the tectonic vs climate interplay on the evolution of equatorial west Africa and SW Africa margins. J Afr Earth Sci 43(1–3):283–300

Sessa JA, Callapez PM, Dinis PA et al (2013) Paleoenvironmental and paleobiogeographical implications of a Middle Pleistocene mollusc assemblage from the marine terraces of Baía das Pipas, southwest Angola. J Paleontol 87(6):1016–1079

Silva GH (1962) Fósseis do Miocénico de Luanda (Angola). Associação Portuguesa para o Progresso das Ciências. Actas do XXVI Congresso Luso-Espanhol (Porto, 22–26 June, 1962), sections II and IV, 3 pp

Silva R, Geirinhas F (2010) Colecções geológicas das antigas Províncias Ultramarinas Portuguesas arquivadas na Litoteca do LNEG. e-Terra 15(4):1–4

Silva R, Pereira P (2014) Redescoberta dos equinodermes fósseis das coleções históricas ultramarinas do LNEG. Comunicações Geológicas 101(Especial III):1379–1382

Silva GH, Soares AF (1962) Contribuição para o conhecimento da fauna miocénica de S. Pedro da Barra e do Farol das Lagostas (Luanda, Angola). Garcia de Orta 9:721–736

Soares AF (1961) Nouvelle espèce de *Chlamys* du Miocène de la région de Luanda (Angola). Memórias e Notícias 51:1–6

Soares AF (1962) Nota sobre alguns lamelibrânquios e gastrópodes do Miocénico de Luanda (Angola). Memórias e Notícias 53:31.35

Soares AF (1965) Contribuição para o estudo dos lamelibrânquios Cretácicos da região de Moçâmedes. Serviços de Geologia e Mina de Angola Boletim 11:137–168

Solé F, Noiret C, Desmares D et al (2018) Reassessment of historical sections from the Paleogene marine margin of the Congo Basin reveals an almost complete absence of Danian deposits. Geosci Front. https://doi.org/10.1016/j.gsf.2018.06.002

Spath LF (1922) On Cretaceous ammonites from Angola, collected by Prof. J.W. Gregory, D.Sc., F.R.S. Trans R Soc Edinb Earth Sci 53:91–160

Spath LF (1931) On cretaceous Ammonoidea from Angola, collected by Pr. J.W. Gregory. Trans Geol Soc Edinb. 53 (1):91–160

Spath LF (1953) The Upper Cretaceous cephalopod fauna of Grahamland. Sci Rep Falkland Islands Depend Surv 3:1–60

Strganac C, Salminen J, Jacobs LL et al (2014) Carbon isotope stratigraphy, magnetostratigraphy, and 40Ar/39Ar age of the Cretaceous South Atlantic coast, Namibe Basin, Angola. J Afr Earth Sci 99:452–462

Strganac C, Jacobs LL, Polcyn M et al (2015a) Stable oxygen isotope chemostratigraphy and paleotemperature regime of mosasaurs at Bentiaba, Angola. Neth J Geosci 94(1):137–143

Strganac C, Jacobs LL, Polcyn MJ et al (2015b) Geological setting and paleoecology of the Upper Cretaceous Bench 19 marine vertebrate bonebed at Bentiaba, Angola. Neth J Geosci 94(1):121–136

Tasch P (1979) Permian and Triassic Conchostraca from the Bowen Basin (with a note on a Carboniferous leaiid from the Drummond Basin), Queensland. Aust Bureau Mineral Resour Geolo Geophys Bull 185:33–44

Tasch P (1987) Fossil Conchostraca of the Southern Hemisphere and continental drift: Paleontology, biostratigraphy, and dispersal. Memoir of the Geological Society of America 165:1–290

Tavares T (2006) Ammonites et échinides de l'Albien du bassin de Benguela (Angola). Systématique, biostratigraphie, paléogéographie et paléoenvironnement. Unpublished PhD Thesis, Université de Bourgogne, Dijon, 389 pp

Tavares T, Meister C, Duarte-Morais ML et al (2007) Albian ammonites of the Benguela Basin (Angola): a biostratigraphic framework. S Afr J Geol 110(1):137–156

Taverne L (2016) New data on the osteoglossid fishes (Teleostei, Osteoglossiformes) from the marine Danian (Paleocene) of Landana (Cabinda Enclave, Angola). Geo-Eco-Trop 40(4):297–304

Teixeira C (1947) Contribuição para o conhecimento geológico do Karroo da África Portuguesa. I-Sobre a flora fóssil do Karroo da região de Téte. Anais da Junta de Investigações do Ultramar, Estudos de Geologia e Paleontologia 2(2):9–28

Teixeira C (1948a) Fósseis vegetais do Karroo de Angola. Boletim da Sociedade Geológica de Portugal 7(1–2):73–77

Teixeira C (1948b) Vegetais fósseis do grés do Quilungo. Anais da Junta de Investigações Coloniais 2:85–92

Teixeira C (1958) Note paléontologique sur le Karroo de la Lunda, Angola. Boletim da Sociedade Geológica de Portugal 12(3):83–92

Teixeira C (1961) Paleontological notes on the Karroo of the Lunda (Angola). Garcia de Orta 9(2):307–311

Van Straelen V (1937) Parapirimela angolensis. Brachyure nouveau du Miocène de l'Angola. Bulletin du Musée Royal d'Histoire Naturelle de Belgique 8(5):1–4

Vasconcelos P (1951) Sur la découverte d'algues fossiles dans les terrains anciens de l'Angola. Int Geol Congr XVIIIth Sess 14:288–293

Walker RT, Telfer M, Kahle RL et al (2016) Rapid mantle-driven uplift along the Angolan margin in the late Quaternary. Nat Geosci 9(12):909–914. https://doi.org/10.1038/NGEO2835

Wood RC (1973) Fossil marine turtle remains from the Paleocene of the Congo. Annales du Musée Royal d'Afrique Centrale, Sciences Geologiques 75:1–28

Zachos JC, Röhl U, Schellenberg SA et al (2005) Rapid acidification of the ocean during the Paleocene-Eocene thermal maximum. Science 308(5728):1611–1615

Part II
Flora and Vegetation: Patterns and Evolutionary Processes

5

Botanical Activity in Angola: Current and Historical Overview

David J. Goyder and Francisco Maiato P. Gonçalves

Abstract Angola is botanically rich and floristically diverse, but is still very unevenly explored with very few collections from the eastern half of the country. We present an overview of historical and current botanical activity in Angola, and point to some areas of future research. Approximately 6850 species are native to Angola and the level of endemism is around 14.8%. An additional 230 naturalised species have been recorded, four of which are regarded as highly invasive. We draw attention to the paucity of IUCN Red List assessments of extinction risk for Angolan vascular plants and note that the endemic aquatic genus *Angolaea* (Podostemaceae), not currently assessed, is at high risk of extinction as a result of dams built on the Cuanza river for hydro-electric power generation. Recent initiatives to document areas of high conservation concern have added many new country and provincial records and are starting to fill geographic gaps in collections coverage.

Keywords Botanical collectors · Botanical diversity · Botanical history · Gossweiler · Invasive species · Welwitsch

D. J. Goyder (✉)
Herbarium, Royal Botanic Gardens, Kew, Richmond, Surrey, UK

National Geographic Okavango Wilderness Project, Wild Bird Trust,
Parktown, Gauteng, South Africa
e-mail: D.Goyder@kew.org

F. M. P. Gonçalves
National Geographic Okavango Wilderness Project, Wild Bird Trust,
Parktown, Gauteng, South Africa

Instituto Superior de Ciências da Educação da Huíla, Lubango, Angola

University of Hamburg, Institute for Plant Science and Microbiology, Hamburg, Germany
e-mail: francisco.maiato@gmail.com

History of Botanical Exploration in Angola

It appears that the earliest extant botanical collections from Angola date from either 1669 (Exell 1939; Martins 1994) or more probably 1696 (Dandy 1958; Exell 1962; Mendonça 1962; Figueiredo et al. 2008), and were made by Mason in the Luanda region, and by John Kirckwood in Cabinda. These reached Hans Sloane whose plant and insect collections formed the core of the British Museum (now the Natural History Museum), London, via James Petiver who encouraged surgeons on English ships to send him natural history collections from their overseas travels. Other Pre-Linnean collections from Angola in the Sloane Herbarium were made by Gladman and William Browne (Fig. 5.1). The earliest known Portuguese collector was the naturalist Joaquim José da Silva, who collected along the Angolan coast and the western escarpment between 1783 and 1804. This material was taken from Lisbon to Paris, where it now resides, in 1808 during the Napoleonic Peninsula War (Mendonça 1962; Figueiredo et al. 2008).

Mendonça (1962) presents a historical account of plant collectors in Angola, giving helpful insights to the itineraries of a number of early expeditions. A more complete list of collectors is given by Figueiredo et al. (2008), which volume also includes a useful listing of references relevant to the study of the flora of Angola.

Fig. 5.1 One of the earliest herbarium specimens collected in Angola, in 1706 or 1707, by W Browne and now housed in the Sloane Herbarium at the Natural History Museum in London. The New World starch crop – cassava *Manihot esculenta* (Euphorbiaceae)

Most eighteenth and early nineteenth century explorers visited only coastal regions of Angola, but by the 1850s, botanists and explorers were starting to document plants from more elevated parts of the interior. Friedrich Welwitsch, who spent 6 years in Angola, amassed over 8000 collections of plants representing around 5000 species, of which around 1000 were new to science (Albuquerque 2008; Albuquerque et al. 2009; Albuquerque and Figueirôa 2018). He spent his first year in Angola in the coastal zone between the mouth of the Rio Sembo ('Quizembo') just north of Ambriz, and the mouth of the Cuanza. In September 1854 he embarked on a three-year excursion, initially following the Bengo River and reaching Golungo Alto (Cuanza-Norte). He based himself eventually at Sange from where he made excursions to Ndalatando ('Cazengo') and the banks of the Luinha. In October 1856 he arrived at Pungo Andongo (Malange) where he was based for the next eight months, making collections from Pedras Negras, Pedras de Guinga and localities along the Cuanza River – the furthest point he reached upstream was Quissonde, south of Malange. After an extended period back in Luanda, he headed south via Benguela to Namibe ('Little Fish Bay') in June 1859, gradually extending his journeys along the coast to Cabo Negro, the port of Pinda (probably Tômbua) and Baía dos Tigres. In October 1859 he headed inland from Namibe, following the Rio Giraul ('Maiombo river') to Bumbo on the slopes of the Serra da Chela. He was based at Lopollo on the Huíla plateau until 1860. In 1866, José Anchieta moved to Angola and was based at Caconda on the Huíla plateau. And by the 1880s, missionaries such as José Maria Antunes and Eugène Dekindt, and collectors such as Francisco Newton and Henry Johnston were also making significant collections from this region.

Three nineteenth century German expeditions to the Congo travelled through Angola – Pechuël-Lösche's 1873 Loango Expedition with Paul Güssfeldt and Hermann Soyaux started from Cabinda; Pogge, Buchner and Wissmann's Cassai Expedition made collections from Malange and the Lundas (Mona Quimbundo, Saurimo, Cuango River) in 1876; while Teucsz and Mechow's Cuango Expedition made collections from Dondo (Cuanza-Norte), Pungo Andongo and Malange (Malange), and the Cuango river (Uíge) in 1879–1881. A fourth German expedition, the Kunene-Sambesi Expedition, left Namibe on 11 August 1899 and travelled east, through present-day Cunene and Cuando Cubango provinces, reaching the Cuando River in March 1900 before returning to Namibe in June of that year. Over 1000 collections were made on this expedition by the botanist Hugo Baum (Warburg 1903; Figueiredo et al. 2009b).

The first half of the twentieth century was dominated by the efforts of Kew-trained Swiss botanist John Gossweiler who in the course of 50 years' work collected in all of Angola's provinces, and amassed over 14,000 collections. His final 2 years' collections, in 1946 and 1948, were from the remote northeast of the country, and formed the basis of Cavaco's Flora of Lunda (Cavaco 1959). Other significant colonial era collectors included Portuguese and British participants of the *Missões Botânicas* such as Luiz Carrisso, Francisco Mendonça, Arthur Exell and Francisco de Sousa (as well as John Gossweiler), whose work formed the basis of early parts of the *Conspectus Florae Angolensis*, and the first vegetation map of Angola

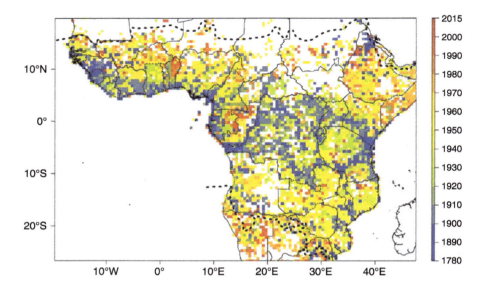

Fig. 5.2 Time lapse of botanical collecting history across tropical Africa. The map represents the date of the first botanical collection made within each 0.5° sampling unit. Dashed lines represent the limits for tropical Africa as defined by Sosef et al. (Used with permission from Sosef et al. 2017: http://rainbio.cesab.org)

(Gossweiler and Mendonça 1939). There are too many other collectors from 1950–1975 to list (see Figueiredo et al. 2008), but two specialist collections are here noted – Hans Hess's aquatic and wetland plants from many of the rivers of western Angola in 1950–1952 are now housed principally in Zurich, and Larry Leach and IC Cannell who travelled up the arid and semi-arid coastal plain between 1967 and 1973, focussed mostly on the succulent flora. After Angolan independence in 1975 and the commencement of the long-running civil war, collection of plants essentially ground to a halt until the end of the twentieth century. Several recent collecting programmes will be described in a later section of this paper. Despite Gossweiler and his successor Brito Teixeira's efforts to survey little known regions of Angola, plant collection coverage and intensity is skewed heavily to the western half of the country, and large parts of Moxico, Cuando Cubango, the Lundas and Uíge are still devoid of collections (Sosef et al. 2017: Fig. 5.2; http://rainbio.cesab.org).

Floristic Diversity and Endemism

Under the leadership of Estrela Figueiredo and Gideon Smith, thirty-two authors from around the world compiled the first checklist of vascular plants for Angola (Figueiredo and Smith 2008; Smith and Figueiredo 2017). A total of 6735 native species were recorded with an additional 226 non-native species. The exotic flora of

Angola was documented by Gossweiler (1948, 1949, 1950). Four of these alien species pose particular threats as they are highly invasive in Angola (Rejmánek et al. 2017). Forty-four additional species have been described or entered onto the International Plant Names Index since publication of Figueiredo and Smith (2008), and inventories in Lunda-Norte (see below) and elsewhere added a further 70 or so species to the Angolan list. So the current estimate of the vascular plants native to Angola is around 6850 species. Current accepted nomenclature for plants can be checked on the African Plants Database (2018), and local plant names in Gossweiler (1953) and Figueiredo and Smith (2012).

Figueiredo et al. (2009a) reported that 997 species (14.8%) are endemic to the country. This percentage is considerably lower than the estimate of 27.3% by Exell and Gonçalves (1973) based on a limited sample of the flora, or studies of individual families of plants where 19% of both Rubiaceae (Figueiredo 2008) and legume species (Soares et al. 2009) were recorded as endemics. Several genera are endemic to Angola, including *Calanda* K.Schum. and *Ganguelia* Robbr. (Rubiaceae); *Carrissoa* Baker f. (Leguminosae); and *Angolaea* Wedd. (Podostemaceae) – the latter now possibly extinct as it was described from the Cambambe rapids on the now heavily dammed Cuanza River.

Legumes (934 spp.), grasses (526 spp.), Compositae (463 spp.) and Rubiaceae (444 spp.) are the most diverse families in the flora, and *Crotalaria* L. and *Euphorbia* L. each have more than 40 Angolan endemic species.

Two of the six tropical African centres of endemism identified by Linder (2001) fall partially or entirely within Angola. A recent analysis of RAINBIO data (Droissart et al. 2018) identifies the western Angolan highlands as a distinct floristic bioregion, although the limited data preclude statements on the remainder of the country. The Huíla plateau consistently stands out as being rich in endemic species (Exell and Gonçalves 1973, Brenan 1978: 472, Linder 2001) and Soares et al. (2009) record 83 endemic legumes from the province. For Rubiaceae, Cabinda has the highest level of diversity with 175 species, but Huíla possesses the most endemics (Figueiredo 2008). Figueiredo (2008) also demonstrates that for Rubiaceae, Huíla is the most intensively collected province. However, our experience is that many of these collections have not necessarily been well studied. Clark et al. (2011) state that the western highlands of Angola comprise the least well-documented stretch of the Great Escarpment of southern Africa.

The western margin of the Huíla Plateau reaches its highest elevation along the Lubango Escarpment of the Serra da Chela and runs in a southwesterly direction from near Tundavala c. 15 km NW of Lubango to Bimbe c. 20 km NW of Humpata. It reaches a height of just over 2200 m and Goyder et al. (in prep.) estimate around 200 species are endemic to this area. However, as other mountains further to the north are surveyed botanically, some of these supposed local endemics may prove to be more widely distributed than currently thought.

Linder's (2001) second area of high species diversity and endemism, the Zambezi-Congo watershed, encompasses eastern Angola, northern Zambia and the Katanga region of the DR Congo. This area has not been well documented in Angola.

Biogeography, Regional Centres of Endemism and Vegetation

With its extremes of landform, climate and rainfall, Angola is host to six of White's (1983) phytochoria, or regional centres of endemism.

Outliers of the Guineo-Congolian forests in Cabinda, Uíge and Cuanza-Norte are progressively smaller in area to the south, ending in the isolated coffee forests of Gabela and Cumbira in Cuanza-Sul. The northward-draining tributaries of the Cuango and Cassai rivers in Uíge and Lunda-Norte have fingers of pure Congolian forest along them. However, much of northern Angola forms a transition zone between Guineo-Congolian vegetation and Zambezian – the latter covers the rest of the country with the exception of the fragmented Afromontane centre of endemism at higher elevations, and the more arid Karoo-Namib and Kalahari-Highveld zones in the southwest.

Geologically, the eastern half of Angola is notable for its deep deposits of Kalahari sand, while to the west crystalline rocks predominate. Marine sediments and recent sands cover the coastal plain (Huntley and Matos 1994; Huntley 2019). The coastal plain is arid in the south due to the cold, upwelling Benguela current, and semi-arid further to the north. Most of the rainfall occurs on the escarpment and the plateau, again with a steady increase to the north. Central Angolan headwaters of major river systems drain into the Okavango (Cuito and Cubango), the Indian Ocean (Cuando, Lungué Bungo and Zambezi) and the Atlantic (Cassai, Cuango, Cuanza and Cunene).

The standard work for vegetation is Barbosa's (1970) *Carta Fitogeográfica de Angola* which recognises 32 vegetation types ranging from desert to moist evergreen and swamp forests. Huntley and Matos (1994) present a concise summary. Barbosa's vegetation map built on the painstaking pioneering work of Gossweiler and Mendonça (1939) – a major contribution that reached a wider audience through the extended English summary by Airy-Shaw (1947).

Angola has a diverse seaweed flora and 169 species have been recorded (Lawson et al. 1975; Anderson et al. 2012). Biogeographically, Angola's marine algae group with those of tropical West Africa, but with a well-developed southern element from around 13°S comprising mainly cooler-water species from the Benguela Marine Province of Namibia and western South Africa.

Recent Botanical Survey Initiatives

In 1968, Angola had only three National Parks (Quiçama, Cameia and Iona) and two Nature reserves (Mupa and Luando), plus a number of forest and game reserves (Teixeira 1968a). Between 1971 and 1975 a programme of field surveys was undertaken to identify areas of high importance for biodiversity conservation (Huntley 1973, 1974; Huntley and Matos 1994). These were supplemented by fieldwork in Huíla, Namibe, Cuanza-Sul and Huambo (Huntley 2009; Mills et al.

2011), and synthesised into an 'Angolan Protected Area Expansion Strategy – APAES' (Huntley 2010). The APAES report was submitted to the Angolan Ministry of Environment in 2010, and formed the basis for the proposals approved by the Angolan *Conselho do Ministros* on 28th April 2011 (GoA 2011).

Much of the recent botanical activity in Angola has focused on the eleven areas highlighted in this conservation planning document. The areas proposed for protection were: Maiombe (Cabinda), Serra do Pingano (Uíge), Lagoa Carumbo (Lunda-Norte), Serra Mbango (Malange), Gabela and Cumbira Forests (Cuanza-Sul), Morro Namba (Cuanza-Sul), Morro Moco (Huambo) Serra da Neve (Namibe), Serra da Chela (Huíla) and Luiana (Cuando Cubango). A listing of post-Independence botanical collectors in Angola is given in Appendix, following the format used for earlier collectors used by Figueiredo et al. (2008).

A collaborative Rapid Biodiversity Assessment and training expedition to the Huíla Plateau and to Iona National Park, with 30 scientific participants from 10 countries and with 15 Angolan students, was convened in 2009. Over 2700 botanical collections were made and deposited in the National Herbarium, Pretoria with duplicates deposited in the ISCED-Huíla Herbarium in Lubango (Huntley 2009).

In northern Angola botanical surveys have been initiated in the moist coffee forests of Serra do Pingano, and more widely in Uíge Province, by a team from Dresden in cooperation with the Universidade Kimpa Vita (Lautenschläger and Neinhuis 2014; Neinhuis and Lautenschläger 2014). These have resulted in a revised list of bryophytes for Angola (Müller 2014; Müller et al. 2018), the description of new species of vascular plant (Abrahamczyk et al. 2016), and ethnobotanical assessments (Göhre et al. 2016; Mawunu et al. 2016; Heinze et al. 2017; Lautenschläger et al. 2018). In total, about 820 species were identified; several of these are new records for Angola.

Lagoa Carumbo and the Luxico, Luele and Lovua valleys were surveyed by a team from Kew, the Ministry of the Environment and Agostinho Neto University, Luanda in 2011, and again in 2013, trebling the known flora of Lunda-Norte as compared to Cavaco (1959) – the combined report documents 752 taxa including 72 additions to the flora of Angola, and 22 potential new species (Darbyshire et al. 2014; Cheek et al. 2015). This part of Lunda-Norte has Congolian swamp forest in the river valleys, moist miombo woodland on the slopes, and Zambezian savanna grasslands on the plateau.

The isolated patch of Guineo-Congolian forest at Cumbira was the subject of a rapid botanical assessment with more than a hundred botanical specimens collected, including new Guineo-Congolian records for Angola and species potentially new to science (Gonçalves and Goyder 2016).

Plants collected from Mount Namba are currently being studied by the Kew/Lubango team – this work may inform studies on the Lubango Escarpment further to the south. Both share a mosaic of Afromontane forest, grassland and miombo woodland habitats, although most of Lubango's woody vegetation is now heavily degraded. Comparisons with the much better preserved vegetation on Mount Namba might inform habitat restoration initiatives in the area.

Serra da Neve and Serra da Chela were visited briefly in 2013 as part of a wider floristic survey of the Angolan Escarpment led by a team from Rhodes University in South Africa, ISCED-Huíla in Lubango, and Kew. One or two new species have been published from these collections (Hind and Goyder 2014), but wider analysis of the flora is still on-going. Through the German-funded Southern African Science Service Centre for Climate Change and Adaptive Land Management (SASSCAL) project, researchers at the Lubango Herbarium are working on vegetation classification of the woodlands of Huíla Province, towards a new vegetation map for the region (Chisingui et al. 2018). A checklist of the Huíla flora is one of the expected early outputs.

In addition to the Protected Areas Expansion Strategy sites mentioned above, three cross-border initiatives have focused on the catchment of the Okavango system in Angola, Namibia and Botswana in recent years. Botswana's flagship wetland ecosystem – the Okavango Delta – is dependent entirely on the two main Angolan tributaries (Cuito and Cubango) for its hydrology. The Southern Africa Regional Environmental Program (SAREP) and OKACOM organised fieldwork in Cuando Cubango in 2013 with botanists from Kew and the University of Botswana. About 350 collections were made from the southeast corner of Angola, as far east as the Cuando river, thus contributing to the documentation of the Luiana proposed protected area. The Future Okavango (TFO) project led by a research team from Hamburg focused on two research sites in Angola (Cusseque, Bié Province; Caiundo, Cuando Cubango Province) both in the more westerly Cubango catchment, one in Namibia (Mashare), and Seronga in Botswana. This project contributed significantly to a better understanding of Angolan miombo and *Baikiaea-Burkea* woodlands in terms of recovery following disturbance caused by shifting cultivation (Wallenfang et al. 2015, Gonçalves et al. 2018, Gonçalves et al. 2017). A checklist of woody species and geoxylic suffrutices in the grasslands of south-central Angola was provided, documenting potential new species and new records for the country (Gonçalves et al. 2016; Revermann et al. 2017, 2018). Further vegetation and ecological studies are published in Oldeland et al. (2013).

The easterly Cuito and Cuanavale catchment has been the focus of the National Geographic Okavango Wilderness Project from 2015 onwards. Surveys in the upper Cubango were initiated in 2017. To date, over 1300 plant collections have been made by a Kew, South African and Angolan team, who have recorded 417 species of vascular plant from the high-rainfall upper Cuito and Cuanavale drainage system, and 176 from the lower rainfall zones further south (e.g., Fig. 5.3). Over 100 new provincial records were reported for Moxico, with a further 24 for Cuando Cubango, underlining how poorly documented and understood this vast and sparsely inhabited part of Angola is, even now (Goyder et al. 2018). Baseline botanical collection data such as these feed into wider biodiversity assessments of the area and provide vital evidence in building a case to protect the headwaters of not only the Okavango system, but other major river systems originating in central Angola (NGOWP 2018).

Fig. 5.3 Some plants collected during recent fieldwork in central and eastern Angola as part of the National Geographic Okavango Wilderness Project. Top to bottom, left to right: *Protea poggei* subsp. *haemantha* (Proteaceae); *Clerodendrum baumii* (Lamiaceae); *Erythrina baumii* (Leguminosae); *Monotes gossweileri* (Dipterocarpaceae); *Gloriosa sessiliflora* (Colchicaceae); *Raphionacme michelii* (Apocynaceae). All photos: David Goyder

Future Botanical Work

Almost every botanical survey made in recent years in Angola has revealed undescribed species and new country or provincial records. Eastern and northern provinces are in most need of collecting programmes and botanical documentation. Most national parks lack basic botanical inventories. To give one example, Teixeira's (1968b) work on plant diversity in Bicuar National Park (Huíla Province) resulted in the recognition of six vegetation types in the park. But recent SASSCAL-funded surveys revealed species unaccounted for by Figueiredo and Smith (2008), underlining the need for more botanical surveys in both existing and newly proposed areas of conservation concern.

Analysis of the collections from recent surveys is starting to reveal little-documented areas of endemism. The Lubango Escarpment is one obvious focus, but so too is the highly leached high-rainfall Kalahari sand system of Moxico Province and adjacent area that has its own peculiar and little-understood flora.

Only 399 species of vascular plant in Angola have been formally assessed for extinction risk through the IUCN Red List system (IUCN 2018), and a mere 36 of these appear in threatened categories. None of the genera listed in an earlier section of this paper as Angolan endemics have been assessed. Much work is needed in this area.

Four Angolan institutions are listed in Index Herbariorum (Thiers, continually updated), LUAI (ex-*Centro Nacional de Investigação Cientifica* (CNIC), Luanda), LUA (*Instituto de Investigação Agronómica*, Huambo), LUBA (*Instituto Superior de Ciencias da Educação*, Lubango), and DIA (*Museu do Dundo*). While the Dundo Museum has been refurbished and reopened to the public in 2012, it appears that the herbarium collections formerly housed there no longer exist. The LUA herbarium contains 40,000 collections. It was evacuated to Luanda in 1995, and has now returned to Huambo, but is in poor condition and funds are needed to employ well-trained young staff to conserve, rehabilitate and work on this important collection. LUAI contains 35,000 collections and LUBA around 50,000. There are ongoing digitisation programmes at both institutions that will make these collections more widely accessible.

Outside of Angola, Portuguese institutes in Coimbra (COI) and Lisbon (LISC, LISU) hold the largest collections of Angolan plants, an estimated 90,000 collections (Figueiredo and César 2008). 8700 of Gossweiler's Angolan collections are housed at COI and these are available online. The collections at LISC are also available digitally, and are now being incorporated into the Lisbon University herbarium LISU. Most other herbaria with significant Angolan holdings have only digitised their type collections, although mass digitisation of entire national collections has made material in the Paris Natural History Museum (P) and Leiden's Naturalis (L, WAG, U) accessible. In the UK, the Natural History Museum (BM) and Royal Botanic Gardens, Kew (K) in London – both of which contain significant Angolan holdings, and Royal Botanic Gardens, Edinburgh (E) have plans to follow suit. In Germany, the collection of Technische Universität Dresden (DR) comprises 2400 specimens, kept separately from the main herbarium. The Future Okavango project has augmented Hamburg's (HBG) Angolan collections by around 2000 numbers.

Once these combined resources are available online, georeferencing the Angolan material should be a priority. Such collections data could then be used in a variety of projects or programmes. Georeferenced specimen data underpins IUCN conservation assessments, for example, and these in turn inform Important Plant Area designations (Darbyshire et al. 2017) and other forms of conservation planning.

Acknowledgements We are delighted to acknowledge the support of the former Minister of Environment, Dr. Fátima Jardim, and the present Minister, Dr. Paula Francisco, at the Ministério do Ambiente, Luanda, in our attempts to provide the botanical evidence for the conservation of Angola's unique flora. We are grateful to Thea Lautenschläger for providing biographical and other information relating to projects in Uíge Province.

Appendix

Post-Independence collectors in Angola. Entries follow a format developed from Figueiredo et al. (2008).

Surname, first names (birth–death); **C:** period when collecting in Angola; **H:** herbaria [abbreviations after Thiers, continuously updated; FC-UAN = Faculdade de Ciências, Universidade Agostinho Neto, Luanda; INBAC = Instituto Nacional da Biodiversidade e Áreas de Conservação of the Ministério do Ambiente, Luanda]; **L:** provinces abbreviated after Figueiredo and Smith 2008: principal localities; **B:** biographical information.

Alcochete, António (1963–)
C: 1991; **H:** K; **L:** CU HI NA; **B:** Angolan botanist, collected with Gerrard, Matos and Newman.

Baragwanath, S.
C: 1994. **H:** PRE.

Barker, Nigel P.(1962-)
C: 2013, 2015, 2017; **H:** GRA, INBAC, K, LUBA, PRE; **L:** CC HI NA: Lubango Escarpment, Mt. Tchivira, Serra da Neve, Mundondo Plateau, Okavango, Cuito and Longa Rivers; **B:** South African Professor of Plant Science at University of Pretoria, formerly at Rhodes University.

Bester, Stoffel Petrus (Pieter) (1969–)
C: 2009, 2015; **H:** GRA, INBAC, K, LUBA, PRE; **L:** CC CU HI NA: Iona, Lubango Escarpment, Bicuar, Okavango, Cuito and Longa Rivers; **B:** South African botanist based at PRE.

Bruyns, Peter Vincent (1957–)
C: 2006, 2007; **H:** BOL, E, K, NBG, PRE; **L:** BE HI NA: Lubango Escarpment and coastal plain; **B:** South African mathematician and botanist with particular interest in succulent plants.

Cardoso, João Francisco (1974–)
C: 2005, 2006; **H:** LISC, LUAI; **L:** HI NA: Serra da Leba, Virei, Caraculo, Cainde; **B:** Agronomist with Agostinho Neto University.

Cheek, Martin Roy (1960–)
C: 2012; **H:** K; **L:** CA; **B:** British botanist at Royal Botanic Gardens Kew, specialist on West African flora.

Clark, Vincent Ralph (1977–)
C: 2013; **H:** GRA, K, LUBA, PRE; **L:** HI NA: Lubango Escarpment, Mt. Tchivira, Serra da Neve; **B:** South African botanist.

Cooper, C.E.
C: 1997; **H:** PRE.

Crawford, Frances Mary (1981–)
C: 2009, 2011; **H:** INBAC, K, PRE; **L:** HI LN NA: Lucapa, Lagoa Carumbo, Iona, Lubango Escarpment; **B:** British botanist, Curator of WIND herbarium, formerly at Royal Botanic Gardens, Kew; collected with Darbyshire and Goyder in LN.

Daniel, José Maria (1943–2015)
C: 1964–2008; **H:** LUBA, LUA, LUAI, L: Collected in all Angolan Provinces; **B:** Angolan botanist at Lubango Herbarium until his retirement; collected with Huntley, Matos and Gonçalves.

Darbyshire, Iain Andrew (1976–)
C: 2011, 2013; **H:** INBAC, K, LISC; **L:** LN: Lucapa, Lagoa Carumbo; **B:** British botanist at Royal Botanic Gardens, Kew; collected with Crawford, Gomes, Goyder & Kodo.

Dexter, Kyle Graham (1980–)
C: 2017–; **H:** E, COLO, LUBA, WIND; **L:** CU HI NA; **B:** Senior Lecturer at University of Edinburgh and Associate Researcher at Royal Botanic Garden Edinburgh.

Ditsch, Barbara (1961–)
C: 2013, 2015 **H:** DR, LUA; **L:** UI: Serra do Pingano, Municipality of Uíge, Kimbele, Damba, Mucaba; **B:** German botanist at Dresden Botanic Garden.

Finckh, Manfred (1963–)
C: 2011–; **H:** HGB, LUBA, WIND; BI CC HA HI MO: Chitembo (Cusseque), Caiundo, Cachingues, Savate, Cuangar, Bicuar National Park, Cameia National Park, Tundavala Observatory under TFO and SASSCAL Projects; **B:** Ecologist at University of Hamburg, Germany.

Francisco, Domingos Mumbundu (1974–)
C: 2008–; **H:** LISC, LUAI, LUBA; **L:** CA CC LA MA NA ZA: Barra do Cuanza, Iona, Cangandala, Quiçama National Parks; **B:** Angolan botanist at Universidade Agostinho Neto, Centro de Botânica, LUAI Herbarium.

Frisby, Arnold.
C: 2016, 2017; **H:** INBAC, K, LUBA, PRE; **L:** BI CC: Cubango and Cuito Rivers; **B:** South African botanist at University of Pretoria.

Gerrard, Jacqueline
C: 1991; **H:** K; **L:** CU HI NA.

Godinho, Elizeth
C: 2013; **H:** INBAC, K, LISC; **L:** LN: Lagoa Carumbo; **B:** Angolan botanist at INBAC; collected with Darbyshire, Goyder and Kodo.

Göhre, Anne (1990–).
C: 2014–2016; **H:** B, BR, BONN, P; **L:** UI: Municipality of Uíge, Kimbele, Damba, Mucaba; **B:** German botanist at Dresden Botanic Garden.

Gomes, Amândio Luís (1971–).
C: 2010–; **H:** FC-UAN, INBAC, K, LISC, LUAI, LUBA; **L:** BE BI BO CC CN CS HA LN ZA: Lucapa, Lagoa Carumbo, Chitembo (Cusseque), Tundavala Observatory under TFO and SASSCAL Projects; **B:** Angolan botanist at Universidade Agostinho Neto, Luanda; collected with Crawford, Darbyshire and Goyder in LN.

Gonçalves, Francisco Maiato Pedro (1982–).
C: 2008–; **H:** HBG, INBAC, K, LUBA; **L:** BI CC CU CS HA HI LA NA MO: Chitembo (Cusseque), Cumbira forest, Mt. Namba, Lubango Escarpment, Okavango headwaters, Huíla Province SASSCAL Project; **B:** Angolan botanist at Lubango Herbarium, ISCED Huíla, Lubango.

Goyder, David John (1959–)
C: 2011–; **H:** GRA, INBAC, K, LUBA, PRE; **L:** BI CC CS HI LN MO NA: Cumbira, Mt. Namba, Serra da Neve, Lubango Escarpment, Mt. Tchivira, Okavango headwaters, Lucapa, Lagoa Carumbo; **B:** British botanist at Royal Botanic Gardens, Kew; collected with Crawford, Darbyshire, Godinho, Gomes and Kodo in LN, with Barker and Clark on the western escarpment, with Gonçalves in CS and Okavango headwaters, with Barker, Bester, Frisby and Janks in CC.

Harris, Timothy (1982–)
C: 2013; **H:** K, LUAI, PSUB, WIND; **L:** CC: Okavango, Cuito and Cuando Rivers; **B:** British botanist; collected with Murray-Hudson.

Heinze, Christin (1993–).
C: 2014–2017; **H:** DR, LUA; **L:** CN: all municipalities; **B:** German botanist at Technische Universität Dresden.

Janks, Matthew.
C: 2015; **H:** GRA, INBAC, LUBA, PRE; **L:** CC: Okavango, Cuito and Longa Rivers; **B:** South African botanist; collected with Barker, Bester & Goyder.

Jürgens, Norbert (1953–)
C: 2008–; **H:** HGB, WIND, LUBA; **L:** CU HI NA; **B:** Professor at Institute for Plant Science and Microbiology, University of Hamburg, Germany.

Kodo, Felipe
C: 2013; **H:** INBAC, K, LISC; **L:** LN: Lagoa Carumbo; **B:** Angolan botanist at INBAC; collected with Darbyshire, Godinho and Goyder.

Lautenschläger, Thea (1980–)
C: 2012–2018; **H:** DR, LUA; **L:** UI: Municipality of Uíge, Mucaba, Maquela do Zombo, Quitexe, Milunga, Sanza Pombo, Kimbele, Ambuila, Songo, Bungo, Bembe, Puri, Negage, Altocauale, Damba; **B:** German botanist at Technische Universität Dresden.

Luís, José Camôngua (1984–)
C: 2015–. **H:** K, LUBA; **L:** CS HI: Lubango Escarpment, Mt. Namba; **B:** Angolan botanist.

Maiato, Francisco
See **Gonçalves, Francisco Maiato Pedro**.

Manning, Stephen D.
C: 1986–1998.

Matos, Elizabeth (Liz), Merle (1938–)
C: 1975–; **B:** British botanist, founder and director of Angola's National Plant Genetic Resources Centre, Agostinho Neto University, Luanda. Retired in 2008.

Mawunu, Monizi (1973–)
C: 2013–2018; **H:** DR, LUA; **L:** UI, whole province; **B:** Angolan botanist at Universidade Kimpa Vita.

Müller, Frank (1966–)
C: 2015; **H:** DR, LUA; **L:** UI: Municipality of Uíge, Songo, Mucaba; **B:** German botanist at Technische Universität Dresden.

Murray-Hudson, Frances
C: 2013; **H:** K, LUAI, PSUB, WIND; **L:** CC: Okavango, Cuito and Cuando Rivers; **B:** Volunteer at Peter Smith University of Botswana Herbarium (PSUB); collected with Harris.

Neinhuis, Christoph (1962–)
C: 2012–2018; **H:** DR, LUA; **L:** UI: Municipality of Uíge, Mucaba, Maquela do Zombo, Quitexe, Milunga, Sanza Pombo; **B:** German botanist at Technische Universität Dresden, director of the Botanical Garden TU Dresden.

Newman, Mark Fleming (1959–)
C: 1991; **H:** K; **L:** CU HI NA; **B:** British botanist at Royal Botanic Garden Edinburgh. In 1991, at the Seed Bank, Royal Botanic Gardens, Kew; collected with Alcochete, Gerrard and Matos, mainly for seeds, with herbarium voucher specimens for identification.

Rejmánek, Marcel (Marek/Marc) (1946–)
C: 2014; **H:** LUBA, STE; **L:** BE BO CN CS HA HI MA NA UI; **B:** Czech botanist based at University of California, Davis, working on biological invasions. Conducted a rapid inventory of invasive plants in Angola in 2014 with Huntley, Roux and Richardson.

Revermann, Rasmus (1979–)
C: 2011–; **H:** HGB, WIND, LUBA; **L:** BI CC HA HI: Chitembo (Cusseque), Caiundo, Cachingues, Savate, Cuangar under TFO and SASSCAL Projects; **B:** Ecologist at University of Hamburg, Germany.

Roux, Jacobus Petrus (Koos) (1954–2013)
C: 2001; **H:** PRE; **B:** South African Pteridophyte specialist.

Tripp, Erin Anne (1979–)
C: 2017–; **H:** COLO, E, LUBA, WIND; **L:** CU HI NA; **B:** Researcher at Colorado Herbarium, University of Colorado.

References

Abrahamczyk S, Janssens S, Xixima L et al (2016) *Impatiens pinganoensis* (Balsaminaceae), a new species from Angola. Phytotaxa 261:240–250

African Plant Database version 3.4.0 (2018) Conservatoire et Jardin Botaniques de la Ville de Genève and South African National Biodiversity Institute, Pretoria. http://www.ville-ge.ch/musinfo/bd/cjb/africa

Airy-Shaw H (1947) The vegetation of Angola. J Ecol 35:23–48

Albuquerque S (2008) Friedrich Welwitsch. Figueiredo E, Smith GF Plants of Angola/Plantas de Angola. *Strelitzia* 22: 2–3

Albuquerque S, Figueirôa S (2018) Depicting the invisible: Welwitsch's map of travellers in Africa. Earth Sci Hist 37(1):109–129

Albuquerque S, Brummitt RK, Figueiredo E (2009) Typification of names based on the Angolan collections of Friedrich Welwitsch. Taxon 58:641–646

Anderson RJ, Bolton JJ, Smit AJ et al (2012) The seaweeds of Angola: the transition between tropical and temperate marine floras on the west coast of southern Africa. Afr J Mar Sci 34:1–13

Barbosa LAG (1970) Carta Fitogeográfica de Angola. Instituto de Investigação Científica de Angola, Luanda

Brenan JPM (1978) Some aspects of the phytogeography of tropical Africa. Ann Mo Bot Gard 65(2):437–478

Cavaco A (1959) Contribution à l'Étude de la Flore de la Lunda d'Après les Récoltes de Gossweiler (1946–1948*)*. Publicações Culturais da Companhia de Diamantes de Angola 42, 230 pp

Cheek M, Lopez Poveda L, Darbyshire I (2015) *Ledermanniella lunda* sp. nov. (Podostemaceae) of Lunda-Norte, Angola. Kew Bull 70:10

Chisingui AV, Gonçalves FMP, Tchamba JJ et al (2018) Vegetation survey of the woodlands of Huíla Province. Biodivers Ecol 6:426–437

Clark VR, Barker NP, Mucina L (2011) The Great Escarpment of southern Africa: a new frontier for biodiversity exploration. Biodivers Conserv 20:2543–2561

Dandy JE (1958) The Sloane Herbarium. An annotated list of the horti sicci composing it; with biographical accounts of the principal contributors. British Museum (Natural History), London

Darbyshire I, Goyder D, Crawford F, et al (2014) Update to the Report on the Rapid Botanical Survey of the Lagoa Carumbo Region, Lunda-Norte Prov., Angola for the Angolan Ministry of the Environment, following further field studies in 2013, incl. Appendix 2: checklist to the flowering plants, gymnosperms and pteridophytes of Lunda-Norte Prov, Angola. Ministério do Ambiente, Luanda

Darbyshire I, Anderson S, Asatryan A et al (2017) Important Plant Areas: revised selection criteria for a global approach to plant conservation. Biodivers Conserv 26:1767–1800

Droissart V, Dauby G, Hardy OJ et al (2018) Beyond trees: biogeographical regionalization of tropical Africa. J Biogeogr 2018:1–15

Exell AW (1939) Notes on the flora of Angola. IV. 1. Collections from Angola in the Sloane Herbarium. J Bot 77:146–147

Exell AW (1962) Pre-Linnean collections in the Sloane Herbarium from Africa south of the Sahara. In: Fernandes A (ed) Comptes Rendus de la IVe Réunion Plénière de l'Association pour l'Étude Taxonomique de la Flore d'Afrique Tropicale (Lisbonne et Coïmbre, 16–23 Septembre, 1960). Junta de Investigações do Ultramar, Lisbon, pp 47–49

Exell AW, Gonçalves ML (1973) A statistical analysis of a sample of the flora of Angola. Garcia de Orta, Série de Botânica 1:105–128

Figueiredo E (2008) The Rubiaceae of Angola. Bot J Linn Soc 156:537–638

Figueiredo E, César J (2008) Herbaria with collections from Angola/Herbários com colecções de Angola. Strelitzia 22:11–12

Figueiredo E, Smith GF (eds) (2008) Plants of Angola/Plantas de Angola. Strelitzia 22:1–279

Figueiredo E, Smith GF (2012) Common names of Angolan plants. Inhlaba Books, Pretoria

Figueiredo E, Matos S, Cardoso JF et al (2008) List of collectors/Lista de colectores. Strelitzia 22:4–11

Figueiredo E, Smith GF, César J (2009a) The flora of Angola: first record of diversity and endemism. Taxon 58:233–236

Figueiredo E, Soares M, Siebert G et al (2009b) The botany of the Cunene-Zambezi Expedition with notes on Hugo Baum (1867-1950). Bothalia 39:185–211

GoA (Government of Angola) (2011) Plano Estratégico da Rede Nacional de Áreas de Conservação de Angola. Direcção Nacional da Biodiversidade, Ministério do Ambiente, Luanda, 35 pp

Göhre A, Toto-Nienguesse AB, Futuro M et al (2016) Plants from disturbed savannah vegetation and their usage by Bakongo tribes in Uíge, Northern Angola. J Ethnobiol Ethnomed 12:42

Gonçalves FMP, Goyder DJ (2016) A brief botanical survey into Kumbira forest, an isolated patch of Guineo-Congolian biome. PhytoKeys 65:1–14

Gonçalves FMP, Tchamba JJ, Goyder DJ (2016) *Schistostephium crataegifolium* (Compositae: Anthemideae), a new generic record for Angola. Bothalia 46:a2029

Gonçalves FMP, Revermann R, Gomes AL, et al (2017) Tree species diversity and composition of Miombo woodlands in south-central Angola, a chronosequence of forest recovery after shifting cultivation. Int J For Res 2017(Article ID 6202093), 13 pp

Gonçalves FMP, Revermann R, Cachissapa MJ, et al (2018) Species diversity, population structure and regeneration of woody species in fallows and mature stands of tropical woodlands of SE Angola. J Forest Res. Published online 13 January 2018

Gossweiler J (1948) Flora exótica de Angola. Nomes vulgares e origem das plantas cultivadas ou sub-espontâneas. Agronomia Angolana 1:121–198

Gossweiler J (1949) Flora exótica de Angola. Nomes vulgares e origem das plantas cultivadas ou sub-espontâneas. Agronomia Angolana 2:173–255

Gossweiler J (1950) Flora exótica de Angola. Nomes vulgares e origem das plantas cultivadas ou sub-espontâneas. Agronomia Angolana 3:143–167

Gossweiler J (1953) Nomes indígenas das plantas de Angola. Agronomia Angolana 7:1–587

Gossweiler J, Mendonça FA (1939) Carta Fitogeográfica de Angola. Ministério das Colónias, Lisboa, 242 pp

Goyder DJ, Barker N, Bester SP et al (2018) The Cuito catchment of the Okavango system: a vascular plant checklist for the Angolan headwaters. PhytoKeys 113:1–31. https://doi.org/10.3897/phytokeys.113.30439

Heinze C, Ditsch B, Congo MF et al (2017) First Ethnobotanical Analysis of Useful Plants in Cuanza Norte North Angola. Res Rev J Bot Sci 6:44

Hind DJN, Goyder DJ (2014) *Stomatanthes tundavalaensis* (Compositae: Eupatorieae:

Eupatoriinae), a new species from Huíla Province, Angola, and a synopsis of the African species of *Stomatanthes*. Kew Bull 69(9545):1–9

Huntley BJ (1973) Proposals for the creation of a strict nature reserve in the Maiombe forest of Cabinda. Report 16. Repartição Técnica da Fauna, Serviços de Veterinária, Luanda, Mimeograph report, 10 pp

Huntley BJ (1974) Ecosystem conservation priorities in Angola. Report 28. Repartição Técnica da Fauna, Serviços de Veterinária, Luanda, Mimeograph report, 22 pp

Huntley BJ (2009) SANBI/ISCED/UAN Angolan biodiversity assessment capacity building project. Report on Pilot Project. Unpublished Report to Ministry of Environment, Luanda, 97 pp, 27 figures

Huntley BJ (2010) Estratégia de Expansão de Rede da Áreas Protegidas da Angola/Proposals for an Angolan Protected Area Expansion Strategy (APAES). Unpublished Report to the Ministry of Environment, Luanda, 28 pp, map

Huntley BJ (2019) Angola in Outline: Physiography, Climate and Patterns of Biodiversity. In: Huntley BJ, Russo V, Lages F, Ferrand N (eds) Biodiversity of Angola. Science & conservation: a modern synthesis. Springer Nature, Cham

Huntley BJ, Matos EM (1994) Botanical diversity and its conservation in Angola. Strelitzia 1:53–74

IUCN (2018) The IUCN Red List of Threatened Species. Ver. 2017-3. http://www.iucnredlist.org. Downloaded on 27 March 2018

Lautenschläger T, Neinhuis C (eds) (2014) Riquezas Naturais de Uíge – uma Breve Introdução Sobre o Estado Atual, a Utilização, a Ameaça e a Preservação da Biodiversidade. Technische Universität Dresden, Dresden

Lautenschläger T, Monizi M, Pedro M et al (2018) First large-scale ethnobotanical survey in the province Uíge, northern Angola. J Ethnobiol Ethnomed 14:51

Lawson GW, John DM, Price JH (1975) The marine algal flora of Angola: its distribution and affinities. Bot J Linn Soc 70:307–324

Linder HP (2001) Plant diversity and endemism in sub-Saharan tropical Africa. J Biogeogr 28:169–182

Martins ES (1994) John Gossweiler. Contribuição da sua obra para o conhecimento da flora angolana. Garcia de Orta, Série de Botânica 12:39–68

Mawunu M, Bongo K, Eduardo A et al (2016) Contribution à la connaissance des produits forestiers non ligneux de la Municipalité d'Ambuila (Uíge, Angola): Les plantes sauvages comestibles [Contribution to the knowledge of no-timber forest products of Ambuila Municipality (Uíge, Angola): The wild edible plants]. Int J Innov Sci Res 26:190–204

Mendonça FA (1962) Botanical collectors in Angola. In: Fernandes A (ed) Comptes Rendus de la IVe Réunion Plénière de l'Association pour l'Étude Taxonomique de la Flore d'Afrique Tropicale (Lisbonne et Coïmbre, 16–23 septembre, 1960). Junta de Investigações do Ultramar, Lisbon, pp 111–121

Mills MSL, Olmos F, Melo M et al (2011) Mount Moco: its importance to the conservation of Swierstra's Francolin *Pternistis swierstrai* and the Afromontane avifauna of Angola. Bird Conserv Int 21:119–133

Müller F (2014) About 150 years after Welwitsch – a first more extensive list of new bryophyte records for Angola. Nova Hedwigia 100:487–505

Müller F, Sollman P, Lautenschläger T (2018) A new synonym of *Weissia jamaicensis* (Pottiaceae, Bryophyta) and an extension of the range of the species from the Neotropics to the Palaeotropics. Plant Fungal Syst 63(1):1–5

Neinhuis C, Lautenschläger T (2014) The potentially natural vegetation in Uíge province and its current status – arguments for a protected area in the Serra do Pingano and adjacent areas. Unpublished Report to Ministry of Environment, Luanda, 64 pp

NGOWP (National Geographic Okavango Wilderness Project) (2018) *Initial findings from exploration of the upper catchments of the Cuito, Cuanavale, and Cuando Rivers, May 2015 to December 2016.* Report prepared for and submitted to the Ministério do Ambiente of the Republic of Angola, the Ministry of Environment, Wildlife and Tourism Botswana, and the Ministry of Environment and Tourism of the Republic of Namibia. Available (as Report 1) from: http://www.wildbirdtrust.com/owp-publications/

Oldeland J, Erb C, Finckh M, Jürgens N (eds) (2013) Environmental assessments in the Okavango region. Biodivers Ecol 5:1–418

Rejmánek M, Huntley BJ, le Roux JJ, Richardson DM (2017) A rapid survey of the invasive plant species in western Angola. Afr J Ecol 55:56–69

Revermann R, Gonçalves FM, Gomes AL et al (2017) Woody species of the miombo woodlands and geoxylic grasslands of the Cusseque area, south-central Angola. Check List 13:2030

Revermann R, Oldenland J, Gonçalves FM et al (2018) Dry tropical forests and woodlands of the Cubango basin in southern Africa – First classification and assessment of their woody species diversity. Phytocoenologia 48:23–50

Smith GF, Figueiredo E (2017) Determining the residence status of widespread plant species: studies in the flora of Angola. Afr J Ecol 55:710–713

Soares M, Abreu J, Nunes H et al (2009) The Leguminosae of Angola: diversity and endemism. Syst Geogr Plants 77:141–212

Sosef MSM, Dauby G, Blach-Overgaard A et al (2017) Exploring the floristic diversity of tropical Africa. BMC Biol 15:15

Teixeira JB (1968a) Angola. In: Hedberg I, Hedberg O (eds.) Conservation of vegetation in Africa south of the Sahara. Proceedings of a symposium held at the 6th plenary meeting of the "Association pour l'Etude Taxonomique de la Flore d'Afrique Tropicale" (A.E.T.F.A.T.) in Uppsala, Sept. 12th–16th, 1966. *Acta Phytogeographica Suecica* 54:193–197

Teixeira JB (1968b) Parque Nacional do Bicuar. Carta da Vegetação (1ª aproximação) e Memória Descritiva. Instituto de Investigação Agronómica de Angola, Nova Lisboa

Thiers B (continuously updated). Index Herbariorum: A global directory of public herbaria and associated staff. New York Botanical Garden's Virtual Herbarium. http://sweetgum.nybg.org/science/ih/

Wallenfang J, Finckh M, Oldeland J et al (2015) Impact of shifting cultivation on dense tropical woodlands in southeast Angola. Trop Conserv Sci 8:863–892

Warburg O (1903) Kunene-Sambesi-Expedition. Kolonial-Wirtschaftliches Komitee, Berlin

White F (1983) The Vegetation of Africa – A Descriptive Memoir to Accompany the UNESCO/AETFAT/UNSO Vegetation Map of Africa. UNESCO, Paris 356 pp

6

Surveying, Mapping and Classifying Angolan Vegetation

Rasmus Revermann and Manfred Finckh

Abstract Spatial information about plant species composition and the distribution of vegetation types is an essential baseline for natural resource management planning. In Angola, the first countrywide vegetation map was elaborated by Gossweiler in 1939. Subsequently, Barbosa published a revised map with much higher detail in 1970 and his work has remained the main reference for the vegetation of Angola until today. However, these early maps were expert drawn and were not based on systematic surveys. Instead, the delimitation of vegetation units was based on many years of field observations and also incorporated results of local studies carried out by other authors. In spite the rich history of the scientific exploration of Angola's vegetation in colonial times, quantitative and plot based studies were rare. After the end of the armed conflict, new vegetation surveys making use of new methodological developments in numerical approaches to vegetation classification in combination with modern remote sensing imagery have provided spatial information of unprecedented detail. However, vast areas of the country still remain seriously understudied. At the same time, sustainable land management strategies are urgently needed due to the increasing pressure on natural resources driven by socio-economic development and global change, thus calling for a new era of vegetation surveys that will enable data-based landuse and conservation planning in Angola.

Keywords Conservation · Landuse planning · Natural resources · Plant communities · Remote sensing

---

R. Revermann (✉) · M. Finckh
Institute for Plant Science and Microbiology, University of Hamburg, Hamburg, Germany
e-mail: rasmus.revermann@gmail.com; manfred.finckh@uni-hamburg.de

Introduction

Knowledge on the spatial distribution of vegetation and its species composition is paramount for any kind of natural resource management and conservation planning. Vegetation serves as habitat for other organismic groups and is the source of energy in the ecosystem. As such, vegetation integrates many ecological processes and reflects patterns of topography, geology, soil, hydrology and climate. Thus, vegetation classification is ideal to provide an aggregated image of the landscape and its ecological communities.

Historical Exploration of Vegetation Patterns in Angola

First reports on the vegetation of Angola were directly linked to the floristic exploration of the country, as outlined by Goyder and Gonçalves (2019). Scientific missions during colonial times in Angola served several purposes: on the one hand they should chart the potential for economic exploitation and development while on the other hand they may also have been used to demonstrate the supremacy of the colonial power (Gago et al. 2016). The expedition by the geographer Jessen (1936) provided a first sketch of the vegetation along the routes of his transects through western Angola. Jessen's work remains a classic as he was among the first to document the landscape and ecosystem properties of the region. However, it is hardly read today as it is only available in German.

The systematic descriptions of the vegetation of Angola started with Gossweiler and Mendonça's (1939) phytogeographical map of Angola. The often-cited English summary by Shaw (1947) contributed much to the recognition of Gossweiler's work internationally. The map is based on the combined structural and ecological approach to vegetation classification developed by Brockmann-Jerosch and Rübel (1912) in Zurich. Thus, in a first level of classification they categorised the vegetation according to woodiness and persistence into the three categories Lignosa (woody), Herbosa (herbaceous) and Deserta (land surfaces without permanent vegetation cover). The next step of the classification included climatic and edaphic factors, as well as leaf traits, leading for instance to five sub-categories of woody vegetation called Pluviilignosa, Laurilignosa, Durilignosa, Ericilignosa, Aestililignosa and Hiemilignosa. Stand structure was the main criterion for the next categories, dividing the afore-mentioned categories between tall forests (-silva) and dense but low forests (-fruticeta) (e.g. Pluviisilva vs. Pluviifruticeta or Durisilva vs. Durifruticeta). Below this third level we finally find floristically defined vegetation units, albeit mostly named after one or two dominant species. Similar structural criteria were used for the sub-classification of the Herbosa and Deserta.

The vegetation map used this rather rigid classification scheme for the 19 main mapping units. However, the resulting map apparently did not fully satisfy the authors, who then applied 29 additional symbols to indicate occurrence of small-scale vegetation units, of transition zones and of species that appeared to be of special interest to the authors – a very nice real world example of dealing with rigid mapping manuals. However, neglecting these small methodological inconsistencies, the map by Gossweiler & Mendonça presented the first overall picture of the vegetation of Angola, a first approach towards a systematic compilation of observations of phytogeographical patterns and a first attempt at ecological interpretation. While many of the mapped polygons seem outdated in times of modern earth observation, the number of observed details in remote parts of Angola is still surprising for today's botanists. The authors were probably the first to report for Angola on invasive species, on seed dispersal by bats, on the morphological plasticity of the genus Syzygium and on many other current scientific topics. Also quite astonishing was the classification of the suffrutex grassland within the woody vegetation types (Ericifruticeta), more than 30 years before White (1976) published the groundbreaking paper on the 'Underground forests of Africa'.

The next important integrating step towards a synthesis on the vegetation units of Angola and their spatial distribution was Grandvaux Barbosa's phytogeographical map of Angola (Barbosa 1970). His work can be seen as a continuation and extension of the Gossweiler approach. The map clearly benefited from several regional studies that had been carried out in the meantime (see below) and of course also from Barbosa's own knowledge gained during several field missions throughout the country and his extensive experience of similar vegetation types found in Mozambique. As ancillary information Barbosa included descriptions of the main soil types and climatic zones of Angola.

The mapping approach adopted by Barbosa was to some degree harmonised with the parallel efforts of the Flora Zambesiaca map and the UNESCO initiative mapping the vegetation of Africa. The first level of classification differentiates the vegetation based on the formation, i.e. deals with the physiognomy of the vegetation such as closed forests, forest savanna mosaics, woodlands etc. and beyond that includes azonal edaphic vegetation units such as mangrove stands and coastal dune vegetation. In the second level of classification vegetation types are distinguished according to dominant species. In total, the map by Barbosa displays 32 main vegetation types and the descriptive text accompanying the map provides details on over 100 subordinate types (for a brief summary in English see Barbosa 1971).

The result was a good overview of the main vegetation types of Angola, in terms of spatial patterns much superior to the first attempt by Gossweiler and Mendonça (1939). Until today the vegetation units of the map by Barbosa (1970) constitute the foundation for the Angolan part of most continental or global scale vegetation maps (see below). However, due to Barbosa's floristic rather than ecological emphasis, the report on the vegetation units did neither contribute much to a better understanding of the ecology of the main vegetation patterns, nor did it make use of a modern classification concept based on plant communities.

Shortly after Barbosa's vegetation map, Diniz (1973) published a monograph on the physical properties of the agricultural zones of Angola. Included in this monograph, are soil and vegetation maps for 36 agricultural zones, albeit in rather fragmented components and without an overall map of the country. The vegetation classification scheme he used is not clearly defined, somewhere in between those of Gossweiler and Barbosa, but sometimes with more detail than Barbosa (1970). The main achievement of Diniz (1973) is that he assembled sound environmental information (with a focus on geology and soils) for all delimited agricultural zones. However, due to the lack of a seamless map and his unclear classification approach his contribution to the knowledge on the vegetation of Angola did not receive much attention in the subsequent scientific literature and due to the violent conflicts following Angola's independence in 1975, Barbosa's work has remained the main reference on the vegetation of Angola.

Integration of the Vegetation Map of Angola Within Continental Scale Maps

The next important step for a better understanding of Angolan vegetation was the UNESCO/AETFAT/UNSO initiative for a Vegetation Map of Africa (UNESCO/ AETFAT/UNSO 1981), compiled and described by White (1983). For Angola the continental map is largely based on the units supplied by Barbosa (1970) but they were subject to further generalisation resulting in only 14 mapping units compared to Barbosa's 32 vegetation types. However, the important achievement of White's map lies in the fact that it inserted Angolan vegetation in a common conceptual and methodological framework with the vegetation of neighbouring countries and the African continent as a whole. As such, the UNESCO (UNESCO/AETFAT/UNSO 1981)/(UNESCO 1981) map and White's (1983) description established the now widely used term 'miombo woodlands' in our scientific and geographic frameworks, allowing thus for the comparison of Angola's ecosystems with similar vegetation types throughout Africa. Although Barbosa and White provided seamless maps covering the entire country, the level of information supporting the mapping units varies strongly and for some units, especially in the more remote eastern parts of the country, barely any details are given. All these early maps are based on expert knowledge and no quantitative data was involved in the process of map making.

The Vegetation Map of Africa then again was the main baseline for the WWF's approximation of the world's terrestrial ecoregions (Olson et al. 2001) in so far as the African continent was concerned. Although without a presentation of a systematic biogeographical database, the map of the terrestrial ecoregions currently constitutes the most used baseline map for strategic conservation planning on a continental and subcontinental scale (e.g. MacKinnon et al. 2016). The availability of modern remote sensing techniques has allowed the generation of continental or global land cover products, i.e. GlobCover, MODIS/Terra Land Cover, GlobLand30 or the map of African ecosystems by Sayre et al. (2013). However, these maps display structural vegetation types only and provide no floristic information.

Regional and Local Studies on Vegetation Composition

Early plot based studies were conducted by Ilse von Nolde on the Planalto de Quela (von Nolde 1938a, b, c). Since the mid-1950s, several local studies based on missions assessing natural resources were carried out at the regional level in Angola. Monteiro studied the forest resources in Moxico (Monteiro 1957), in the northern Maiombe and Dembos forests (Monteiro 1962, 1965a, b, 1967), and in Bié (Monteiro 1970a) contributing to our knowledge on species composition in the respective forest types. Monteiro's (1970a, b) work in Bié needs to be highlighted as for Angola he implemented new methods in mapping the vegetation. His map of the woody vegetation of the province of Bié is not drawn based on pure observations but is based on quantitative vegetation plot data. He collected data on species composition in 144 vegetation relevés sized 30 m × 30 m that were subject to a vegetation classification based on vegetation tables. The mapping process was guided by aerial photography, quite an advanced approach for its time. Menezes (1965, 1971) undertook phytosociological studies and produced local vegetation maps in pastoral ecosystems of the Cunene Province. Teixeira elaborated vegetation maps for two of the main protected areas of Angola, the Quiçama and Bicuar national parks (Teixeira et al. 1967; Teixeira 1968). A few years later, Huntley produced a much more detailed map of the Quiçama National Park in 1972 at a scale of 1:100000 depicting 28 plant communities (Huntley 1972). Aguiar and Diniz (1972) mapped the vegetation of the western plateau of Cela. Coelho explored the potential of forestry in Cuando Cubango and elaborated a classification of the lower Cubango Basin into 32 forestry zones (Coelho 1964, 1967). Santos (1982) used a transect approach, so called 'itinerários florísticos', in order to generate an expert-drawn vegetation map for Cuando Cubango Province (Fig. 6.1).

Modern Approaches to Vegetation Mapping and Classification

This early period of vegetation mapping and classification was followed by the absence of any such activities for the coming decades due to the long-lasting armed conflict in the country. During this period significant methodological advances were made in vegetation ecology and phytosociology as well as in remote sensing techniques. The advent of computers allowed the development of new methodological tools to semi-automatically classify large amounts of multivariate vegetation plot data based on objective criteria. As such, vegetation classification moved away from the subjective assignments of vegetation types to more formalised data analysis. Similarly, remote sensing imagery became readily available often at no cost and in unprecedented temporal and spatial resolution. Thus, new numerical methods together with modern remote sensing products have the potential to provide a much more detailed and objective picture of vegetation and plant diversity patterns than expert drawn maps and arbitrarily assigned vegetation types of earlier times (Fig. 6.2).

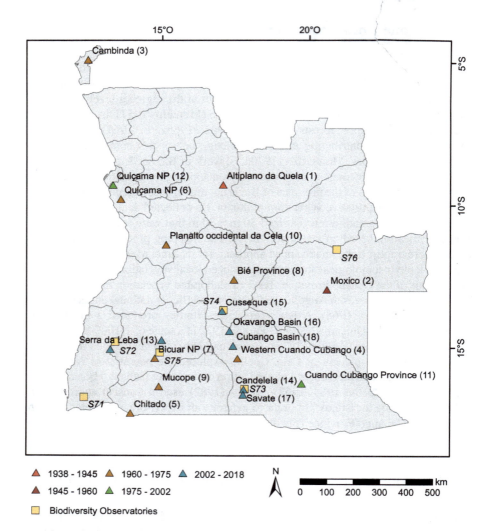

Fig. 6.1 Location of regional and local studies on vegetation composition, vegetation classification or vegetation mapping approaches according to the year the study was published. The country-wide maps by Gossweiler and Mendonça (1939), Barbosa (1970) and Diniz (1973) are not depicted. (1) von Nolde 1938a, b, c (2) Monteiro 1957 (3) Monteiro 1962 (4) Coelho 1964 (5) Menezes 1965 (6) Teixeira et al. 1967, Huntley 1972 unpublished (7) Teixeira 1968 (8) Monteiro 1970a (9) Menezes 1971 (10) Diniz and Aguiar (1968) (11) dos Santos 1982 (12) De Bruyn and Eberle 2001 (13) Cardoso et al. 2006 (14) Revermann and Finckh 2013a (15) Revermann et al. 2013, Schneibel et al. 2013, Gonçalves et al. 2017 (16) Revermann and Finckh 2013b, Stellmes et al. 2013 (17) Wallenfang et al. 2015 (18) Revermann 2016, Revermann et al. 2018a (19) Chisingui et al. 2018. Furthermore, the six biodiversity observatories installed by the SASSCAL project are shown: Espinheira (S71), Tundavala (S72), Candelela (S73), Cusseque (S74), Bicuar National Park (S75), Cameia National Park (S76)

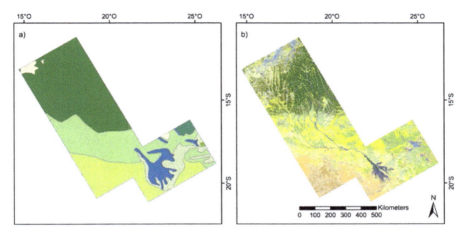

Fig. 6.2 Maps for the Okavango Basin located in southeast Angola and extending into Namibia and northern Botswana. (**a**) shows the ecoregions as defined by Olson et al (2001) which are largely based on the vegetation maps of Barbosa (1970) and White (1983), (**b**) Vegetation map produced for the same area by The Future Okavango project based on unsupervised classification of land surface phenology metrics derived from 16–day MODIS EVI time series from the years 2000–2011 (Stellmes et al. 2013) and interpreted using the information of vegetation plots stored in the vegetation database of the Okavango Basin (Revermann and Finckh 2013b; Revermann et al. 2016a). For an explanation of the vegetation units depicted in the maps please refer to the original publications

Recent years have seen increasing activity in the investigation of vegetation patterns at the local and regional scale. During the years 1995–2002 some vegetation surveys were carried out in the Quiçama National Park south of Luanda, for which the map elaborated by Huntley in the year 1972 served as a baseline. The activities aimed at gathering data for the re-establishment of the national park and to develop management strategies (Jeffery et al. 1996). De Bruyn and Eberle (2001) studied a small fenced of area in the north of the park where they collected 74 relevés and identified four plant communities including eight subcommunities. Additional quantitative data was collected to investigate grazing and browsing capacities. Cardoso et al. (2006) studied the vegetation communities along the steep altitudinal gradient of the Serra da Leba near Lubango.

Within The Future Okavango (TFO, www.future-okavango.org) project detailed investigations have been carried out in the Okavango (Cubango) River Basin. The project team assembled a vegetation database containing vegetation relevé data on all terrestrial vegetation types within the Okavango Basin (Revermann et al. 2016a). The plot design followed the standards implemented for woodland vegetation in the southern neighbour countries, i.e. a nested plot design of one small 10 m × 10 m plot in a large 20 m × 50 m plot (Strohbach 2001; Jürgens et al. 2012). Based on this data, classifications for local study sites based on numerical classification approaches have been published (Revermann and Finckh 2013a; Wallenfang et al. 2015) and a first classification of the terrestrial vegetation of the entire Cubango Basin was elaborated (Revermann et al. 2018a).

The vegetation database of the Okavango Basin was also the foundation to produce a first vegetation map based on quantitative ground data for the Okavango Basin (Fig. 6.2b Revermann and Finckh 2013b; Stellmes et al. 2013) and allowed modelling the α-diversity of vascular plants for the same region (Revermann et al. 2016b).

Based on vegetation relevés various studies have investigated the impact of land use on vegetation (Revermann et al. 2017) and studied the regeneration of the vegetation after land use had ceased (Wallenfang et al. 2015; Gonçalves et al. 2017, 2018).

Presently a number of vegetation classification and mapping initiatives are underway in the framework of the research project SASSCAL (Southern African Science Service Centre for Climate Change and Adaptive Land Management). For a compilation of project outcomes see Revermann et al. 2018b, e.g. in the Huíla Province (Chisingui et al. 2018) and along the coastal plain from the Cunene River to Benguela including Iona National Park (Jürgens et al. in prep.). The same project includes six newly implemented biodiversity observatories (http://www.sasscalobservationnet.org/), depicted on Fig. 6.1. The standardised monitoring of the 1 km² sites (Jürgens et al. 2012) will allow the long term monitoring of changes in plant species composition and plant diversity. Zigelski et al. (2018) present first analyses of the data gathered on such a biodiversity observatory in the Cameia National Park.

Outlook: A Call for a New Vegetation Survey of Angola

Vegetation and natural resources in general are under strong pressure from the increasing demands of a growing population and the transition from traditional lifestyles to modern consumerism (cf. Pröpper et al. 2015). The main drivers of deforestation and degradation of woodlands and the general loss of pristine vegetation cover in Angola are the clearing of new fields for shifting cultivation, industrialised agricultural schemes and the production of charcoal (Cabral et al. 2010; Hansen et al. 2013; Schneibel et al. 2013, 2016, 2018; Röder et al. 2015; Wallenfang et al. 2015; Mendelsohn 2019). Without adequate knowledge of the spatial distribution and extent of vegetation types, their species composition and the environmental drivers of vegetation patterns (climate, geology, soils, landuse) sound landuse management is not feasible. Thus, a nationwide vegetation survey based on quantitative, plot level data is urgently needed. Combined with remote sensing data and ecological modelling tools an accurate vegetation map can be produced serving the needs of conservationists, planners, entrepreneurs and scientists alike. A successful vegetation survey however relies on good taxonomic knowledge, current flora compendia and plant identification guides. Functioning and strengthened herbaria are also of great importance building the capacity of the future generation of field ecologists and environmental scientists.

References

Aguiar FQB, Diniz AC (1972) *Carta de Vegetação do Planalto Ocidental da Cela: Estudo Interpretativo*. Série Científica N° 26. Instituto de Investigação Agronómica de Angola, Nova Lisboa

Barbosa LAG (1970) Carta Fitogeográfica de Angola. Instituto de Investigação Científica de Angola, Luanda, 323 pp

Barbosa LAG (1971) Phytogeographical map of Angola. Mitteilungen der Botanischen Staatssammlung München 10:114–115

Brockmann-Jerosch H, Rübel E (1912) Die Einteilung der Pflanzengesellschaften nach okologisch-physiognomischen Gesichtspunkten. Engelmann, Leipzig, 72 pp

Cabral AIR, Vasconcelos MJ, Oom D et al (2010) Spatial dynamics and quantification of deforestation in the central-plateau woodlands of Angola (1990–2009). Appl Geogr 31(3):1185–1193

Cardoso J, Duarte M, Costa E et al (2006) Communidades vegetais da Serra da Leba. In: Moreira I (ed) Angola: Agricultura, Recursos Naturais e Desenvolvimento. ISA Press, Lisbon, pp 205–223

Chisingui AV, Gonçalves FMP, Tchamba JJ et al (2018) Vegetation survey of the woodlands of Huíla province. In: Revermann R, Krewenka KM, Schmiedel U et al (eds) Climate change and adaptive land management in southern Africa – assessments, changes, challenges, and solutions, Biodivers & Ecol, vol 6. Klaus Hess Publishers, Göttingen & Windhoek, pp 426–437

Coelho H (1964) Contribuição para o Conhecimento da Composição Florística e Possibilidades de uma Zona Compreendida entre os rios Cubango, Cueio e Quatir. Agronomia Angolana 20:49–82

Coelho H (1967) Zonagem Florestal do Distrito do Cuando-Cubango. Primeiros elementos. Agronomia Angolana 26:3–28

De Bruyn PJN, Eberle D (2001) An ecological study of the plant communities of the fenced sector of the Quiçama National Park, Angola, with Management Recommendations. B.Sc. (Hons) Thesis. University of Pretoria, Pretoria

Diniz AC (1973) Características mesológicas de Angola. Missão de Inquéritos Agrícolas de Angola, Nova Lisboa, 482 pp

Diniz AC, Aguiar FB (1968) Regiões Naturais de Angola. Série Científica N° 2. Instituto de Investigação Agronómica de Angola, Nova Lisboa, 6 pp + 1 map

Gago MM, Macedo M, Castelo C (2016) Surveying Angola, São Tomé and Timor: experts and transnational practices. In: Serrão JV, Freire D, Fernández L (eds) Old and new worlds: the global challenges of rural history, Conference eBook. ISCTE – Instituto Universitário de Lisboa. Centro de Investigação e Estudos de Sociologia, Lisboa

Gonçalves FMP, Revermann R, Gomes AL, et al. (2017) Tree species diversity and composition of Miombo woodlands in south-central Angola, a chronosequence of forest recovery after shifting cultivation. Int J For Res 2017(6202093), 13 pp.

Gonçalves FMP, Revermann R, Cachissapa MJ, et al (2018) Species diversity, population structure and regeneration of woody species in fallows and mature stands of tropical woodlands of SE Angola. J For Res. Published online 13 January 2018

Gossweiler J, Mendonça FA (1939) Carta Fitogeográfica de Angola. Ministério das Colónias, Lisbon, 242 pp

Goyder DJ, Gonçalves FMP (2019) The flora of Angola: collectors, richness and endemism. In: Huntley BJ, Russo V, Lages F, Ferrand N (eds) Biodiversity of Angola. Science & conservation: a modern synthesis. Springer Nature, Cham

Hansen MC, Potapov PV, Moore R et al (2013) High-resolution global maps of 21st-century forest cover change. Science 342(6160):850–853

Huntley BJ (1972) Parque Nacional da Quiçama. Carta da Vegetação, 1° Aproximação Julho 1972. Ecologist's report 22. Repartição Técnica da Fauna, Serviços de Veterinária, Luanda, Mimeograph report

Jeffery RF, van der Waal C, Radloff F (1996) An ecological evaluation with management guidelines for the re-establishment of the Quiçama National Park, Angola. B.Sc. (Hons) thesis. University of Pretoria, Pretoria

Jessen O (1936) Reisen und Forschungen in Angola. Dietrich Reimer Verlag, Berlin, 397 pp
Jürgens N, Schmiedel U, Haarmeyer DH et al (2012) The BIOTA biodiversity observatories in Africa-a standardized framework for large-scale environmental monitoring. Environ Monit Assess 184(2):655–678
MacKinnon J, Aveling C, Olivier R et al (2016) Inputs for an EU strategic approach to wildlife conservation in Africa – regional analysis. European Commission, Directorate-General for International Cooperation And Development, Brussels, 494 pp
Mendelsohn JM (2019) Landscape changes in Angola. In: Huntley BJ, Russo V, Lages F, Ferrand N (eds) Biodiversity of Angola. Science & conservation: a modern synthesis. Springer Nature, Cham
Menezes JA (1965) Estudo fitosociológico e características das pastagens da região do Chitado. Boletim do Instituto de Investigação Científica de Angola 2(2):137–181
Menezes JA (1971) Estudo fitoecológico da região de Mucope e carta da vegetação. Boletim Instituto de Investigação Científica de Angola 8(2):7–54
Monteiro RFR (1957) Aspectos da exploração florestal no distrito do Moxico. Garcia de Orta 5(1):129–146
Monteiro RFR (1962) Le massif forestier du Mayombe angolais. Revue Bois et Forêts des Tropiques 82:3–17
Monteiro RFR (1965a) A formação florestal dos Dembos. Boletim do Instituto de Investigação Científica de Angola 2(1):71–82
Monteiro RFR (1965b) Correlação entre as florestas do Maiombe e dos Dembos. Indicação de factores predominantes. Boletim do Instituto de Investigação Científica de Angola 1(2):257–265
Monteiro RFR (1967) Essências Florestais de Angola. Estudo das Suas Madeiras. Espécies do Maiombe. Instituto de Investigação Científica de Angola, Luanda
Monteiro RFR (1970a) Estudo da Flora e da Vegetação das Florestas Abertas do Plantalto do Bié. Instituto de Investigação Científica de Angola, Luanda, 352 pp
Monteiro RFR (1970b) Alguns Elementos de Interese Ecológico da Flora Lenhosa do Planalto do Bié (Angola). Instituto de Investigação Científica de Angola, Luanda, 166 pp
Olson DM, Dinerstein E, Wikramanayake ED et al (2001) Terrestrial ecoregions of the world: a new map of life on earth. Bioscience 51(11):933–938
Pröpper M, Gröngröft A, Finckh M et al (2015) The future Okavango – findings, scenarios and recommendations for action. Research project final synthesis report 2010-2015. Hamburg, 190 pp
Revermann R (2016) Analysis of vegetation and plant diversity patterns in the Okavango basin at different spatial scales – integration of field based methods, remote sensing information and ecological modelling. PhD thesis, University of Hamburg Hamburg, 295 pp
Revermann R, Finckh M (2013a) Caiundo – vegetation. Biodivers Ecol 5:91–96
Revermann R, Finckh M (2013b) Okavango basin – vegetation. Biodivers Ecol 5:29–35
Revermann R, Gomes A, Gonçalves FM et al (2013) Cusseque – vegetation. Biodivers Ecol 5:59–63
Revermann R, Gomes AL, Gonçalves FM et al (2016a) Vegetation database of the Okavango basin. Phytocoenologia 46(1):103–104
Revermann R, Finckh M, Stellmes M et al (2016b) Linking land surface phenology and vegetation-plot databases to model terrestrial plant alpha diversity of the Okavango Basin. Remote Sens 8:370
Revermann R, Wallenfang J, Oldeland J et al (2017) Species richness and evenness respond to diverging land-use patterns – a cross-border study of dry tropical woodlands in southern Africa. Afr J Ecol 55:152–161
Revermann R, Oldeland J, Gonçalvess FM et al (2018a) Dry tropical forests and woodlands of the Cubango Basin in southern Africa – first classification and assessment of their woody species diversity. Phytocoenologia 48(1):23–50
Revermann R, Krewenka KM, Schmiedel U et al (eds) (2018b) Climate change and adaptive land management in southern Africa – assessments, changes, challenges, and solutions. Biodivers Ecol 6:1–497
Röder A, Pröpper M, Stellmes M et al (2015) Assessing urban growth and rural land use transformations in a cross-border situation in northern Namibia and southern Angola. Land Use Policy 42:340–354

Santos RM (1982) Itinerários Florísticos e Carta da Vegetacão do Cuando Cubango. Instituto de Investigação Científica Tropical, Lisbon, 265 pp

Sayre R, Comer P, Hak J et al (2013) A new map of standardized terrestrial ecosystems of Africa. Association of American Geographers, Washington, DC, 24 pp

Schneibel A, Stellmes M, Revermann R et al (2013) Agricultural expansion during the post-civil war period in southern Angola based on bi-temporal Landsat data. Biodivers Ecol 5:311–320

Schneibel A, Stellmes M, Röder A et al (2016) Evaluating the trade-off between food and timber resulting from the conversion of Miombo forests to agricultural land in Angola using multi-temporal Landsat data. Sci Total Environ 548-549:390–401

Schneibel A, Röder A, Stellmes M et al (2018) Long-term land use change analysis in south-central Angola. Assessing the trade-off between major ecosystem services with remote sensing data. Biodivers Ecol 6:360–367

Shaw HKA (1947) The vegetation of Angola. J Ecol 35(1):23–48

Stellmes M, Frantz D, Finckh M et al (2013) Okavango basin – earth observation. Biodivers Ecol 5:23–27

Strohbach BJ (2001) Vegetation survey of Namibia. J Namibia Sci Soc 49:93–124

Teixeira JB (1968) Parque Nacional do Bicuar. Carta da vegetação (1ª aproximação) e Memória Descritiva. Instituto de Investigação Agronómica de Angola, Nova Lisboa

Teixeira JB, Matos GC, Sousa JNB (1967) Parque Nacional da Quiçama. Carta da Vegetação e Memória Descritiva. Instituto de Investigação Agronómica de Angola, Nova Lisboa

UNESCO/AETFAT/UNSO (1981) Vegetation map of Africa – scale 1:5 000 000. In: White F (ed) UNESCO, Paris

von Nolde I (1938a) Probeflächen verschiedener Savannenformationen im Hochland von Quela in Angola. Notizblatt des Botanischen Gartens und Museums zu Berlin-Dahlem 14:298–311

von Nolde I (1938b) Probeflächen verschiedener Waldformationen aus dem Hochland von Quela in Angola. Notizblatt des Botanischen Gartens und Museums zu Berlin-Dahlem 14:483–486

von Nolde I (1938c) Botanische Studie über das Hochland von Quela in Angola. Feddes Repertorium Beihefte 101:35

Wallenfang J, Finckh M, Oldeland J et al (2015) Impact of shifting cultivation on dense tropical woodlands in southeast Angola. Trop Conserva Sci 8(4):863–892

White F (1976) The underground forests of Africa: a preliminary review. Gard Bull Singapore 11:57–71

White F (1983) The vegetation of Africa – a descriptive memoir to accompany the Unesco/AETFAT/UNSO vegetation map of Africa. UNESCO, Paris, 356 pp

Zigelski P, Lages F, Finckh M (2018) Seasonal changes of biodiversity patterns and habitat conditions in a flooded savanna – the Cameia national park biodiversity observatory in the upper Zambezi catchment, Angola. In: Revermann R, Krewenka KM, Schmiedel U et al (eds) Climate change and adaptive land management in southern Africa – assessments, changes, challenges, and solutions, Biodivers Ecol, vol 6. Klaus Hess Publishers, Göttingen & Windhoek, pp 438–447

Geoxylic Suffrutices and Suffrutex-Grasslands in Angola

Paulina Zigelski, Amândio Gomes and Manfred Finckh

Abstract A small-scale mosaic of miombo woodlands and open, seasonally inundated grasslands is a typical aspect of the Zambezian phytochorion that extends into the eastern and central parts of Angola. The grasslands are home to so-called 'underground trees' or *geoxylic suffrutices*, a life form with massive underground wooden structures. Some (but not all) of the *geoxylic suffrutices* occur also in open woodland types. These iconic dwarf shrubs evolved in many plant families under similar environmental pressures, converting the Zambezian phytochorion into a unique evolutionary laboratory. In this chapter we assemble the current knowledge on distribution, diversity, ecology and evolutionary history of geoxylic suffrutices and suffrutex-grasslands in Angola and highlight their conservation values and challenges.

Keywords Endemism · Geoxyles · Miombo · Phytochorion · Underground forests · Vegetation

Introduction

Open grassy vegetation is a common aspect of Angolan landscapes and is a characteristic part of the Zambezian phytochorion. Grasses are the most conspicuous element of these landscapes towards the end of the rainy season, whereas at the onset of the rainy season many woody species of so called *geoxylic suffrutices* or 'underground trees' (Davy 1922; White 1976) dominate the aspect of the vegetation. Thus, in vast areas of central and eastern Angola, the open 'grasslands' are de-facto co-dominated by grasses and geoxylic suffrutices. Closely intertwined with miombo

---

P. Zigelski (✉) · M. Finckh
Institute for Plant Science and Microbiology, University of Hamburg, Hamburg, Germany
e-mail: paulina.meller@gmx.de; paulina.meller@studium.uni-hamburg.de;
manfred.finckh@uni-hamburg.de

A. Gomes
Institute for Plant Science and Microbiology, University of Hamburg, Hamburg, Germany

Faculty of Sciences, Agostinho Neto University, Luanda, Angola
e-mail: amandiogomes2@hotmail.com

woodlands and with wetlands, suffrutex-grasslands constitute one of the main and most particular ecosystem types of Angola. According to Mayaux et al. (2004), they cover at least 70,080 km^2 or 5.6% of the Angolan territory (not including the small scale woodland suffrutex-grassland mosaics of the central Angolan plateau).

The geoxylic suffrutex life form is marked by proportionally massive underground woody organs, in literature often termed as *lignotuber, xylopodia* or *woody rhizomes*. Annual shoots sprout readily from the buds on these perennial woody organs, bearing leaves, inflorescences and fruits before they die back after the end f the rainy season. Coexistence of grasses and suffrutices is made possible by cupation of different ecological niches together with phase-delayed activity peri- ; (i.e. main assimilation/flowering/fruiting time) that reduces competition.

Exploration of Geoxylic Grasslands

The first authors who indicated the distribution and ecological particularity of suffrutex-grasslands in Angola were Gossweiler and Mendonça (1939), who classified them as heathland-like woodlands ('Ericilignosa'). They already noted the main differentiation between the *Cryptosepalum* spp. dominated suffrutex communities ('Anharas de Ongote') on ferralitic and psammoferralitic soils and the vegetation types characterised by *Parinari capensis* and the Apocynaceae *Landolphia thollonii* and *L. camptoloba* on leached sandy soils ('Chanas da Borracha'). They had also already observed the strong thermic oscillations of which at least the 'Anharas de Ongote' are subject (see below) and commented on the generative cycle of *Cryptosepalum maraviense* from flowering to fruiting in the dry season (and thus, being inverse to the generative cycle of the C4-grasses).

Using a different mapping and classification approach, typical suffrutex-grasslands mostly on sandy soils were again mapped and described by Barbosa (1970) as 'Chanas da Borracha' (alluding to the presences of species of the genus *Landolphia*), 'Chanas da Cameia', and 'Anharas do Alto'. The *Cryptosepalum* spp. dominated 'Anharas de Ongote' on ferralitic soils are described (but not depicted on the map) as being inserted in the main miombo types of the Angolan plateau. However, he describes the typical spatial pattern, i.e. how they appear close to the headwaters of the small tributaries and then follow the watercourses in narrow or broad fringes downstream. Gossweiler and Mendonça (1939) as well as Barbosa (1970), treated these ecosystems as particular site specific plant communities closely linked to woodland ecosystems, and not as grass-dominated savannas.

White (1983), however, mapped and described only the sandy 'Chanas' as 'Kalahari and dambo-edge suffrutex grassland' in the context of the 'Zambezian edaphic grassland', but did not refer to the 'Anharas de Ongote' which constitute a key (but small scale) element of the miombo ecosystems of the Angolan Plateau. Even in his prominent suffrutex review, White (1976) focuses solely on the 'Chanas' in the range of the Zambezi Graben and neither mentions (psammo-) ferralitic

'Anharas', nor lists their dominant key species *Cryptosepalum maraviense* and *C. exfoliatum* ssp. *suffruticans* in his suffrutex list. He certainly recognises a transition zone between Zambezian and Guineo-Congolian floras that spans over central and northern Angola (where the 'Anharas' are included) (White 1983). However, he did not recognise the importance and floristic singularity of the ferralitic suffrutex-grasslands dominated by *Cryptosepalum* spp.

Suffrutex Flora and Endemism

The suffrutex life form appears in many different floristic groups and obviously evolved convergently. A similar center of geoxyle diversity has been reported from the Brazilian Cerrado. Today, 198 species from 40 families are listed for the western Zambezian phytochorion (White 1976; Maurin et al. 2014, own data), but an even higher number is expected as floristic exploration of the region is still poor and new species might be found (see Goyder and Gonçalves 2019). In some cases suffrutices are considered a dwarf variety or subspecies of a closely related tree species (e.g. *Gymnosporia senegalensis* var. *stuhlmanniana*, *Syzygium guineense* ssp. *huillense*) and hence classified as such and not as one species, although the genetic relatedness between tree and dwarf form is rarely investigated. On the other hand, not all dwarf forms are obligate suffrutices; some can facultatively outgrow the dwarf state if protected from environmental stressors (White 1976), for instance *Oldfieldia dacty-lophylla* or *Syzygium guineense* ssp. *macrocarpum* (Zigelski et al. 2018).

Within the suffrutex communities of the Zambezian phytochorion, the Rubiaceae have the highest number of described taxa (46), followed by Anacardiaceae (22) and Lamiaceae (14). Table 7.1 lists all families with known geoxylic suffrutex taxa occurring in Angola and gives examples of common geoxyles for each family. Furthermore, Fig. 7.1 shows some examples and aspects of suffrutex species given in Table 7.1. The unique Zambezian geoxylic flora with a high number of endemic species (Brenan 1978; White 1983; Frost 1996) is a consequence of challenging environmental conditions, as illustrated further below. According to Figueiredo and Smith's catalogue of Angolan plants (2008) and our list of suffrutices (Table 7.1), 121 of the 198 suffrutex species occurring in the Zambezian phytochorion are known from Angola (61%). Of these 121 species 12 are endemic to Angola (10%).

Environmental Conditions of Suffrutex-Grasslands Through the Year

The substrate strongly influences the species composition of the suffrutex-grasslands. In Angola geoxylic suffrutices occur on (a) well-drained arenosols which are found as seasonally flooded savannas in the Zambezi Graben of the

Geoxylic Suffrutices and Suffrutex-Grasslands in Angola

Table 7.1 List of plant families with geoxylic suffrutices in the Zambezian phytochorion

Plant family	N°	Species common in Angola	Angolan endemics
Rubiaceae	46	*Pygmaeothamnus zeyheri* (Sond.) Robyns, *Pachystigma pygmaeum* (Schltr.) Robyns	2, e.g. *Leptactina prostrata*
Anacardiaceae	22	*Lannea edulis* (Sond.) Engl., *Rhus arenaria* Engl.	3, e.g. *Lannea gossweileri*
Lamiaceae	14	*Clerodendrum ternatum* Schinz, *Vitex madiensis* ssp. *milanjensis* (Britten) F.White	
Fabaceae-Papilionioideae	13	*Erythrina baumii* Harms, *Abrus melanospermum* ssp. *suffruticosus* Hassk.	3, e.g. *Adenodolichos mendesii*
Proteaceae	11	*Protea micans* ssp. *trichophylla* (Engl. & Gilg) Chisumpa & Brummitt	1, *Protea paludosa* (Hiern) Engl.
Ochnaceae	9	*Ochna arenaria* De Wild. & T. Durand, *Ochna manikensis* De Wild.	
Passifloraceae	7	*Paropsia brazzaeana* Baill.	
Fabaceae-Detarioideae	6	*Cryptosepalum maraviense* Oliv., *C. exfoliatum* ssp. *suffruticans* (P.A.Duvign.)	
Apocynaceae	5	*Chamaeclitandra henriquesiana* (Hallier f.) Pichon	1, *Landolphia gossweileri*
Ebenaceae	5	*Diospyros chamaethamnus* Mildbr, *Euclea crispa* (Thunb.) Gürke	
Celastraceae	4	*Gymnosporia senegalensis* var. *stuhlmanniana* Loes.	
Dichapetalaceae	4	*Dichapetalum cymosum* (Hook.) Engl.	
Fabaceae-Caesalpinioideae	4	*Entada arenaria* Schinz	
Myrtaceae	4	*Syzygium guineense* ssp. *huillense*, (Hiern) F. White *Eugenia malangensis* (O.Hoffm.) Nied.	
Tiliaceae	4	*Grewia herbaceae* Hiern	
Combretaceae	3	*Combretum platypetalum* Welw. ex M. A. Lawson	2, e.g. *Combretum argyrotrichum*
Euphorbiaceae	3	*Sclerocroton oblongifolius* (Müll.Arg.) Kruijt & Roebers	
Loganiaceae	3	*Strychnos gossweileri* Exell	
Annonaceae	2	*Annona stenophylla* ssp. *nana* (Exell) N. Robson	
Apiaceae	2	*Steganotaenia hockii* (C. Norman) C. Norman	
Chrysobalanaceae	2	*Parinari capensis* Harv., *Magnistipula sapinii* De Wild.	
Meliaceae	2	*Trichilia quadrivalvis* C.DC.	
Moraceae	2	*Ficus pygmaea* Welw. ex Hiern	
Myricaceae	2	*Morella serrata* (Lam.) Killick	
Phyllanthaceae	2	*Phyllanthus welwitschianus* Müll.Arg.	
Ranunculaceae	2	*Clematis villosa* DC.	

(continued)

Table 7.1 (continued)

Plant family	N°	Species common in Angola	Angolan endemics
Achariaceae	1	*Caloncoba suffruticosa* (Milne-Redh.) Exell & Sleumer	
Anisophyllaceae	1	*Anisophyllea quangensis* Engl. ex Henriq.	
Clusiaceae	1	*Garcinia buchneri* Engl.	
Dilleniaceae	1	*Tetracera masuiana* De Wild. & T. Durand	
Fabaceae-Caesalpinioideae	1	*Bauhinia mendoncae* Torre & Hillc.	
Hypericaceae	1	*Psorosperum mechowii* Engl.	
Ixonanthaceae	1	*Phyllocosmus lemaireanus* (De Wild. & T. Durand) T. Durand & H. Durand	
Lecythidaceae	1	*Napoleonaea gossweileri* Baker f.	
Linaceae	1	*Hugonia gossweileri* Baker f. & Exell	
Malpighiaceae	1	*Sphedamnocarpus angolensis* (A. Juss.) Planch. ex Oliv.	
Malvaceae	1	*Hibiscus rhodanthus* Gürke	
Melastomaceae	1	*Heterotis canescens* (E. Mey. ex Graham) Jacq.-Fél.	
Picrodendraceae	1	*Oldfieldia dactylophylla* (Welw. ex Oliv.) J.Léonard	
Rhamnaceae	1	*Ziziphus zeyheriana* Sond.	
Urticaceae	1	*Pouzolzia parasitica* (Forssk.) Schweinf.	

N°: overall number of Suffrutex species in the Zambezian phytochorion; examples of species occurring in Angola are given for each family. Compilation of families and species according to White (1976), Maurin et al. (2014) and own data

Moxíco province or as sandy alluvial deposits on fossil river terraces along the valleys of the southern slopes of the Angolan plateau (Fig. 7.2a); (b) on psammo–ferralitic plinthisols as they frequently occur on the Bíe Plateau in central Angola. The suffrutex-grasslands on ferralitic soils mostly occur on mid- and foot-slopes and are embedded within a matrix of miombo woodland (Fig. 7.2b).

Environmental conditions in suffrutex-grasslands change dramatically throughout the year. The most perceived stresses are man-made fires in the dry season (May–October) which are mostly deployed to induce resprouting for livestock fodder or to facilitate hunting (Hall 1984). Depending on fire intensity, which in turn depends mostly on fuel load, ambient temperature and wind (Govender et al 2006), such fires can completely burn unprotected aboveground biomass.

Another abiotic stress occurring mostly in the early dry season (June–August) is nocturnal frost, peaking immediately before sunrise. At this time of year masses of cold dry air from southern latitudes intrude into south-central Africa (Tyson and Preston-Whyte 2000). As depressions accumulate confluent cold air, the undulating topography of the Angolan highlands facilitates frequent radiation frost especially in valleys (Revermann and Finckh 2013; Finckh et al. 2016). Up to 44 frost events per dry season (with a minimum temperature of −7.5 °C) were recorded by Finckh et al. (2016), with a temperature span of up to 40 degrees within 12 h. Most woody species from tropical background (including geoxylic suffrutices) are sensitive to frost, their leaves wilt or their shoots die-off entirely.

The geoxylic suffrutex species seem to be triggered by the destruction of their shoots by frost and/or fire, as they readily resprout after these disturbances and in most cases already start flowering in the dry season. The suffrutices therefore have often already finished their generative cycle when the grasses start to cover them.

The suffrutex-grasslands of the sandy plains in eastern Angola are furthermore subject to seasonal flooding in the late rainy and early dry season (January–May), leading, for example, in the Cameia National Park to standing water up to 0.5 m deep. Whereas grass species dominate the sites which are inundated for several months, suffrutex species seem to avoid fully waterlogged sites and grow patchily on slightly elevated termite mounds (Fig. 7.2a) or other well drained sites.

The dominant grass species seem to profit from inundation. Their tufts develop massively in the middle of the rainy season and they flower and bear fruits throughout the flooding season (own observations).

Knowledge Gaps on the Evolution of the Geoxylic Suffrutices and the Formation of Suffrutex-Grasslands

A common observation within suffrutex ecosystems is the resemblance (Meerts 2017) and assumed close relatedness of suffrutex species to tree species that occur in forests and woodlands. The indigenous people (e.g. the Chokwe in eastern Angola) in many cases recognise the similarity and relatedness and use similar local names for such pairs, for instance Muhaua and Mupaua for the tree and suffrutex forms of *Syzygium guineense* Willd. DC. The striking fact that the suffrutex life form was developed by several plant families independently and at roughly the same time (Maurin et al. 2014) indicates a common driver that triggered its convergent evolution.

Grassy biomes emerged in Africa in the late Miocene approximately 10 mya (Cerling et al. 1997; Keeley and Rundel 2005; Herbert et al. 2016). This period is characterised by global climatic fluctuations which led to cooler, drier conditions, to a drop of atmospheric CO_2 concentrations and particularly to pronounced precipitation seasonality (i.e. wet and dry seasons) in southern Africa (Pagani et al. 1999). As a consequence, humid tropical forests retreated to more favorable sites further north and were replaced by more open dry and seasonal tropical forest ecosystems like the miombo (Bonnefille 2011). In parts where miombo landscapes prevail today, canopies were disrupted and allowed the establishment of open ecosystems embedded in woodland matrices. These open ecosystems were then rapidly occupied by light-demanding C4-grasses and the evolving geoxylic suffrutices.

It is still an open discussion why open suffrutex-grasslands are able to persist within the woodlands (or vice versa). It is however likely that rainfall seasonality and the above described abiotic stresses that characterise the suffrutex-grasslands play a major role in their establishment and maintenance (Sankaran et al. 2005; Staver et al. 2011).

Savanna ecologists tend to see fire as the main driver for grassland formation. On the one hand frequent fires prevent tree establishment if saplings cannot outgrow the reach of the flames and are destroyed therein. For woodlands in eastern South Africa, a fire free time period of at least 5 years is necessary for many tree species to escape the 'fire trap' (Sankaran et al. 2004; Gignoux et al. 2009). This time window, allowing for successful reestablishment of trees, is rarely achieved in Angolan grasslands, at least nowadays (Schneibel et al. 2013; Stellmes et al. 2013). C4-savanna grasses, however, respond positively to periodic burning and esprout within weeks (Bond and Keeley 2005), thus being able to colonise seasonally burnt sites.

Forest ecologists, on the other hand, attribute the frequent short duration frost events in the dry season for preventing tree recruitment in the open areas (Finckh et al. 2016). As the list of suffrutices (Table 7.1) shows, mainly (but not exclusively) tropical families or genera evolved suffrutex life forms. Frost is deleterious to most tropical tree taxa, as they have not developed physiological adaptations to this 'un-tropical' stress factor, thus showing little or no frost tolerance (Sakai and Larcher 2012). As the suffrutex-grasslands are typically situated in particularly frost prone sites (depressions), tree taxa that are not adapted to frost are being filtered out of such environments.

In any case, a promising strategy to cope with seasonally returning thermic stress (by frost or fire) is to protect sensitive organs (buds) by hiding them underground. Tree species relocated their woody biomass and regenerative buds belowground at the expense of growth height and were thus able to cope with frost and fire prone sites (White 1976; Maurin et al. 2014; Finckh et al. 2016). Even shallow soil depths of less than 10 cm are sufficient to alleviate thermic stresses (Revermann and Finckh 2013). The high number of tropical genera and families that contribute to the suffrutex flora show how successful this strategy is for frost sensitive and fire suscep-tible taxa, in order to survive the adverse conditions of the open grasslands.

Concomitantly other evolutionary advantages of the geoxylic life form have been discussed, for instance poor edaphic conditions, as favoured by White (1976). He considered the low nutrient status of the leached and locally seasonal waterlogged soils on Kalahari sands as a likely cause for the lack of regular trees and the suffru-tication of them as means of compensation. However, trees as well as suffrutices often grow on the same or similarly poor soils, with comparable physical and chemi-cal properties (Gröngröft et al. 2013); forests and grasslands are not separated by edaphic boundaries but follow topographic rather than edaphic logics.

The waterlogging argument on the other hand would imply that the woody underground organs show adaptations to inundation, for instance aerenchymatic tis-sue or adventitious roots (Parolin 2008). Anatomical analyses of the rootstocks of four common suffrutex species however did not provide any support for aerenchy-matic tissue nor other adaptations to inundation (Sanguino 2015). Moreover, in sea-sonally flooded savannas suffrutices avoid inundated sites. This is even the case for

Fig. 7.1 Common Angolan suffrutex species. (**a**) *Ochna arenaria* (Ochnaceae), fruiting and growing on sandy sediments of the Bíe Plateau. (**b**) *Syzygium guineense* ssp. *huillense* (Myrtaceae) flowering in the dry season and growing on sandy soils of the Bíe Plateau. (**c**) *Lannea edulis* (Anacardiaceae), bearing edible fruits, growing on Kalahari sands in southeast Angola. (**d**) *Hibiscus rodanthus* (Malvaceae), growing on Kalahari sands in southeast Angola and flowering in the rainy season. (**e**) *Landolphia gossweileri* (Apocynaceae), typical element of the 'Chanas da Borracha', growing on sandy soils of the Bíe Plateau and bearing edible fruits. (**f**) *Phyllanthus welwitschianus* (Phyllanthaceae), growing on sandy soils of the Bíe Plateau and flowering in the rainy season. (**g**) *Cryptosepalum exfoliatum* ssp. *suffruticans* (Fabaceae – Detarioideae) with excavated rootstocks, typical element of the 'Anharas de Ongote', growing on psammoferralitic soils of the Bíe Plateau. (**h**) *Parinari capensis* (Chrysobalanaceae), typical element of the 'Chanas da Borracha', growing on slightly elevated termite mounds in flooded savannas of the Cameia National Park, Moxico Province

Fig. 7.2 Typical geoxylic suffrutex grasslands of Angola. (**a**) 'Chanas da Cameia' in the Cameia National Park, Moxíco Province, during dry season in June. The slightly elevated termite mounds provide habitat for several geoxyle species that avoid the low-lying areas that are waterlogged from January to May. (**b**) 'Anharas de Ongote' in the Sovi Valley on the southern slopes of the Bié Plateau, in August. The mid- and footslopes are dominated by suffrutex-grassland with the characteristic reddish and green patches of the fresh leaves of *Cryptosepalum maraviense*, whereas the wetlands in the drainage lines are covered mostly by Cyperaceae (background, in dark green)

Syzygium guineense ssp. *huillense*, a suffrutex closely related to a tree species that grows along and in rivers and floodplains (Coates Palgrave 2002; Meerts and Hasson 2016).

To summarise, so far the main environmental driver for the astonishing radiation of geoxylic suffrutices has not been conclusively identified. The emergence of the suffrutex grassland at the end of the Pliocene and the peak of radiation at the beginning of the Pleistocene is clearly related to climatic seasonality and pronounced dry seasons. Dry seasons, however, did not only provide the necessary dry fuel for fire but also provided the atmospheric conditions for nocturnal frost events – the seasonality argument, thus, does not tip the balance toward fire or frost.

Conservation Value and Conservation Challenges

Various studies recognise the high floristic singularity of the Zambezian phytochorion and suffrutex-grasslands with its unique life forms contribute prominently to its high number of endemic species (Clayton and Cope 1980; White 1983). The high degree of suffrutex-grassland endemics within the Zambezian phytochorion as well as within Angola is a consequence of a unique setting of environmental drivers like nutrient poor soils, frequent frosts and fires or precipitation seasonality in a small-scale heterogeneous landscape (Linder 2001). Thus, the Zambezian phytochorion can be seen as an evolutionary laboratory that promoted the evolution of many specialised plant species, e.g. suffrutices, orchids and grasses.

Suffrutex-grasslands are sometimes misunderstood as 'degraded forests', over looking their naturalness. Through this misconception they are listed as sites for reforestation in order to recover presumably lost forests and to sequestrate atmospheric CO_2 (Parr et al. 2014). However, the well-intentioned act of reforestation would in fact destroy biodiverse natural ecosystems (Bond 2016). A lack of understanding, however, frustrates the development of appropriate conservation measures for the suffrutex grasslands today and in the future. The rebuilding process in Angola also has risks, happening at a rapid pace and shaping the landscape to human demands with limited consideration for sustainable management (Pröpper et al. 2015). Flooded savannas in the Moxíco Province for instance are targeted for large-scale agro-industrial development (ANGOP 2017). Not even National Parks offer adequate protection to ecosystems in this area, as the first rice schemes emerged during 2016 within the limits of Cameia National Park (own observation). Deficiencies in communication and cooperation between different ministries and governance levels aggravate such problems.

Outlook

Many questions still remain to be answered around the enigmatic life form of the geoxylic suffrutices. In order to efficiently safeguard suffrutex-grasslands, we need to understand the evolutionary drivers and evolutionary processes shaping these ecosystems. For instance, a thorough understanding of the evolutionary drivers and the response of suffrutices to them would help to assess how current environmental conditions affect the Zambezian ecosystems and how landscape shaping processes work. Moreover, investigations about genetic patterns of suffrutices and close tree-relatives would give insight to speciation processes, means of propagation (clonal or sexual) and evolutionary history. Also, ecophysiological or morphological measurements would contribute another perspective from which to assess how suffrutices react to environmental stresses and change processes. All these facets are currently the subjects of incipient research.

References

ANGOP (2017) Cameia prepara mais de mil hectares para cultivar arroz. Agência Angola Press, Moxico, 16.07.2017. http://www.angop.ao/angola/pt_pt/noticias/economia/2017/6/28/Moxico-Cameia-prepara-mais-mil-hectares-para-cultivar-arroz,be825477-10c7-472d-b6e3-cb1ed5d25974.html. Accessed 22 Aug 2017

Barbosa LAG (1970) Carta fitogeográfica de Angola. Instituto de Investigação Científica de Angola, Luanda

Bond WJ (2016) Ancient grasslands at risk. Science 351(6269):120–122

Bond W, Keeley J (2005) Fire as a global 'herbivore': the ecology and evolution of flammable ecosystems. Trends Ecol Evol 20(7):387–394

Bonnefille R (2011) Rainforest responses to past climatic changes in tropical Africa. In: Tropical rainforest responses to climatic change. Springer, Berlin/Heidelberg, pp 125–184

Brenan JP (1978) Some aspects of the phytogeography of tropical Africa. Ann Mo Bot Gard 65(2):437–478

Cerling TE, Harris JM, MacFadden BJ et al (1997) Global vegetation change through the Miocene/Pliocene boundary. Nature 389(6647):153–158

Clayton WD, Cope TA (1980) The chorology of Old World species of Gramineae. Kew Bull 35(1):135–171

Coates Palgrave K (2002) Trees of southern Africa, 3rd edn. New edition revised and updated by Meg Coates Palgrave, Struik, Cape Town, pp 833–836

Davy JB (1922) The suffrutescent habit as an adaptation to environment. J Ecol 10(2):211–219

Figueiredo E, Smith G (eds) (2008) Plants of Angola: Plantas de Angola. Strelitzia 22:1–279

Finckh M, Revermann R, Aidar MP (2016) Climate refugees going underground–a response to Maurin et al. (2014). New Phytol 209(3):904–909

Frost P (1996) The ecology of miombo woodlands. In: Campbell B (ed) The Miombo in transition: woodlands and welfare in Africa. CIFOR, Jakarta, pp 11–57

Gignoux J, Lahoreau G, Julliard R et al (2009) Establishment and early persistence of tree seedlings in an annually burned savanna. J Ecol 97(3):484–495

Gossweiler J, Mendonça FA (1939) Carta fitogeografica de Angola. Ministério das Colónias, Lisboa, p 242

Govender N, Trollope WS, Van Wilgen BW (2006) The effect of fire season, fire frequency, rainfall and management on fire intensity in savanna vegetation in South Africa. J Appl Ecol 43(4):748–758

Goyder DJ, Gonçalves FPM (2019) The Flora of Angola: collectors, richness and endemism. In: Huntley BJ, Russo V, Lages F, Ferrand N (eds) Biodiversity of Angola. Science & Conservation: a modern synthesis. Springer

Gröngröft A, Luther-Mosebach J, Landschreiber L et al (2013) Cusseque – soils. Biodivers Ecol 5:51–54

Hall M (1984) Man's historical and traditional use of fire in southern Africa. In: de Booysen PV, Tainton NM (eds) Ecological effects of fire in South African ecosystems. Springer Berlin/Heidelberg, pp 40–52

Herbert TD, Lawrence KT, Tzanova A et al (2016) Late Miocene global cooling and the rise of modern ecosystems. Nat Geosci 9(11):843–847

Keeley JE, Rundel PW (2005) Fire and the Miocene expansion of C4 grasslands: miocene C4 grassland expansion. Ecol Lett 8(7):683–690

Linder HP (2001) Plant diversity and endemism in sub-Saharan tropical Africa. J Biogeogr 28(2):169–182

Maurin O, Davies TJ, Burrows JE et al (2014) Savanna fire and the origins of the 'underground forests' of Africa. New Phytol 204(1):201–214

Mayaux P, Bartholomé E, Fritz S et al (2004) A new land-cover map of Africa for the year 2000. J Biogeogr 31(6):861–877

Meerts P (2017) Geoxylic suffrutices of African savannas: short but remarkably similar to trees. J Trop Ecol 33(4):1–4

Meerts PJ, Hasson M (2016) Arbres et arbustes du Haut-Katanga. National Botanic Garden of Belgium

Pagani M, Freeman KH, Arthur MA (1999) Late Miocene atmospheric CO2 concentrations and the expansion of C4 grasses. Science 285(5429):876–879

Parolin P (2008) Submerged in darkness: adaptations to prolonged submergence by woody species of the Amazonian floodplains. Ann Bot 103(2):359–376

Parr CL, Lehmann CER, Bond WJ et al (2014) Tropical grassy biomes: misunderstood, neglected, and under threat. Trends Ecol Evol 29(4):205–213

Pröpper M, Gröngröft A, Finckh M et al (2015) The future Okavango: findings, scenarios and recommendations for action: research project final synthesis report 2010–2015. University of Hamburg-Biocentre Klein Flottbek, pp 53–129

Revermann R, Finckh M (2013) Cusseque – micro-climatic conditions. Biodivers Ecol 5:47–50

Sakai A, Larcher W (2012) Frost survival of plants: responses and adaptation to freezing stress, vol 62. Springer, New York, pp 138–173

Sanguino G (2015) Wood anatomy and adaptation strategies of suffrutescent shrubs in South-Central Angola. MSc. thesis, Universität Hamburg, Hamburg, 68 pp

Sankaran M, Ratnam J, Hanan NP (2004) Tree-grass coexistence in savannas revisited – insights from an examination of assumptions and mechanisms invoked in existing models. Ecol Lett 7(6):480–490

Sankaran M, Hanan NP, Scholes RJ et al (2005) Determinants of woody cover in African savannas. Nature 438(7069):846–849

Schneibel A, Stellmes M, Frantz D et al (2013) Cusseque – earth observation. Biodivers Ecol 5:55–57

Staver AC, Archibald S, Levin SA (2011) The global extent and determinants of savanna and forest as alternative biome states. Science 334(6053):230–232

Stellmes M, Frantz D, Finckh M et al (2013) Okavango basin – earth observation. Biodivers Ecol 5:23–27

Tyson PD, Preston-Whyte RA (2000) Weather and climate of Southern Africa, 2nd edn. Oxford University Press Southern Africa, Cape Town Chapter 10–12

White F (1976) The underground forests of Africa: a preliminary review. Gard Bull Singap 29:57–71

White F (1983) The Zambezian regional centre of endemism. In: White F (ed) The vegetation of Africa – a descriptive memoir to accompany the UNESCO/AETFAT/UNSO vegetation map of Africa. UNESCO, Paris, 356 pp

Zigelski P, Lages F, Finckh M (2018) Seasonal changes of biodiversity patterns and habitat conditions in a flooded savanna – the Cameia National Park Biodiversity Observatory in the Upper Zambezi catchment, Angola. Biodivers Ecol 6:438–447

8

Changes in Vegetation, Soils and Water Quality at Landscape Scales

John M. Mendelsohn

Abstract Landscape changes in Angola are dominated by woodland and forest losses due to clearing for crops, bush fires (which convert woodland into shrubland) and the harvesting of fuel (as wood and charcoal) and timber. Rates of clearing for small-scale dryland crops are high over much of Angola as a result of poor soil fertility. Erosion is also a severe problem, which has caused widespread losses of topsoil, soils nutrients and ground water. Rates of erosion are greatest in areas with steep slopes, sparse plant cover and high numbers of people, as well as around diamond mines in Lunda-Norte. Patterns of river flow and water quality have been changed, largely as a result of soil erosion and plant cover loss, as well as large irrigation schemes and dams. High rates of urban growth and the production of untreated urban waste have led to large concentrations of contamination around towns. Further research is needed, for example to assess the environmental impacts of the fishing and petroleum industries offshore, the effects of large volumes of urban waste being washed into and down major rivers to the sea, and landscape changes in an around areas of highland forests and grasslands that support populations of rare and endemic species.

Keywords Bushmeat · Charcoal · Deforestation · Fire · Land transformation · Mining impacts · River flows · Shifting cultivation · Soil erosion · Urbanisation

Introduction

Angola is a developing country, its development occurring in multiple ways in different areas of the country and affecting a variety of natural resources. Some changes and developments are likely to accelerate as the country seeks to diversify its economy and reduce dependence on revenues from oil and diamonds. It is also likely that the changes will contribute to such global trends as loss of biodiversity and land degradation.

J. M. Mendelsohn (✉)
RAISON (Research & Information Services of Namibia), Windhoek, Namibia
e-mail: john@raison.com.na

This brief review provides perspectives and information on changes to Angola's terrestrial landscapes, particularly in the southern half of the country. There are three sections to the chapter, the first of which describes the major kinds of landscape change. The second is an account of conditions that drive changes, both ultimately and proximately. Finally, areas most affected by major changes are identified in the third part of the paper.

Major Changes

Woodland and Forest Loss

Losses of woodland are by far the most obvious and conspicuous of changes in Angola. Much of this has been due to clearing for small-scale crop farming, particularly of dry-land crops, and large-scale commercial agriculture (including relatively small areas of exotic tree plantations). Other losses have come from the harvesting of charcoal, wood fuel, timber production (both for commercial and domestic uses), and runaway bush fires. On a smaller scale, swathes of riverine forest have been removed to give miners access to alluvial diamonds in rivers in Lunda-Norte.

As a result of all these losses, large areas of forest and savanna are now grasslands or shrublands. For example, the greater part of Huambo and Angola's central *planalto* was originally wooded, and 78.4% of the province of Huambo was covered in miombo woodland in 2002. In 13 years that figure had dropped in 2015 to 48.3%, amounting to the loss of some 1.265 million ha, 63.2% of which was converted from forest to crop land (Palacios et al. 2015). Similar losses in western Cuando Cubango, eastern Huíla and eastern Huambo have been documented by Schneibel et al. (2013), and elsewhere in Huíla and the Cuvelai drainage in Cunene (Mendelsohn and Mendelsohn 2018).

A countrywide perspective on the loss of forest or tree canopy cover is presented in Fig. 8.1. Several relevant features are visible in this image. First is the open, deforested expanse stretching southwest to northeast across western Huíla, southwestern Huambo and western Bié. Much of this area of highlands was cleared for crops between the 1950s and 1970s, although grasslands (*anharas do alto*) probably always dominated high altitude areas of the central *planalto* above about 1900 m above sea level. Substantial areas were cleared at the same time in parts of Cuanza-Norte, Cuanza-Sul and Malange, but their boundaries are not easily defined.

Second is the clearing of woodlands around urban areas. Many had already been cleared of tree cover by 2000, after which clearings expanded as trees were removed progressively further from the town centres, a trend illustrated by Schneibel et al. (2018). Examples of recent clearings between 2000 and 2015 are conspicuous as 'red bands' around Dundo, Menongue, Luena, Malange, Cafunfo, Cubal and Caimbambo in Fig. 8.1. Much of the deforestation is clear-felling for dryland fields by residents, while other trees are removed for charcoal production, wood fuel and timber.

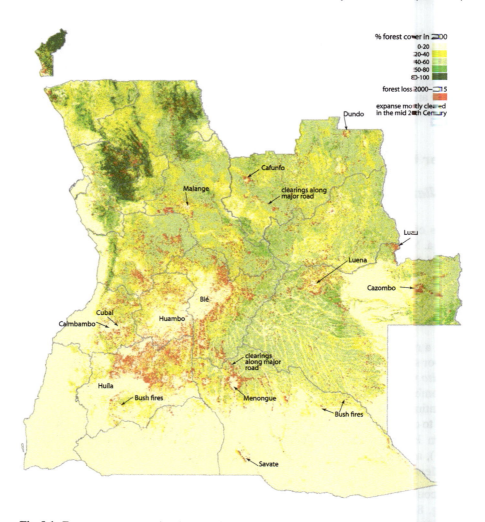

Fig. 8.1 Forest or tree canopy loss between 2000 and 2015 derived from data described by Hansen et al. (2013), updated and available from http://earthenginepartners.appspot.com/science-2013-global-forest. Percentage forest cover in the year 2000 is shown in shades of green. Red areas are those which, by 2015, had lost all the forest or canopy cover that still remained in 2000. (Source: Hansen/UMD/Google/NASA)

Third, the concentration of clearings along major roads where many rural families choose to settle is visible, but requires closer inspection of Fig. 8.1. Most losses of tree cover here are also due to clear-felling for dryland crops. Local residents also produce charcoal on a large scale, particularly along roads frequently travelled by trucks that can transport large volumes of charcoal to urban markets. However, the effects of charcoal – and timber – harvesting are seldom visible in satellite images of tree or canopy cover because harvesters typically remove only larger, taller trees, leaving smaller trees and shrubs which present a seemingly intact canopy of woodland when viewed from high above. After some years of regrowth, harvesters return to fell those bigger individual trees that produce good charcoal.

Timber has been harvested on a substantial scale for many years. Most of it has been used for the construction of domestic homes, palisades and fences, or sold as exported hardwood. The harvesting of selected species and large individual trees has evidently increased substantially in recent years, and further increases are to be expected (ANGOP 2017). Conversely, the use of poles for houses, palisades and fences may be declining, at least in certain areas where people increasingly build with home-made or bought bricks, and fence with wire (Calunga et al. 2015).

Trees were evidently harvested in large numbers to fuel railway engines running between Benguela and Huambo, and perhaps elsewhere, in the early twentieth century (Silva 2008). There are also reports of Zambezi teak *Baikiaea plurijuga* and *Marquesia macroura* timber being used for sleepers on the *Caminho do Ferro Moçâmedes* (CFM) and *Caminho de Ferro Benguela* (CFB) lines, respectively, while indigenous woodlands were cleared to make way for the many eucalyptus plantations established along the CFB line.

Bush fires have major effects on woodlands, particularly in limiting the growth of trees and shrubs in savannas. Indeed, fires maintain the 'balance' between grass and trees that characterise savannas. However, hot, intense runaway fires set by people are seemingly more frequent than before. The fiercest of fires kill all plants, old sizeable trees being burnt and scarred year after year until they eventually die. Large areas have thus been converted from woodland and forest into shrubland, particularly in southern Angola (Fig. 8.2). Much of Cuando Cubango and parts of Moxico are mosaics of open woodland separated along sharp margins from dense woodland and forest. As a probable result of fire, the edges of the dense cover are smoothed and often rounded, in some cases creating circular patches of forest (Fig. 8.3).

Soil Loss (Bulk and Nutrients)

At least three areas appear to have lost large volumes of soil and soil nutrients. The first is the central *planalto* and surrounding higher areas of ferralsol soils. In the catchment of the Cunene River, erosion has been greatest in areas that are densely populated, extensively cultivated with dryland crops, largely cleared of plant cover and that have at least moderate slopes (Fig. 8.4). The catchments of other major rivers (Cuando, Queve, Quicombo, Catumbela, Guvrire and Coporolo, for example) that drain the central catchment are likewise eroded, particularly where slopes are steep and plant cover is sparse. Similar, more concentrated effects are seen in cities where inadequate management of storm water has led to the formation of erosion gullies, many of them damaging urban roads, houses and other infrastructure.

Second is in Lunda-Norte where open-cast mining leads to considerable volumes of soil (probably also ferralsols) being washed into rivers that flow north into the Congo Basin (Fig. 8.5; see Ferreira-Baptista et al. 2018).

Fig. 8.2 An example of woodlands converted by repeated hot fires into shrublands in Bicuar National Park. The fires normally start in the grassy drainage lines (*mulolas*) from where they spread into the surrounding woodlands. With the same areas being burnt by fierce fires every few years, large areas of woodland (dark greenish zones) have progressively been turned into shrublands (pale areas). These satellite images from Google Earth (LandSat/Copernicus) were taken between 1984 and 2016, and viewed from about 15.3 South, 14.4 East. The red line marks the western border of Bicuar National Park. (From Mendelsohn and Mendelsohn 2018)

Changes in Vegetation, Soils and Water Quality at Landscape...

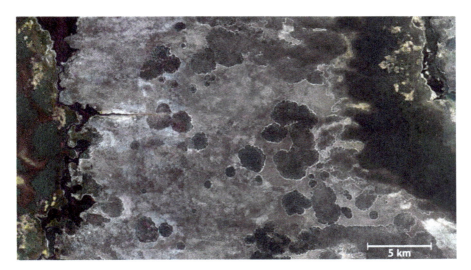

Fig. 8.3 Patches of open woodland (pale grey areas) and dense miombo forest (dark green) between the Longa and Sovi rivers in Cuando Cubango. The forest margins have probably been sharpened and smoothed by bush fires. Isolated blocks are so rounded and reminiscent of the Namib Desert's fairy circles that they may be called 'fairy forests'. (The image was taken from Google Earth (LandSat/Copernicus) as viewed from about 15.4 South, 18.9 East)

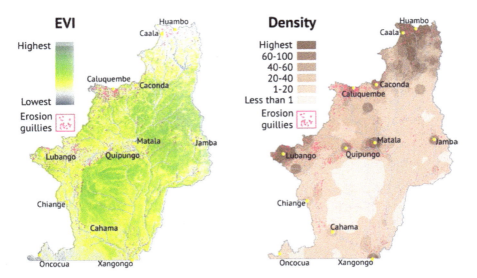

Fig. 8.4 The distribution of erosion gullies in relation to vegetation cover (Enhanced Vegetation Index – EVI) and population density in the catchment of the Cunene River between Huambo in the north and Xangongo in the south. (Adapted from Mendelsohn and Mendelsohn 2018)

Fig. 8.5 Mining impacts on Angolan rivers. Top left and right: The confluence of the clear Cassai River and the turbid Lubembe River carrying suspended sediments from open-cast diamond mining in Lunda-Norte. The confluence is in the DRC about 80 km north of Angola's border. The left photograph was taken on 30 May 2007, while the right image from Google Earth was taken 10 years later on 21 May 2017, viewed from 6.62 South, 21.07 East. Bottom left: The confluence of the Calonga and Cunene rivers at Quiteve (16.02 South, 15.20 East), showing the volumes of eroded sediments from upstream in the Cunene catchment. By contrast, the clear waters of the Calonga mainly come from areas where arenosols predominate, where few people live and where large areas of woodland have not been cleared for dryland agriculture. Bottom right: Erosion from open-cast mining along the Luachimo River 22 km north of Lucapa. (The image from Google Earth was taken in May 2017 as viewed from 8.23 South, 20.77 East)

There is a likely net loss of certain soil nutrients in the third area, which is where bush fires are frequent and/or intense, predominantly so in Cuando Cubango, Moxico and the Lunda provinces (Figs. 8.6 and 8.7). Fires often result in the loss of nitrogen, phosphorus and organic carbon, although cooler fires also facilitate the release of nutrients from plant matter into the soil (Jain et al. 2008). A study comparing open and dense woodland near Savate (see Fig. 8.1), found much lower nutrient levels in open than dense woodland soils (Wallenfang et al. 2015). This stark difference was probably a consequence of the open areas being burnt often and intensely, while the dense woodlands were seldom burnt (Stellmes et al. 2013)

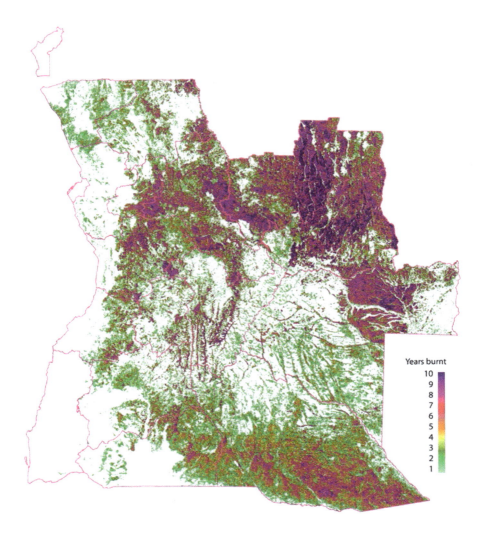

Fig. 8.6 The frequency of fires expressed as the number of years each area of 500 by 500 m burnt between 2000 and 2010. (From Archibald et al. (2010) and data available at http://wamis.meraka.org.za/products/firefrequency-map)

Water Flows and Quality

Discharges and the quality of water have changed significantly in certain rivers, and in a number of ways. The most obvious changes are in the heavy sediment loads which impair the functioning of aquatic animals and plants that require well-lit waters, and reduce the capacity of dams. For example, eroded sediments washed down the Cunene River have evidently accumulated in Gove and Matala dams to such an extent that their production of hydro-electricity has declined (António 2017).

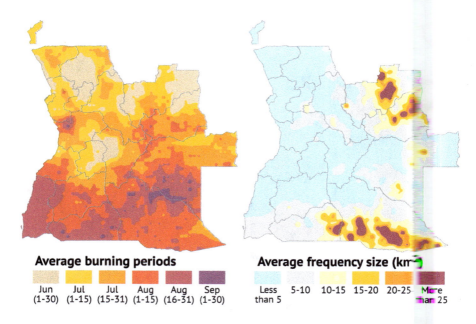

Fig. 8.7 Left: the seasonality of fires, reflected by the average period of the year when fires were recorded. Right: the average size of fires. (From Archibald et al. (2010) and data available at http://wamis.meraka.org.za/products/firefrequency-map)

River flows, soil moisture levels and groundwater recharge have been affected by losses of plant cover. Sheets of surface flows after heavy rain have increased in bare areas, causing higher river flows and probabilities of flooding, especially in seasons with above average rainfall. For example, the clearing of plant cover in the catchment of the Guvrire River around Caimbambo and Cubal (Fig. 8.1) is considered to have increased the risk and frequency of flooding at the river mouth in the city of Benguela (Development Workshop 2016).

A different impact of plant cover loss and erosion may affect the Cuvelai. Many residents there believe that surface flows down the floodplains (*chanas*) are now slower and wider than before because eroded sediments deposited in the shallow channels have further reduced their depths and slopes (Calunga et al. 2015).

Reductions in plant cover result in lower volumes of rain water being trapped or impeded, thus reducing seepage into the top soil to replenish soil moisture and recharge local aquifers. With lower soil moisture, seepage to sustain river flows during the dry season also declines. This is a likely – and at least partial – explanation for flows of the Cunene River at Ruacana dropping to less than 10 cubic metres/second in September 2017. Such low levels were only recorded previously during extreme drought years in 1993–1994 and 1994–1995 (Mendelsohn and Mendelsohn 2018).

Contamination

Quantitative assessments of the magnitude of environmental contamination from urban waste are apparently not available for Angola. However, substantial volumes of waste are generated, particularly in Luanda (now with more than seven million residents) and other major cities with populations approaching a million or more people, such as Cabinda, Lubango, Lobito, Huambo and Benguela. Solid waste is not collected in many middle to low-income *bairros*, which also lack sewage systems. The resulting volumes and concentrations of untreated waste from these large cities have significant impacts on human and environmental health (Development Workshop 2016).

Drivers of Landscape Change

Population Growth and Natural Resource Exploitation

As elsewhere in the world, but particularly in developing countries, most changes have been driven by rising demands for natural resources to meet the needs of Angola's growing population and increasing consumption per capita. The country's population rose from about 6 million people in 1970 to almost 26 million in 2014, which amounts to an annual growth rate of 3.4%. Over the same period a high proportion of the population shifted from rural to urban areas and their accompanying economies. Most urban residents live in low income areas where they generally lack piped water supplies, electricity, secure tenure, sewage systems and solid waste collection services. For example, 85% of Luanda residents live in such areas, with similar percentages in Lobito (90%), Cabinda (86%) and Benguela (92%) (Development Workshop 2016). Similar conditions and proportions hold in Huambo, Malange, Cuito and Ndalatando.

Urban consumption patterns differ from those in rural areas, but one difference of particular interest concerns the use of fuels for cooking. Rural homes generally use wood collected from around their homes, whereas the majority of urban people use purchased charcoal because alternative fuels are more expensive or not available in towns. Their supplies of charcoal all come from rural areas, in particular from poor families that harvest and then sell bags of charcoal along roads leading to major urban areas. The supply of charcoal is therefore an informal one that generates incomes for many rural homes. For both consumers and suppliers this seems to be an ideal market, providing affordable fuel for urban consumers and incomes for rural families, that often being their only monetary income. Similar market arrangements hold for supplies of bush meat from rural suppliers to urban consumers.

Another new, sometimes surprising economic link between urban and rural areas involves investments in cattle by wealthier town folk. Their cattle (and sometimes goats and sheep) are placed in rural areas where they are normally tended by relatives and held as savings or capital, with the best returns coming from large stock numbers (Gomes 2012). Owners are thus encouraged to have as many animals as possible, which places added pressures on forage, water and the limited resources available to poor rural residents (who seldom have other incomes).

Food Production

An abundance of wealth, much of it derived from the boom in oil revenues, has provided resources to develop large-scale agricultural projects, often with limited or no environmental impact assessments. For example, several new irrigation schemes have been developed along the Cunene River. If and when the farms are fully developed, downstream stretches of the Cunene could be dry for much of the year. Elsewhere, tens of thousands of hectares of woodland and forest have been cleared in recent years, one example being the Angola Biocom project which has 30,116 ha allocated to produce sugar, ethanol fuel and electricity south of Malange (Angola Biocom 2017).

Clearing for small-scale dryland crop production has caused much of the loss of woodland and forest in Angola. The *rate* at which trees are cleared is however driven by four related, but arguably separate factors. First is the need to feed a growing number of rural residents. Second is the need for farmers to abandon their fields after several years of use and to clear new fields (which will produce better yields than those that have had their supplies of nutrients exhausted). Third is the general low-input/low-output crop production strategy adopted and adapted for dryland agriculture, which means that fertilisers are seldom used to replenish soil nutrients. Fourth is the poor quality of soils available for dryland farming (Ucuassapi and Dias 2006; Asanzi et al. 2006; Wallenfang et al. 2015). Indeed, the relative lack of nutrients and moisture in soils is arguably the most important factor driving the rapid rate at which Angolan woodlands and forests are cleared, as well as the very slow rate of recovery.

Grassland and Woodland Fires

Much of Angola's vegetation has been moulded by frequent fire. This is particularly true for savanna woodlands, the grasslands of the *anharas de ongote* in the central highlands and grassy *chanas da borrachas* in the Lunda provinces. Most woody plants in the latter habitats and many in open woodlands are geoxylic suffrutices, their growth forms adapted to survive frequent hot fires (see Zigelski et al. 2019).

Fire therefore has major impacts on Angola's vegetation, and any changes in fire regimes are likely to result in landscape changes. Against that background and the widely held assumption that burning has increased in frequency, the following information is provided on fires in Angola.

Fires are recorded most frequently in grasslands of the Lunda provinces, Malange, and the Bulozi Floodplains in Moxico, and in open, savanna woodlands in Cuando Cubango (Fig. 8.6). Additionally, fires are frequent in highland grasslands (*anharas do alto*) distributed between Serra da Chela (near Lubango) in the south, the Benguela, Huambo and Huíla highlands and higher elevations in northern Cuanza-Sul and southwestern Malange.

Almost all fires are in the dry season between late April and early November. However, those in northern Angola and the central highlands burn considerably earlier than in the south (Fig. 8.7). Grass fuel is likely to contain more moisture in June than later in August, with the likely result that the earlier fires are cooler, less intense and probably less damaging to vegetation than later, hotter burns in southern Angola. That seems true for the large fires in the tall grasses that comprise the widespread *chanas da borrachas* in the Lunda provinces (Huntley 2017). Similar trends were found by Stellmes et al. (2013) within the river basins of the Cubango and Cuito Rivers where fires in the northern catchment areas were both earlier and less intense than those in the south. That trend also roughly corresponded to land cover, with dense miombo woodland in the northern areas and open savanna woodland, often called *Baikiaea-Burkea* woodland in the southern zones.

Areas of Major and Widespread Landscape Change

Figure 8.8 provides perspectives on the distribution of the major landscape changes described in this chapter. The majority of changes are in and around the central plateau (*planalto*) and to the north in parts of Cuanza-Norte, Bengo and Uíge. The effects of fire are probably most severe in Cuando Cubango, although the large (but probably cooler) fires that are so frequent in Lunda provinces may too have major effects on those extensive grasslands.

The landscape changes around towns are limited to the 18 provincial capitals shown as dark red circles in Fig. 8.8. But landscape changes around many other large towns need to be recorded.

Future Needs for Research and Documentation

Large volumes of waste are washed into the Atlantic, both close to major coastal cities – such as Luanda, Benguela and Cabinda – and down large rivers that drain large areas of the country, such as the Cunene, Cuanza and Queve Rivers. As far as is known, the volumes, nature and impacts of the waste have not been assessed. The same is true for impacts on populations of fish and other marine animals which are harvested from Angolan and foreign vessels that operate offshore, where their activities and impacts are not monitored.

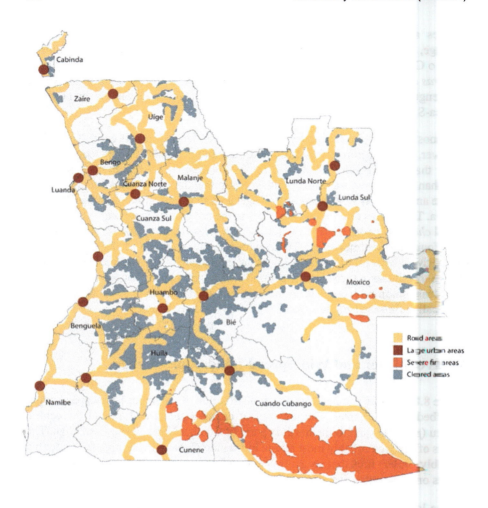

Fig. 8.8 Areas in Angola where substantial landscape changes have occurred in recent years as a result of woodland and forest clearing, and of bush fires, and in areas around main roads and towns where major changes have occurred, or are likely to occur. Woodland and forest clearings are large contiguous areas shown in Fig. 8.1. Severe fire areas are those where fires burnt in five or more years between 2000 and 2010, where fires normally burn in August and September when grass is driest and where fires are normally large (> 20 km^2), as derived from data shown in Figs. 8.2 and 8.7. Zones where people settle, farm and harvest wood are usually within 10 km either side of roads or within 15 km of major towns

The construction of very large dams on the Cuanza River is likely to have affected the functioning of that river. However, I am not aware of assessments of those effects, either by individual dams or the cumulative impounding of large volumes of water.

There are other activities and areas of concern, for example the offshore impacts of exploration and exploitation by the petroleum industry; large-scale logging in Cabinda and more recently Moxico and Cuando Cubango; pollution of river water used for washing and other domestic uses, especially where rivers flow through large towns; and pesticide contamination from crop farming, particularly from big commercial farms where large volumes of agricultural poisons are applied.

Finally, the fragmentary patchwork of mini-landscapes that support many species and which deserve special conservation measures, requires more study and documentation. These include the forests of the Escarpment Zone *(Faixa subplanaltica)* and Marginal Mountain Chain (*Cadeia Marginal de Montanhas*). Considerable numbers of rare and endemic plants and animal species are concentrated in these highland forests, many of which are small, covering no more than a few hundred hectares (Huntley and Matos 1994; Cáceres et al. 2014). The forests have shrunk, and continue to do so as a result of clearing for crops, harvesting of timber and charcoal, and grassland fires that kill trees on the forest edges. None of the forests are legally protected, and all are surrounded by substantial numbers of rural residents. Some forests are privately owned and their owners should be encouraged to manage them for conservation. Likewise, private ownership and management could be encouraged for the protection of other forests and areas of special value.

References

Angola Biocom (2017) http://www.biocom-angola.com/en/company. Accessed 30 Nov 2017
ANGOP (2017) http://www.angop.ao/angola/pt_pt/noticias/economia/2017/5/24/Angola-Mais-228-mil-madeira-serao-explorados-este-ano,40579b4d-10d3-4bed-b2e5-2751edb213eb.html
António PS (2017) Ponto de Situação Albufeira do Gove 2012–2017. Relatório de PRODEL – Empresa Pública de Produção de Electricidade, Luanda
Archibald S, Scholes R, Roy D et al (2010) Southern African fire regimes as revealed by remote sensing. Int J Wildland Fire 19:861–878
Asanzi C, Kiala D, Cesar J et al (2006) Food production in the Planalto of southern Angola. Soil Sci 171:81–820
Cáceres A, Melo M, Barlow J et al (2014) Threatened birds of the Angolan Central Escarpment: distribution and response to habitat change at Kumbira Forest. Oryx 49:727–734
Calunga P, Haludilu T, Mendelsohn J et al (2015) Vulnerabilidade na Bacia do Cuvelai/Vulnerability in the Cuvelai Basin, Angola. Development Workshop, Luanda
Development Workshop (2016) Water resource management under changing climate in Angola's coastal settlements. Project Number: 107025–001. Final technical report to the International Development Research Centre (IDRC), Canada
Ferreira-Baptista L, Manuel J, Aguiar PF et al (2018) Impact of mining on the environment and water resources in northeastern Angola. Biodivers Ecol 6:155–159
Gomes AF (2012) O Gado na Agricultura Familiar Praticada no Sudoeste de Angola – Meios de Vida e Vulnerabilidade dos Grupos Domésticos Pastoralistas e Agro-pastoralistas. PhD thesis, Technical University of Lisbon, Lisbon
Hansen MC, Potapov PV, Moore R et al (2013) High-resolution global maps of 21st-century forest cover change. Science 342:850–853

Huntley BJ (2017) Wildlife at war in Angola: the rise and fall of an African Eden. Protea Book House, Pretoria, 432 pp

Huntley BJ, Matos EM (1994) Botanical diversity and its conservation in Angola. Strelitzia 1:53–74

Jain TB, Gould W, Graham RT et al (2008) A soil burn severity index for understanding soil-fire relations in tropical forests. Ambio 37:563–568

Mendelsohn JM, Mendelsohn S (2018) Sudoeste de Angola: um Retrato da Terra e da Vida. South West Angola: a Portrait of Land and Life. Raison, Windhoek

Palacios G, Lara-Gomez M, Márquez A et al (2015) Spatial dynamic and quantification of deforestation and degradation in Miombo Forest of Huambo Province (Angola) during the period 2002–2015. SASSCAL Proceedings, Huambo, 182 pp

Schneibel A, Stellmes M, Revermann R et al (2013) Agricultural expansion during the post-civil war period in southern Angola based on bi-temporal Landsat data. Biodivers Ecol 2:311–319

Schneibel A, Röder A, Stellmes M et al (2018) Long-term land use change analysis in south-central Angola. Assessing the trade-off between major ecosystem services with remote sensing data. Biodivers Ecol 6:360–367

Silva ERS (2008) Companhia do Caminho de Ferro de Benguela: uma História Sucinta da sua Formação e Desenvolvimento. Lisbon. https://sites.google.com/site/cfbumahistoriasucinta/

Stellmes M, Frantz D, Finckh M et al (2013) Fire frequency, fire seasonality and fire intensity within the Okavango region derived from MODIS fire products. Biodivers Ecol 5:351–362

Ucuassapi AP, Dias JCS (2006) Acerca da fertilidade dos solos de Angola. In: Moreira I (ed) Angola: Agricultura, Recursos Naturais e Desenvolvimento. ISA Press, Lisboa, pp 477–495

Wallenfang J, Finckh M, Oldeland J et al (2015) Impact of shifting cultivation on dense tropical woodlands in southeast Angola. Trop Conserv Sci 8:863–892

Zigelski P, Gomes A, Finckh M (2019) Suffrutex dominated ecosystems in Angola. In: Huntley BJ, Russo V, Lages F, Ferrand N (eds) Biodiversity of Angola. Science & conservation A modern synthesis. Springer, Cham

Part III
Environmental Indicators and Diversity among Dragonflies and Butterflies

9

Odonata of Angola: Researches and Biogeography

Jens Kipping, Viola Clausnitzer, Sara R. F. Fernandes Elizalde and Klaas-Douwe B. Dijkstra

Abstract Prior to 2012, only 158 species of Odonata were known from Angola. Surveys in 2012 and 2013 added 76 species and further additions in 2016 brought the national total to 236 species. This was published earlier in 2017 as the checklist of the dragonflies and damselflies (Odonata) of Angola by the same authors (Kipping et al. Afr Invertebr 58 (I):65–91. https://africaninvertebrates.pensoft.net/article/11382/, 2017) on which this chapter is based. Records obtained in 2017 and 2018 and a survey by two of the authors in December 2017 led to the discovery of 25 additional species, of which several are undescribed. We provide a revised checklist here comprising 260 species and discuss the history of research, the biogeography of the fauna with endemism and the potential for further discoveries. The national total is likely to be above 300 species. This would make Angola one of the richest countries for Odonata in Africa.

Keywords Africa · Biogeography · Checklist · Conservation · Endemism

---

J. Kipping (✉)
BioCart Ökologische Gutachten, Taucha/Leipzig, Germany
e-mail: biocartkipping@email.de

V. Clausnitzer
Senckenberg Museum for Natural History, Görlitz, Görlitz, Germany
e-mail: Viola.Clausnitzer@senckenberg.de

S. R. F. Fernandes Elizalde
SASSCAL – BID GBIF, Instituto de Investigação Agronómica, Huambo, Angola
e-mail: kikas.sara@gmail.com

K.-D. B. Dijkstra
Naturalis Biodiversity Center, Leiden, The Netherlands
e-mail: kd.dijkstra@naturalis.nl

Introduction

Given the country's size, diverse landscapes, climatic regimes and habitats Angola is likely to be one of the richest in Odonata species in Africa. However, Angola's biodiversity is very poorly known, with comparatively limited research before independence in 1975 halting altogether in the three decades of civil war and unrest that followed. Research coverage is also limited for Odonata, with much of the north and east never surveyed at all (Clausnitzer et al. 2012). The potentially very species-rich highland catchments of the Congo, Cuanza, Cubango (Okavango) and Zambezi rivers are almost unknown and may hold many undescribed species. The whole Angolan part of the extensively marshy Cuando River and almost the entire Cuito River system are also largely unsurveyed.

History of Odonata Research in Angola

Research on Odonata began in July 1928, when the Swiss zoologist Albert Monard embarked on the first of his two expeditions to Angola, which lasted until February 1929. Monard was a curator at the Natural Museum of La-Chaux-de-Fonds in Switzerland with a broad interest in nature who mainly collected vertebrates and plants. Ris (1931) identified 27 and described four species from Monard's first expedition.

With the death of Ris, Monard submitted the Odonata from his second expedition (April 1932 to October 1933) to Cynthia Longfield at the British Museum (now the Natural History Museum) in London, who had published several records obtained by Karl Jordan from Mount Moco in 1934 (Longfield 1936). Longfield (1947) identified 77 species from Monard's new material and described 13 new species and two new genera. She also dealt with the Odonata held at the Dundo Museum in northern Angola, first revising the genus *Orthetrum* based on the long series available (Longfield 1955) and later listing 61 species from the collection, including three new species (Longfield 1959).

Elliott CG Pinhey (1961a, b) described five new species of Gomphidae from northern Angola received from António de Barros Machado of the Dundo Museum. While Longfield (1959) stated that the Dundo collection "shows the usual scarcity of the genera Gomphidae", Pinhey (1961a) noted it "was particularly notable for the number of Gomphids." Possibly Machado split the material between the two authors. It is uncertain whether the material was collected in Dundo or only held there, as most records lack details on collector, date and precise locality. However, Pinhey (1961b) did detail collecting in localities around Dundo, suggesting that all material came from this part of Lunda-Norte Province. The collector was probably Machado himself. No-one has worked on this collection since and its state is thus unknown.

Elliot Pinhey was curator at the National Museum of Zimbabwe from 1955 until 1975 and while he collected intensively in adjacent countries, he only visited Angola twice (Vick et al. 2001). In April and May 1963 Pinhey participated in an expedition to northwestern Zambia, also visiting an area east of Caianda and the Lutchigena River in Angola directly adjacent to the Ikelenge Pedicle of Zambia, where he recorded 26 species (Pinhey 1964, 1974, 1984). His second excursion into Angola went to an area between Luanda and the Duque de Bragança Falls on the Lucala River (now known as Calandula Falls) in October 1964 with records of 32 species (Pinhey 1965).

Pinhey further treated the material of three collectors, describing a species in honour of each of them. Edward S Ross of the California Academy of Sciences collected between Cuchi and Dondo in 1957 and 1958 (Pinhey 1966), the American expert of mammal behaviour Richard Estes in central Angola in 1970 (Pinhey 1971a), and Ivan Bampton around the Serra da Chela and Tundavala in 1973 (Pinhey 1975). In the 1975 paper he also repeated records from Pinhey (1964, 1965) and Longfield (1947), and provided a gazetteer, causing confusion about the precise locality of some sites. The correct historic collecting sites could be verified with the gazetteer of Mendes et al. (2013).

Various collectors gathered about 1000 specimens in the collection of the *Instituto de Investigação Agronómica* in Huambo between 1950 and 1974. These records were never published but this will be done shortly by Sara F Elizalde and David Elizalde as a GBIF dataset.

After Angola's independence in 1975 there was a long break in field research, with only a few records by various collectors. Namely in the two decades between 1980 and 2000 not a single record of Odonata is available. Some years after the end of the civil war a renaissance of research began, resulting in a growing number of records (Fig. 9.1). All localities with available Odonata records distinguished in the three periods (a) pre-independence 1928–1974, (b) after independence 1975–2001 and (c) after the end of civil war 2002-today are shown in Fig. 9.2.

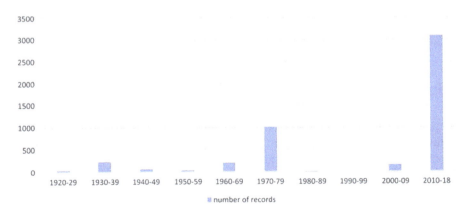

Fig. 9.1 Number of Odonata records from Angola over past decades

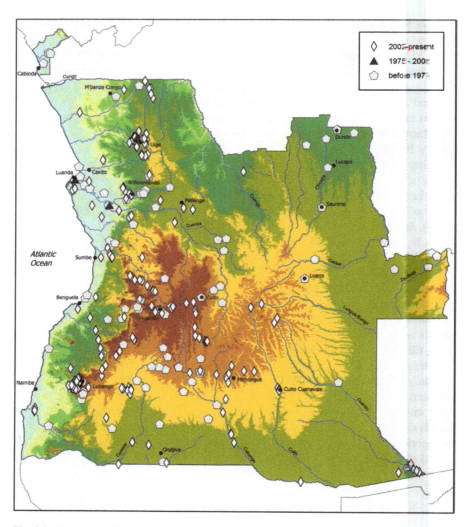

Fig. 9.2 Records of Odonata from Angola before 1975, before 2002 and up to 2018

Origin of Recent Data

In January 2009, an expedition led by Brian Huntley visited the Serra da Chela in southwestern Angola and the Namib Desert to the south. During that survey Warwick Tarboton collected and photographed Odonata around Humpata (7 field days).

Jens Kipping surveyed the upper catchment of the Okavango (Cubango) River on the SAREP (Southern African Regional Environmental Program) Expedition from 5 to 22 May 2012 (18 field days). A second SAREP survey visited southeastern Angola with the Cubango and Cuando River floodplains in April 2013.

Viola Clausnitzer and K-D B Dijkstra in collaboration with the Universidade Kimpa Vita (Uíge) and the Technical University of Dresden (Germany) surveyed around Uíge, Negage and Ndalatando in northern Angola in the wet season from 13 November to 1 December 2012 (19 days). Dijkstra revisited this area in the dry season, from 26 September to 5 October 2013 (10 days).

From 27 November to 10 December 2016 (14 field days), Manfred Haacks and colleagues of SASSCAL (Southern African Science Service Centre for Climate Change and Adaptive Land Management) visited Bicuar NP and a few other places in southern Angola.

Sara F Elizalde and David Elizalde, Chris Hines, André Günther, Raik Moritz and Jens Kipping surveyed the Serra da Chela around Lubango and the mountain range stretching from Huambo northwards to Gabela from 30 November to 19 December 2017 (20 field days).

Sara F Elizalde, Chris Hines, Rogério Ferreira and other experts provided many photographic records from 2016 to 2018.

The National Geographic Okavango Wilderness Project (NGOWP 2018) gathered scattered data on Odonata which has not yet been fully considered, except for some field photographs and exceptional records provided by John Mendelsohn.

Apart from the field surveys the authors also examined the Angolan collections and type material in the Natural History Museum in London, the National History Museum of Zimbabwe in Bulawayo (Dijkstra 2007a, b), the Royal Museum for Central Africa in Tervuren, Belgium and the *Instituto de Investigação Agronómica* in Huambo, Angola. All records are kept in the Odonata Database of Africa – ODA (Kipping et al. 2009) and mapped per species on African Dragonflies and Damselflies Online – ADDO (visit http://addo.adu.org.za/ also for further information about all mentioned species).

Odonata Species Recorded in Angola

From all the historic sources mentioned above, 152 species of Odonata were known to occur in Angola until 2009. Some of the formerly published species had to be deleted from the country list in the light of new taxonomic knowledge and after careful validation of all records (see Kipping et al. 2017).

In 2009 Warwick Tarboton recorded 47 species of Odonata at the Serra da Chela of which five were recorded in Angola for the first time and one was new to science (Tarboton 2009, Dijkstra et al. 2015). The first SAREP Expedition in 2012 yielded 87 species, 17 of them new to the country list and two new to science (Kipping 2012, Dijkstra et al. 2015). One additional species new for the country came from a second SAREP Expedition in April 2013 of which all collected specimens were examined. The first expedition to Uíge, Negage and Ndalatando resulted in 138 species, of which 43 were recorded for the first time in Angola and five were new to science. The second visit produced 86 species, adding another 15 to the national list. With the surveys from 2009 to 2013 and a careful review of the historic data, the known odonate fauna of Angola had increased from 152 species in the year 2009 to 234 species in 2013: an increase of about one-third with only 54 days in the field. Two species were added in 2016 by photographs made by Chris Hines and specimens from the collection of the *Instituto de Investigação Agronómica* in Huambo provided by Sara F Elizalde. The state of knowledge at the end of 2016 was published as the checklist of the dragonflies and damselflies of Angola by Kipping et al. in early 2017 (free download: https://africaninvertebrates.pensoft.net/article/11382/).

The SASSCAL expedition in November–December 2016 recorded 44 species, amongst them one new species for Angola. The latest survey in December 2017 yielded 88 species of which 10 were new for the country list, amongst them probably three species new to science. A further 14 species new for Angola were recorded only in 2017 and early 2018 by Chris Hines and colleagues mostly in northern Angola.

The updated checklist of the Odonata of Angola, now of 260 species, is provided in Appendix 1. The ODA database now holds about 4900 Angolan records from more than 400 localities. All species of Appendix 1 are reliably recorded from Angolan territory. Footnotes will give further information about the 25 additional species to the updated country list and one species that was deleted from it.

There are 15 more species listed in Appendix 2 that are known to occur at rivers bordering the country with Namibia and Zambia. The Namibian bank of the Okavango River is very well surveyed (Suhling and Martens 2007, 2014) and most of the mentioned species derive from this river. These species were technically not found on the Angolan riverbank and therefore not included in the checklist. But naturally they belong to the country's fauna.

Composition

Angola's rich dragonfly fauna expresses its geographic position, size and diversity. Its territory, especially in the north, falls within a region with an estimated highly diverse fauna (Fig. 9.3). Dijkstra et al. (2011) observed that roughly half of tropical African species occur predominantly within the extensive lowland forests of western and central Africa, a quarter is associated with the eastern and southern part dominated by highlands, while the remaining quarter occurs in open habitats throughout much of the Afrotropics. Indeed, about half of Angola's species are widespread across the continent and its exceptional diversity can be attributed to two major sources. Almost 30% are confined to forest habitats in the north, mostly below 1000 m altitude. Nine species confined to the Lower Guinea, the forest area that stretches between the Congo Basin and Atlantic Ocean from Cameroon to Gabon and western Congo, reaching their southern limit in northwestern Angola. Nearly 20% favour the swamps, grasslands, miombo woodlands and gallery forests that stretch eastwards, mostly above 1000 m asl. This fauna is concentrated in Katanga and northern Zambia but has now been proven to extend across to the

Odonata of Angola: Researches and Biogeography

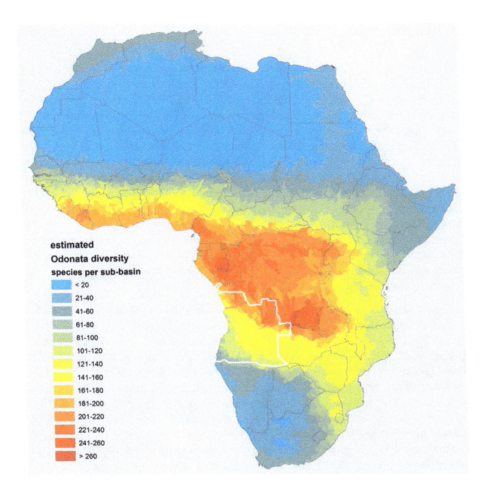

Fig. 9.3 Spatial estimation of Odonata diversity in continental Africa, based on the summation of the inferred ranges of all 770 species known; mapped as the number of species per Hydro1K basin. (Adapted from Clausnitzer et al. 2012). Angola is outlined in white

Angolan upland. This was confirmed by the discovery of *Orthetrum kafwi* at two localities in Cuanza-Sul in December 2017. The species was until then only known from Upemba NP in southern DRC. A number of palustrine species, e.g. *Anax bangweuluensis, Pinheyagrion angolicum, Pseudagrion deningi* and *P. rufostigma*, prefer larger marsh areas and swamps as in the Okavango Delta of Botswana (see Kipping 2010) and spreading north into Bié highlands with the headwaters of that river system. The discovery of *Trithemis integra* near Uíge is also of special interest, as it had seemed to be endemic to the Albertine Rift, being known previously only from western Tanzania and Uganda and eastern DRC.

Endemism

Seventeen valid species and several recently discovered undescribed species have so far only been found in Angola (Figs. 10.4, 10.5 and 10.6 for examples). With the exception of two known only from their alleged type localities in far northeastern Angola, all are limited to the central plateau: the type locality for *Platycypha rubriventris* is questionable as it may be that of *Pseudagrion dundoense*, which could also be a river species from the very poorly sampled southern Congo Basin. No endemics have been found below 1200 m asl in the east, although some drop down to about 500 m west of the escarpment. While the proportion of endemics (7%) is lower than for Ethiopia (12 endemics; 11%) and South Africa (30 endemics; 18%), countries that also enclose distinct highland areas, this still ranks Angola as one of Africa's greatest centres of endemism for Odonata, rivalling the highlands of Cameroon (13 endemics) and the Albertine Rift, Eastern Arc and Katanga. Moreover, the number is expected to increase, as almost two-fifths were described since exploration was reinitiated and undescribed species of *Platycypha*, *Paragomphus* and *Tetrathemis* are already known to us.

Only *Platycypha* presents an endemic radiation. While *Chlorocypha* (the family's other large Afrotropical genus) has diversified with almost 30 species largely in the forested lowlands of west and central Africa, *Platycypha* is ecologically more diverse, with species adapted to open, submontane and lake habitats as well. The Angolan endemics are found mainly between 1300 and 1800 m altitude in open habitats. The widespread *P. angolensis* replaces the common *P. caligata*, which extends from South Africa to Ethiopia but only peripherally into Angola. *Platycypha bamptoni* is probably confined to Serra da Chela; a similar undescribed species appears more widespread. *Platycypha crocea* is typical of very small streams in the Bié highlands and escarpment mountains whereas the other two inhabit larger streams and rivers. A local radiation of a group that has otherwise diversified in the highlands to the east, and forests to the north, fits the overall affinities of Angola's endemic Odonata both geographically and ecologically (Kipping et al. 2017) (Fig. 9.4).

The four endemic *Pseudagrion* species have separate origins but similar links: the nearest relatives of *P. angolense* and *P. estesi* appear to be the rainforest species *P. grilloti* Legrand, 1987 and *P. kibalense* respectively. The former is limited to Congo and Gabon, but the latter extends to Cameroon and Uganda. *P. sarepi* is closely related to *P. fisheri* and *P. greeni*, both of which extend from Angola into Zambia. While these species belong to the genus's A-group, the B-group species *P. dundoense* is known only from Dundo and may not be endemic at all (see above).

Notogomphus kimpavita is the sister-species of *N. praetorius* found in highlands across southern Africa (including Angola), while *Eleuthemis eogaster* is nearest to an unnamed species from Gabon (Dijkstra et al. 2015).

Molecular data for *Umma femina* and *Onychogomphus rossii* are not available yet. By its unusual habitus and colouration *Umma femina* (see Fig. 9.5) is a very distinct member of the genus. It is definitely the odonate flagship species of the

Fig. 9.4 Photographs of some of Angola's (near) endemic dragonflies and damselflies. (**a**) Sarep Sprite (*Pseudagrion sarepi*), (**b**) Blue Wisp (*Agriocnemis angolensis*) that just extends into Namibia and Zambia, (**c**) Angola Claspertail (*Onychogomphus rossii*), (**d**) Angola Longleg (*Notogomphus kimpavita*), (**e**) Sunrise Firebelly (*Eleuthemis eogaster*), (**f**) Angola Micmac (*Micromacromia flava*). (All males, photographs a–d by J Kipping, e–f by K-DB Dijkstra)

Angolan highlands and probably also most threatened. The morphology of *O. rossii* is close to other pale *Onychogomphus* species from the open plateaus stretching from Angola to Zambia and Katanga.

Thus, like the majority of Angola's Odonata, most endemics probably originated quite recently and proximally from the forests to the north and open habitats to the east. However, some affinities are unresolved and potentially more distant: *Agriocnemis toto* and especially *A. canuango* have no obvious close relatives (Dijkstra et al. 2015), while the near-endemic *A. angolensis* and *A. bumhilli* are probably related to each other but even more distinct overall (Kipping et al. 2017).

Fig. 9.5 Photographs of some of Angola's endemic damselflies. (**a**) Angola Blue Jewel (*Platycypha crocea*), (**b**) Highland Blue Jewel (*Platycypha bamptoni*), (**c**) undescribed Blue Jewel (*Platycypha* sp. nov.), (**d**) Angola Dancing Jewel (*Platycypha angolensis*), (**e**) Angola Sparklewing (*Umma femina*), (**f**) Stout Threadtail (*Elattoneura tarbotonorum*), (**g**) Angola Sprite (*Pseudagrion angolense*), (**h**) Estes's Sprite (*Pseudagrion estesi*). (All males, photographs by J Kipping)

These data suggest that Angola may be the centre of diversification of this genus, which includes Africa's smallest damselflies. The morphologically very distinct *Aciagrion rarum* is only known from very few specimens from Lunda Sul Province in the northeast and molecular data is not available yet.

Micromacromia flava is morphologically nearest *M. miraculosa* (Förster 1906), known only from the East Usambara Mountains of north-eastern Tanzania and the only one of four *Micromacromia* species adapted to non-forest habitats, being strongly pruinose with maturity. *Elattoneura tarbotonorum* may be closest to *E. frenulata* of southwestern South Africa (Dijkstra et al. 2015): after its discovery at the Serra da Chela in 2009 it was found more widespread in December 2017 along the mountain range stretching north into Cuanza-Sul.

Potential for Discovery

If we compare the tallies for the well-studied neighbouring countries of Zambia and Namibia, the total number of species in Angola should lie somewhat above 300, meaning that less than 80% of the fauna is currently known. All Odonata expeditions in modern times surveyed areas that are easily accessible.

Additions can be expected throughout the country, but especially in the remote regions on the eastern and particularly northern border, as species diversity is expected to be extraordinarily high in the transition to the Congolian rainforest (Fig. 9.3). The province of Lunda-Norte with only 92 recorded species and 162 records should be the richest area for discovery, around Dundo where exploration began in the 1950s. Generally, all the northern and eastern provinces are largely unsurveyed and the discrepancy between the amount of available data, the number of known species and the expected diversity is extremely high. This applies also to the provinces of Lunda-Sul (10 species, 11 records), Zaire (17 species, 21 records), Malanje (35 species, 152 records) and Moxico (46 species, 51 records). An exception is Uíge where recent surveys increased the number of known species to 145 from 820 records gathered.

The central highlands can also yield more surprises, like the discovery of additional endemic species, with three areas being especially notable. Firstly, despite having most records, the north-south directed mountain range that lies entirely above 1600 m asl and includes the Serra do Chilengue, Serra da Chela and Angola's highest peak at Mount Moco (2620 m asl) is poorly sampled as the large gaps in Fig. 9.2 illustrates.

Secondly, except for its extreme northern and southern ends, the western escarpment has only been surveyed recently, which already led to the discovery of an undescribed *Paragomphus* species from Cumbira Forest. Even more easily accessible provinces such as Bengo and Cuanza-Sul will prove to be much richer in species than currently known. The potential of these mountains is illustrated by the discovery of a spectacular and unique but unknown species by Chris Hines and Rogério Ferreira in May 2018 (Fig. 9.6). Two males at a stream that flows off the

Fig. 9.6 An undescribed species that probably belongs to *Trithemis*, although the extensive markings and dense veins in the wings are unusual even for that highly diverse genus. Two males were observed at a stream running off Namba Mountains in Cuanza-Sul. (Photograph by R Ferreira)

Namba Mountains in Cuanza-Sul were photographed but not collected. These mountains reach over 2000 m in altitude and harbour larger pockets of Afromontane forest than Mt. Moco (Mills et al. 2013). They are known for their plant endemism and fieldwork there will definitely lead to the discovery of more endemic Odonata.

Thirdly, an extensive plateau at 1200–1600 m altitude stretches east from the Bié Highlands. Except for its southern edge, this area shared between Bié and Moxico Provinces, which is almost as large as Uganda (or the United Kingdom), has almost no records. A few collections from the NGS Okavango Wilderness Project suggest that more new species for the country and for science can be expected here. These deep Kalahari sands are the 'watertower' of Angola and its neighbours, incorporating the headwaters of the Cuito, Cuando, Chicapa, Cuango, Cuanza and large tributaries of the Congo and Zambezi such as the Cassai and Lungué-Bungo. The sources of the vast catchments of the Congo, Cuanza, Okavango and Zambezi meet in a small area between Munhango and Cangonga. Watersheds are prone to endemism (Dijkstra et al. 2011) and this region is the top priority for further research.

Studying insect collections of Angola's museums will be also a valuable source of more records and possibly even to get insight into past conditions in the light of the recent landscape change. Of special interest is the Dundo Museum that holds many interesting specimens of Odonata, some of which have been published. This remarkable collection has not been studied since independence but survived the civil war. There is also material dispersed over several museums in Europe and probably also in private collections, mainly in Portugal.

New species are most likely to be found among genera prone to narrow (highland) ranges, i.e. with known Angolan endemics like *Platycypha* and *Pseudagrion*, but also *Agriocnemis*, *Elattoneura*, *Notogomphus* and *Paragomphus*. Also possible

Odonata of Angola: Researches and Biogeography

is the discovery of endemics in genera that are well represented across the country and continent, and that have highland endemics elsewhere but not in Angola, such as *Africallagma*, *Neodythemis* and *Orthetrum*. However, given the biogeographic diversity of Angola's fauna and endemics, we could expect greater surprises. Among forest genera with no known Angolan endemic, *Allocnemis* seems most likely to reveal one, e.g. on the escarpment. The presence (or local endemism) of distinctly Lower Guinean genera like *Neurolestes*, *Africocypha*, *Pentaphlebia* and *Stenocnemis* seems less likely, but the Lower Guinean *Stenocypha gracilis* (Karsch 1899) has four endemic relatives in the Albertine Rift and the sister-taxon of the Upper and Lower Guinean *Tragogomphus* is *Nepogomphoides stuhlmanni* (Karsch 1899) in the Eastern Arc, suggesting an Angolan taxon is possible.

Some typical African highland genera are notably absent from Angola. *Atoconeura* is most likely to be present above 1400 m asl, being found in Zambia, Katanga, the Lower Guinea and Albertine Rift. However, its absence also from South Africa suggests historical factors may have been limiting, e.g. that the highlands were too harsh in cooler periods and too isolated when habitats were suitable (Dijkstra 2006).

This might not apply to *Proischnura*, present in South Africa as well as Cameroon and the Albertine Rift. However, that genus is absent from Katanga and northern Zambia, which lies lower and thus possibly provided no stepping-stone to the mountains of Angola. Kipping et al. (2017) also noted the absence of *Zosteraeschna* and *Pinheyschna*, which have a similar range (although the latter does occur in Katanga and northern Zambia), but isolated populations of *Z. minuscula* (McLachlan 1895) and *P. subpupillata* (McLachlan 1896) were discovered in the Serra da Chela in southern Angola in December 2017.

Conservation

Our findings show that Angola's wealth of aquatic habitats harbours a rich freshwater fauna. Although large areas are relatively untouched, Angola's rapid economic and population growth will have a tremendous impact on the environment and thus human well-being in the future. In the light of this, Angola's development should consider (1) the establishment of sewage works in cities and larger villages; (2) a stop to deforestation, especially along stream courses; (3) restoration of deforested water catchments; (4) village-level awareness campaigns for sustainable use of freshwater sources, e.g. no detergents and waste dumping in rivers; (5) biodiversity surveys and monitoring to feed into a national conservation plan.

With the exception of four species, all endemics are currently considered Data Deficient for the IUCN Red List of Threatened Species. *Platycypha angolensis*, *Pseudagrion angolense* and *Micromacromia flava* are Near Threatened because, while they seem fairly widespread, their dependence on relatively natural habitats may put them at risk as human development progresses. Only *Umma femina* is now listed as threatened. It is currently known from only a few sites in the fairly densely populated highlands around Lubango and seems to inhabit exclusively the smaller and cooler highland streams. There is much development in this densely settled region and increasing pressure on those habitats by grazing, deforestation and urbanisation. As it seems to prefer cool mountain streams we can assume additional risks from climate change and it is therefore thought to be Vulnerable to extinction. More research on all endemic species' statuses and ecology is urgently required.

Angola has an exceptional fauna of dragonflies and damselflies, as well as many valuable rivers and wetlands. Odonata are excellent indicators of the health and biodiversity of both the freshwater and terrestrial realm. As the biological survey of Angola advances, they should be a priority taxon.

Acknowledgements We are grateful to Her Excellency Madame Minister of Environment Dr. Paula C Francisco Coelho (MINAMB) for making the SAREP survey in southern Angola possible, to Dr. Chris Brooks of SAREP for the preparation and organization of the 2012 survey, to Marta Alexandre Zumbo (MINAMB), Maria Helena Loa (MINAMB), Julius Bravo (MINAMB), Francisco de Almeida (INIP), Manuel Domingos (INIP) and Gabriel Cabinda (Agriculture and Rural Development and Fisheries) for their help in organization and management on the 2012 tour, and to Vince Shacks and Werner Conradie for collecting specimens on the second SAREP survey in 2013. We thank Alvaro Bruno Toto Nienguesso, the driving force behind biodiversity research in Uíge Province, Angola, Prof Dr. Neinhuis and Dr. Thea Lautenschlaeger from TU Dresden for inviting us to the field survey in Uíge province. Part of the fieldwork in Angola was supported by a travel fund from the German Academic Exchange Service (DAAD). We thank Dr. Aristófanes Pontes, director of the *Instituto Nacional da Biodiversidade e Áreas de Conservação* (INBAC) for the support during the December 2017 expedition, by providing the necessary permits. Chris Hines provided many valuable photographic records. Further records were provided by Warwick Tarboton, Dr. Manfred Haacks (SASSCAL), John Mendelsohn (RAISON) and Rogério Ferreira. The latter also gave permission to use his wonderful photograph.

Appendices

Appendix 1

Checklist of Odonata recorded from Angola.

\# – see Taxonomic comments in Kipping et al. (2017); \#\# – see footnotes following this table;

(V) Validation of species: "1!" new national record made by the authors; "1!!" new national record made by the authors and addendum to Kipping et al. (2017); "1" records obtained by authors and confirming existing records; "2" specimens kept in collections (identification confirmed or primary types); "3" literature records, regarded as reliable because specimens were described well or location agrees with known biogeographic pattern; "4!!" new national record made by other persons and addendum to Kipping et al. (2017); ** – range restricted to Angola; * – range restricted to Angola with very few exceptions (see Endemism in the discussion).

(RL) Global conservation status according to the IUCN Red List of Threatened Species (2016): CR (Critically Endangered), DD (Data-Deficient), EN (Endangered), NT (Near-Threatened), VU (Vulnerable), LC (Least Concern), NE (Not Evaluated)

Odonata of Angola: Researches and Biogeography 149

Scientific name	English name	V	RL
Lestidae			
Lestes amicus (Martin, 1910)	Yellow-winged Spreadwing	1	LC
Lestes dissimulans (Fraser, 1955)	Cryptic Spreadwing	1	LC
Lestes pallidus (Rambur, 1842)	Pallid Spreadwing	1	LC
Lestes pinheyi (Fraser, 1955)	Pinhey's Spreadwing	1	LC
Lestes plagiatus (Burmeister, 1839)	Highland Spreadwing	1	LC
Lestes tridens (McLachlan, 1895)	Spotted Spreadwing	1	LC
Lestes virgatus (Burmeister, 1839)	Smoky Spreadwing	3	LC
Calopterygidae			
Phaon camerunensis (Sjöstedt, 1900)	Emerald Demoiselle	1!	LC
Phaon iridipennis (Burmeister, 1839)	Glistening Demoiselle	1	LC
Sapho orichalcea (McLachlan, 1869)#	Mountain Bluewing	1!	LC
Umma electa (Longfield, 1933)	Metallic Sparklewing	1	LC
Umma femina (Longfield, 1947)	Angola Sparklewing	1**	VU
Umma longistigma (Selys, 1869)	Bare-bellied Sparklewing	1	LC
Umma mesostigma (Selys, 1879)	Hairy-bellied Sparklewing	1!	LC
Chlorocyphidae			
Chlorocypha aphrodite (Le Roi, 1915)##	Blue Jewel	4!!	LC
Chlorocypha cancellata (Selys, 1879)	Exquisite Jewel	1!	LC
Chlorocypha curta (Hagen in Selys, 1853)	Blue-tipped Jewel	1!	LC
Chlorocypha cyanifrons (Selys, 1873)	Blue-fronted Jewel	1!	LC
Chlorocypha fabamacula (Pinhey, 1961)	Spotted Jewel	1	LC
Chlorocypha victoriae (Förster, 1914)	Victoria's Jewel	1	LC
Platycypha angolensis (Longfield, 1959)	Angola Dancing Jewel	1**	NT
Platycypha bamptoni (Pinhey, 1975)#	Highland Blue Jewel	1**	NE
Platycypha cf. *bamptoni* (Pinhey, 1975)#	(near Highland Blue Jewel)	1!**	NE
Platycypha caligata (Selys, 1853)#	Common Dancing Jewel	2	LC
Platycypha crocea (Longfield, 1947)#	Angola Blue Jewel	1**	LC
Platycypha rubriventris (Pinhey, 1975)#	Red-bellied Blue Jewel	2**	DD
Platycypha rufitibia (Pinhey, 1961)	Beautiful Jewel	1	LC
Platycnemididae			
Allocnemis nigripes (Selys, 1886)	Rainbow Yellowwing	1	LC
Allocnemis pauli (Longfield, 1936)	Orange-tipped Yellowwing	1!	LC
Copera congolensis (Martin, 1908)	Congo Featherleg	1!	LC
Elattoneura acuta (Kimmins, 1938)	Red Threadtail	1!	LC
Elattoneura cellularis (Grünberg, 1902)#	Zambezi Threadtail	3	LC
Elattoneura cf. *glauca* (Selys, 1860)#	(near Common Threadtail)	1	LC
Elattoneura lliba (Legrand, 1985)	Eastern Stream Threadtail	1!	LC
Elattoneura tarbotonorum (Dijkstra, 2015)#	Stout Threadtail	1**	DD
Mesocnemis singularis (Karsch, 1891)##	Common Riverjack	1!!	LC
Mesocnemis cf. *singularis* (Karsch, 1891)#	(near Common Riverjack)	1!	NE
Coenagrionidae			
Aciagrion africanum (Martin, 1908)	Blue Slim	1	LC
Aciagrion macrootithenae (Pinhey, 1972)	Awl-tipped Slim	3	DD

(continued)

Scientific name	English name	V	RL
Aciagrion nodosum (Pinhey, 1964)	Cryptic Slim	1!	LC
Aciagrion rarum (Longfield, 1947)	Tiny Slim	2*=	DD
Aciagrion steeleae (Kimmins, 1955)	Swamp Slim	3	LC
Aciagrion zambiense (Pinhey, 1972)	Zambia Slim	3	DD
Africallagma fractum (Ris, 1921)	Slender Bluet	1	LC
Africallagma glaucum (Burmeister, 1839)	Swamp Bluet	1	LC
Africallagma sinuatum (Ris, 1921)##	Peak Bluet	4!!	LC
Africallagma subtile (Ris, 1921)##	Fragile Bluet	1!!	LC
Africallagma vaginale (Sjöstedt, 1917)	Forest Bluet	1!	LC
Agriocnemis angolensis (Longfield, 1947)	Blue Wisp	1*	LC
Agriocnemis bumhilli (Kipping, Suhling & Martens, 2012)	Bumhill Wisp	1=	LC
Agriocnemis canuango (Dijkstra, 2015)	Bog Wisp	1=*	DD
Agriocnemis exilis (Selys, 1872)	Little Wisp	1	LC
Agriocnemis forcipata (Le Roi, 1915)	Greater Pincer-tailed Wisp	1	LC
Agriocnemis gratiosa (Gerstäcker, 1891)##	Gracious Wisp	4=	LC
Agriocnemis cf. *maclachlani* (Selys, 1877)#	(near Forest Wisp)	1!	LC
Agriocnemis pinheyi (Balinsky, 1963)##	Pinhey's Wisp	1!!	LC
Agriocnemis ruberrima (Balinsky, 1961)	Orange Wisp	1!	LC
Agriocnemis toto (Dijkstra, 2015)	Bruno's Wisp	1=*	DD
Agriocnemis victoria (Fraser, 1928)	Lesser Pincer-tailed Wisp	1	LC
Azuragrion nigridorsum (Selys, 1876)	Sailing Bluet	1	LC
Ceriagrion annulatum (Fraser, 1955)	Green-eyed Citril	1!	LC
Ceriagrion bakeri (Fraser, 1941)	Blue-fronted Citril	3	LC
Ceriagrion corallinum (Campion, 1914)	Green-fronted Citril	1	LC
Ceriagrion glabrum (Burmeister, 1839)	Common Citril	1	LC
Ceriagrion junceum (Dijkstra & Kipping, 2015)	Spikerush Citril	1!	LC
Ceriagrion platystigma (Fraser, 1941)	Variable Citril	1	LC
Ceriagrion sakejii (Pinhey, 1963)	Cream-sided Citril	1!	LC
Ceriagrion suave (Ris, 1921)	Plain Citril	1	LC
Ceriagrion whellani (Longfield, 1952)	Yellow-faced Citril	1!	LC
Ischnura senegalensis (Rambur, 1842)	Tropical Bluetail	1	LC
Pinheyagrion angolicum (Pinhey, 1966)	Pinhey's Bluet	1	LC
Pseudagrion (A) *angolense* (Selys, 1876)	Angola Sprite	=*	NT
Pseudagrion (A) *coeruleipunctum* (Pinhey, 1964)	Pretty Sprite	=	LC
Pseudagrion (A) *estesi* (Pinhey, 1971)	Estes's Sprite	=*	LC
Pseudagrion (A) *fisheri* (Pinhey, 1961)	Dark-tailed Sprite	=	LC
Pseudagrion (A) *greeni* (Pinhey, 1961)	Clasper-tailed Sprite	1	LC
Pseudagrion (A) *inconspicuum* (Ris, 1931)	Little Sprite	1	LC
Pseudagrion (A) *kersteni* (Gerstäcker, 1869)	Powder-faced Sprite	1	LC
Pseudagrion (A) *kibalense* (Longfield, 1959)	Forest Sprite	1	LC
Pseudagrion (A) *makabusiense* (Pinhey, 1950)	Green-striped Sprite	=	LC
Pseudagrion (A) *melanicterum* (Selys, 1876)	Farmbush Sprite	1	LC
Pseudagrion (A) *salisburyense* (Ris, 1921)	Slate Sprite	1	LC

(continued)

Odonata of Angola: Researches and Biogeography 151

Scientific name	English name	V	RL
Pseudagrion (A) *sarepi* (Kipping & Dijkstra, 2015)	Sarep Sprite	1!**	DD
Pseudagrion (A) *serrulatum* (Karsch, 1894)	Superb Sprite	1!	LC
Pseudagrion (A) *simonae* (Legrand, 1987)	Wide-striped Sprite	1!	LC
Pseudagrion (A) *simplicilaminatum* (Carletti & Terzani, 1997)##	Blue Slim Sprite	4!!	LC
Pseudagrion (B) *acaciae* (Förster, 1906)	Acacia Sprite	1	LC
Pseudagrion (B) *camerunense* (Karsch, 1899)##	Yellow-fronted Sprite	4!!	LC
Pseudagrion (B) *coeleste* (Longfield, 1947)	Catshead Sprite	1	LC
Pseudagrion (B) *deningi* (Pinhey, 1961)	Dark Sprite	1!	LC
Pseudagrion (B) *dundoense* (Longfield, 1959)	Dundo Sprite	2**	DD
Pseudagrion (B) *glaucescens* (Selys, 1876)	Blue-green Sprite	1	LC
Pseudagrion (B) *hamoni* (Fraser, 1955)	Swarthy Sprite	1!	LC
Pseudagrion (B) *helenae* (Balinsky, 1964)	Little Blue Sprite	1!	LC
Pseudagrion (B) *isidromorai* (Compte Sart, 1967)	Large Blue Sprite	1!	LC
Pseudagrion (B) *massaicum* (Sjöstedt, 1909)	Masai Sprite	1	LC
Pseudagrion (B) *rufostigma* (Longfield, 1947)	Ruby Sprite	1	LC
Pseudagrion (B) *sjoestedti* (Förster, 1906)	Variable Sprite	1	LC
Pseudagrion (B) *sublacteum* (Karsch, 1893)	Cherry-eye Sprite	1	LC
Aeshnidae			
Afroaeschna scotias (Pinhey, 1952)	Shadow Hawker	1!	LC
Anaciaeschna triangulifera (McLachlan, 1896)##	Evening Hawker	4!!	LC
Anax bangweuluensis (Kimmins, 1955)##	Swamp Emperor	4!!	NT
Anax congoliath (Fraser, 1953)	Dark Emperor	1!	LC
Anax ephippiger (Burmeister, 1839)	Vagrant Emperor	1	LC
Anax imperator (Leach, 1815)	Blue Emperor	1	LC
Anax speratus (Hagen, 1867)	Eastern Orange Emperor	1	LC
Anax tristis (Hagen, 1867)	Black Emperor	1	LC
Gynacantha (A) *sextans* (McLachlan, 1896)	Dark-rayed Duskhawker	3	LC
Gynacantha (A) *vesiculata* (Karsch, 1891)	Lesser Girdled Duskhawker	3	LC
Gynacantha (B) *bullata* (Karsch, 1891)	Black-kneed Duskhawker	1	LC
Gynacantha (B) *manderica* (Grünberg, 1902)	Little Duskhawker	3	LC
Heliaeschna cynthiae (Fraser, 1939)##	Blade-tipped Duskhawker	4!!	LC
Heliaeschna fuliginosa (Karsch, 1893)	Black-banded Duskhawker	1	LC
Heliaeschna ugandica (McLachlan, 1896)	Uganda Duskhawker	3	LC
Pinheyschna subpupillata (McLachlan, 1896)##	Stream Hawker	1!!	LC
Zosteraeschna minuscula (McLachlan, 1895)##	Friendly Hawker	1!!	LC
Gomphidae			
Crenigomphus cf. *cornutus* (Pinhey, 1956)#	(near Horned Talontail)	1!	LC
Diastatomma selysi (Schouteden, 1934)	Common Hoetail	3	LC
Diastatomma soror (Schouteden, 1934)	Painted Hoetail	3	LC
Gomphidia quarrei (Schouteden, 1934)	Southern Fingertail	3	LC
Ictinogomphus dundoensis (Pinhey, 1961)	Swamp Tigertail	1	LC
Ictinogomphus ferox (Rambur, 1842)	Common Tigertail	1	LC
Ictinogomphus regisalberti (Schouteden, 1934)	Congo Tigertail	3	LC

(continued)

Scientific name	English name	V	RL
Lestinogomphus calcaratus (Dijkstra, 2015)	Spurred Fairytail	1!	LC
Libyogomphus tenaculatus (Fraser, 1926)	Large Horntail	1!	LC
Mastigogomphus chapini (Klots, 1944)[#]	Western Snorkeltail	2	LC
Mastigogomphus dissimilis (Cammaerts, 2004)[##]	Southern Snorkeltail	2	LC
Microgomphus cf. *nyassicus* (Grünberg, 1902)[#]	(near Eastern Scissortail)	1!	LC
Neurogomphus alius (Cammaerts, 2004)	Large Siphontail	1!	LC
Notogomphus kimpavita (Dijkstra & Clausnitzer, 2015)	Angola Longleg	1!**	DD
Notogomphus praetorius (Selys, 1878)	Yellowjack Longleg	2	LC
Notogomphus spinosus (Karsch, 1890)	Jungle Longleg	1!	LC
Onychogomphus rossii (Pinhey, 1966)	Angola Claspertail	1**	DD
Onychogomphus cf. *styx* (Pinhey, 1961)[#]	(near Northern Dark Claspertail)	1!	LC
Paragomphus abnormis (Karsch, 1890)	Humdrum Hooktail	1!	LC
Paragomphus cognatus (Rambur, 1842)	Rock Hooktail	1!!	LC
Paragomphus cf. *darwalli* (Dijkstra, Mézière & Papazian, 2015)[#]	(near Darwall's Hooktail)	1!	DD
Paragomphus genei (Selys, 1841)	Common Hooktail	1	LC
Paragomphus machadoi (Pinhey, 1961)	Forest Hooktail	2	LC
Paragomphus cf. *nigroviridis* (Cammaerts, 1969)[#]	(near Black-and-green Hooktail)	1!	LC
Paragomphus sabicus (Pinhey, 1950)[##]	Flapper Hooktail	1!!	LC
Paragomphus sp. nov. [##]	(Hooktail, undescribed species)	1!!**	NE
Phyllogomphus annulus (Klots, 1944)	Crested Leaftail	1	LC
Phyllogomphus selysi (Schouteden, 1933)	Bold Leaftail	3	LC
Macromiidae			
Phyllomacromia aureozona (Pinhey, 1966)	Golden-banded Cruiser	1!	LC
Phyllomacromia contumax (Selys, 1879)	Two-banded Cruiser	1!	LC
Phyllomacromia hervei (Legrand, 1980)	River Cruiser	1!	LC
Phyllomacromia melania (Selys, 1871)	Sombre Cruiser	1	LC
Phyllomacromia overlaeti (Schouteden, 1934)	Clubbed Cruiser	3	LC
Phyllomacromia paula (Karsch, 1892)	Greater Double-spined Cruiser	3	LC
Phyllomacromia picta (Hagen in Selys, 1871)	Darting Cruiser	3	LC
Phyllomacromia unifasciata (Fraser, 1954)	Golden-eyed Cruiser	3	LC
Libellulidae			
Acisoma inflatum (Selys, 1882)	Stout Pintail	1	LC
Acisoma trifidum (Kirby, 1889)	Pied Pintail	1	LC
Aethiothemis bequaerti (Ris, 1919)	Skimmer-like Flasher	1	LC
Aethiothemis ellioti (Lieftinck, 1969)	Plump Flasher	1!	LC
Aethiothemis mediofasciata (Ris, 1931)[#]	Orange Flasher	2	LC
Aethiothemis solitaria (Martin, 1908)	Pearly Flasher	1	LC
Aethriamanta rezia (Kirby, 1889)	Pygmy Basker	1	LC
Brachythemis lacustris (Kirby, 1889)	Red Groundling	1	LC

(continued)

Odonata of Angola: Researches and Biogeography

Scientific name	English name	V	RL
Brachythemis leucosticta (Burmeister, 1839)	Southern Banded Groundling	1	LC
Bradinopyga strachani (Kirby, 1900)##	Red Rockdweller	1!!	LC
Chalcostephia flavifrons (Kirby, 1889)	Inspector	1!	LC
Crocothemis brevistigma (Pinhey, 1961)	Spotted Scarlet	1!	LC
Crocothemis divisa (Baumann, 1898)	Rock Scarlet	1	LC
Crocothemis erythraea (Brullé, 1832)	Broad Scarlet	1	LC
Crocothemis sanguinolenta (Burmeister, 1839)	Little Scarlet	1	LC
Cyanothemis simpsoni (Ris, 1915)	Bluebolt	1!	LC
Diplacodes deminuta (Lieftinck, 1969)	Little Percher	1	LC
Diplacodes lefebvrii (Rambur, 1842)	Black Percher	1	LC
Diplacodes luminans (Karsch, 1893)	Barbet Percher	1	LC
Diplacodes pumila (Dijkstra, 2006)	Dwarf Percher	1!	LC
Eleuthemis eogaster (Dijkstra, 2015)	Sunrise Firebelly	1!**	DD
Eleuthemis libera (Dijkstra & Kipping, 2015)	Free Firebelly	1!	DD
Hadrothemis camarensis (Kirby, 1889)	Saddled Jungleskimmer	3	LC
Hadrothemis coacta (Karsch, 1891)	Robust Jungleskimmer	1!	LC
Hadrothemis defecta (Karsch, 1891)	Scarlet Jungleskimmer	3	LC
Hemistigma albipunctum (Rambur, 1842)	African Piedspot	1	LC
Malgassophlebia bispina (Fraser, 1958)	Ringed Leaftipper	1!	LC
Micromacromia camerunica (Karsch, 1890)	Stream Micmac	1!	LC
Micromacromia flava (Longfield, 1947)	Angola Micmac	1**	NT
Neodythemis afra (Ris, 1909)	Seepage Junglewatcher	1!	LC
Neodythemis klingi (Karsch, 1890)	Stream Junglewatcher	1!	LC
Nesciothemis cf. *farinosa* (Förster, 1898)#	(near Eastern Blacktail)	1	LC
Nesciothemis fitzgeraldi (Longfield, 1955)	Lesser Peppertail	1!	LC
Notiothemis jonesi (Ris, 1919)##	Eastern Forestwatcher	1!!	LC
Notiothemis robertsi (Fraser, 1944)	Western Forestwatcher	1!	LC
Olpogastra lugubris (Karsch, 1895)	Bottletail	1	LC
Orthetrum abbotti (Calvert, 1892)	Little Skimmer	1	LC
Orthetrum austeni (Kirby, 1900)	Giant Skimmer	1	LC
Orthetrum brachiale (Palisot de Beauvois, 1817)	Banded Skimmer	1	LC
Orthetrum caffrum (Burmeister, 1839)	Two-striped Skimmer	1	LC
Orthetrum chrysostigma (Burmeister, 1839)	Epaulet Skimmer	1	LC
Orthetrum guineense (Ris, 1910)	Guinea Skimmer	1	LC
Orthetrum hintzi (Schmidt, 1951)	Dark-shouldered Skimmer	1	LC
Orthetrum icteromelas (Ris, 1910)	Spectacled Skimmer	1	LC
Orthetrum julia (Kirby, 1900)	Julia Skimmer	1	LC
Orthetrum kafwi (Dijkstra, 2015)##	Bog Skimmer	1!!	DD
Orthetrum machadoi (Longfield, 1955)	Highland Skimmer	1	LC
Orthetrum macrostigma (Longfield, 1947)	Sharkfin Skimmer	1	LC
Orthetrum microstigma (Ris, 1911)	Farmbush Skimmer	1	LC
Orthetrum monardi (Schmidt, 1951)	Woodland Skimmer	1	LC
Orthetrum robustum (Balinsky, 1965)	Robust Skimmer	1!	LC

(continued)

Scientific name	English name	V	RL
Orthetrum saegeri (Pinhey, 1966)	Eastern Mushroom Skimmer	1!	LC
Orthetrum stemmale (Burmeister, 1839)	Bold Skimmer	1	LC
Orthetrum trinacria (Selys, 1841)	Long Skimmer	1	LC
Oxythemis phoenicosceles (Ris, 1910)	Pepperpants	1!	LC
Palpopleura albifrons (Legrand, 1979)	Pale-faced Widow	1!	LC
Palpopleura deceptor (Calvert, 1899)	Deceptive Widow	3	LC
Palpopleura jucunda (Rambur, 1842)	Yellow-veined Widow	1	LC
Palpopleura lucia (Drury, 1773)	Lucia Widow	1	LC
Palpopleura portia (Drury, 1773)	Portia Widow	1	LC
Pantala flavescens (Fabricius, 1798)	Wandering Glider	1	LC
Porpax asperipes (Karsch, 1896)	Powdered Pricklyleg	1	LC
Porpax risi (Pinhey, 1958)	Highland Pricklyleg	1	LC
Rhyothemis fenestrina (Rambur, 1842)	Skylight Flutterer	1	LC
Rhyothemis mariposa (Ris, 1913)	Butterfly Flutterer	2	LC
Rhyothemis cf. *notata* (Fabricius, 1781)##	(near Veiled Flutterer)	4!!	LC
Rhyothemis semihyalina (Desjardins, 1832)	Phantom Flutterer	1!	LC
Sympetrum fonscolombii (Selys, 1840)	Nomad	1	LC
Tetrathemis camerunensis (Sjöstedt, 1900)	Forest Elf	2	LC
Tetrathemis fraseri (Legrand, 1977)	Treefall Elf	1!	LC
Tetrathemis polleni (Selys, 1869)	Black-splashed Elf	2	LC
Tetrathemis sp. nov. ##	(Elf, undescribed species)	4!!**	NE
Thermochoria equivocata (Kirby, 1889)	Dash-winged Piedface	1!	LC
Tholymis tillarga (Fabricius, 1798)	Twister	1	LC
Tramea basilaris (Palisot de Beauvois, 1817)	Keyhole Glider	1	LC
Trithemis aconita (Lieftinck, 1969)	Halfshade Dropwing	1!	LC
Trithemis aenea (Pinhey, 1961)##	Bronze Dropwing	4!!	LC
Trithemis annulata (Palisot de Beauvois, 1807)	Violet Dropwing	1	LC
Trithemis anomala (Pinhey, 1956)	Striped Dropwing	1!	LC
Trithemis apicalis (Fraser, 1954)	Furtive Dropwing	1!	LC
Trithemis arteriosa (Burmeister, 1839)	Red-veined Dropwing	1	LC
Trithemis basitincta (Ris, 1912)	Jungle Dropwing	1!	LC
Trithemis dichroa (Karsch, 1893)	Black Dropwing	1	LC
Trithemis dorsalis (Rambur, 1842)	Highland Dropwing	1	LC
Trithemis cf. *dubia* (Fraser, 1954)#	(near Sleek Dropwing)	1!	DD
Trithemis furva (Karsch, 1899)	Navy Dropwing	1	LC
Trithemis imitata (Pinhey, 1961)#	Northern Fluttering Dropwing	1!	LC
Trithemis integra (Dijkstra, 2007)	Albertine Dropwing	1!	LC
Trithemis kirbyi (Selys, 1891)	Orange-winged Dropwing	1	LC
Trithemis leakeyi (Pinhey, 1956)	Mealy Dropwing	1!	LC
Trithemis monardi (Ris, 1931)#	Southern Fluttering Dropwing	1	LC
Trithemis nuptialis (Karsch, 1894)	Hairy-legged Dropwing	1	LC
Trithemis palustris (Damm & Hadrys, 2009)#	Marsh Dropwing	1!	LC

(continued)

Scientific name	English name	V	RL
Trithemis pluvialis (Förster, 1906)	Russet Dropwing	1	LC
Trithemis pruinata (Karsch, 1899)	Cobalt Dropwing	1!	LC
Trithemis stictica (Burmeister, 1839)	Jaunty Dropwing	1	LC
Trithemis werneri (Ris, 1912)	Elegant Dropwing	3	LC
Trithemis sp. nov. (Fig 9.6)##	(Dropwing, undescribed species)	4!!**	NE
Urothemis assignata (Selys, 1872)	Red Basker	1	LC
Urothemis edwardsii (Selys, 1849)	Blue Basker	1	LC
Urothemis venata (Dijkstra & Mézière, 2015)##	Red-veined Basker	4 !!	LC
Zygonoides fuelleborni (Grünberg, 1902)	Southern Riverking	3	LC
Zygonyx denticulatus (Dijkstra & Kipping, 2015)	Pale Cascader	1!	LC
Zygonyx eusebia (Ris, 1912)	Imperial Cascader	3	LC
Zygonyx flavicosta (Sjöstedt, 1900)##	Ensign Cascader	1	LC
Zygonyx natalensis (Martin, 1900)	Blue Cascader	1	LC
Zygonyx regisalberti (Schouteden, 1934)	Regal Cascader	1	LC
Zygonyx torridus (Kirby, 1889)	Ringed Cascader	1	LC

Notes on new country records (by J Kipping and S F Elizalde unless stated otherwise)

Chlorocypha aphrodite – male photographed by C Hines near Lucala north of Uíge in June 2017.

Mesocnemis singularis – first record of true *M. singularis* (see Kipping et al. 2017) from the Angolan bank of the Cunene River in December 2017.

Africallagma sinuatum – single male photographed by C Hines near Cambondo, Cuanza-Norte Province in February 2017.

Africallagma subtile – several collected at marshy floodplains of the Yevedula River, 20 km northwest of Caconda, Benguela Province in December 2017.

Agriocnemis gratiosa – several collected by M Haacks from Bicuar NP, Huíla Province in December 2016.

Agriocnemis pinheyi – several collected at a marsh northwest of Caconda, Benguela Province in December 2017.

Pseudagrion (A) *simplicilaminatum* – male photographed by C Hines near Lucala north of Uíge in June 2017.

Pseudagrion (B) *camerunense* – male photographed by C Hines in Cuanza River floodplains south of Luanda in January 2018.

Anaciaeschna triangulifera – female photographed by C Hines in Cuanza River floodplains south of Luanda in June 2017. Westernmost record; nearest locality is Ikelenge in northwestern Zambia, about 1200 km to the east.

Anax bangweuluensis – teneral male photographed by J Mendelsohn at Lake Saliakembo, Moxico Province in October 2017. The Cuito River links to the nearest known population in the Okavango Delta, Botswana, about 750 km away.

Heliaeschna cynthiae – female and two males recorded by C Hines at the Rio Nzadi and near Quicunga in Uíge Province in June 2017.

Pinheyschna subpupillata – many observed and collected at Tchiamena River near Lubango and Neve River near Humpata on Serra da Chela, Huíla Province in

December 2017. Presumably isolated population; widespread in South Africa, with another isolated population on border of Mozambique and Zimbabwe. With the new finding of this species a former record of a female *P. rileyi* (Calvert, 1892) from Tundavala (Pinhey 1975) became more doubtful and the species is therefore deleted from the country list.

Zosteraeschna minuscula – male collected at the Tchiamena River near Lubango on Serra da Chela, Huíla Province in December 2017. Northernmost record; widespread in South Africa but with scattered records in Namibia and eastern Botswana.

Mastigogomphus dissimilis – the *Instituto de Investigação Agronómica* in Huambo has one male from Nova Sintra (Catabola), Bié Province from October 1973, coll. L Amorim.

Paragomphus cognatus – presence in Angola was uncertain due to lack of reliable material (Kipping et al. 2017), but several males collected at Tchiamena, Leba and Neve Rivers in the Serra da Chela in December 2017.

Paragomphus sabicus – common at the Rio Coporolo, north of Chongoroi, Benguela Province in December 2017.

Paragomphus sp. nov. – two males collected at the Uiri River near Conda, Cuanza-Sul Province in December 2017 belong to an undescribed species similar to *P. cognatus* but darker and with stouter paraprocts and more curved cerci.

Bradinopyga strachani – the *Instituto de Investigação Agronómica* in Huambo has three males from Ndalatando, Cuanza-Norte Province from March 1973, coll. U Passos. Several also collected at Rio Mussenju, south of Quilengues, Benguela Province in December 2017 and photographed by R Ferreira at Calandula Falls, Lunda-Norte Province in June 2018.

Notiothemis jonesi – male was collected in Lubango, Huíla Province in December 2017.

Orthetrum kafwi – several males and females collected at boggy streams and bogs in the highlands around Cassongue, Cuanza-Sul Province in December 2017. Previously know only from the type locality in the Upemba National Park in Katanga, which lies 1400 km to the east.

Rhyothemis cf. *notata* – male photographed by J Mendelsohn at Sacangombe near the Cuito River source in Moxico Province in November 2011. The black markings in the forewings reach only to the nodus and in the hindwings halfway the nodus and pterostigma, which is much less than even the palest variation of *R. notata* illustrated by Dijkstra & Clausnitzer (2014). The habitat is open, while true *R. notata* favour rainforest conditions. This species therefore needs to be verified with specimens.

Tetrathemis sp. nov. – several males photographed by C Hines in dry forest near Cambondo, Cuanza-Norte Province in March 2017. Differs from *T. fraseri* by the smoky wings and shape of the very hairy cerci.

Trithemis aenea – photographed by C Hines near Lucala north of Uíge in June 2017.

Trithemis sp. nov. – see Fig. 9.6 and main text.

Urothemis venata – photographed by Carel van der Merwe in the Cuango area, Cuanza-Norte Province in May 2017.

Appendix 2

Odonata recorded from rivers bordering Angola that most likely also occur in Angola.

Scientific name	English name	Nearest occurrence
Coenagrionidae		
Pseudagrion (A) *spernatum* (Selys, 1881)	Upland Sprite	At Jimbe and other rivers in Ikelenge Pedicle of north-western Zambia.
Pseudagrion (B) *assegaii* (Pinhey, 1950)	Assegai Sprite	Cuando River in Namibian Caprivi Strip.
Pseudagrion (B) *sudanicum* (Le Roi, 1915)	Blue-sided Sprite	Okavango and Cuando Rivers in Namibian Caprivi Strip.
Gomphidae		
Crenigomphus kavangoensis (Suhling & Marais, 2010)	Kavango Talontail	Okavango River in Namibia.
Lestinogomphus angustus (Martin, 1911)	Common Fairytail	Cunene, Okavango and Cuando Rivers in northern Namibia.
Lestinogomphus silkeae (Kipping, 2010)	Silke's Fairytail	One locality on the southern bank of the Okavango River near Rundu, Namibia.
Paragomphus cataractae (Pinhey, 1963)	Cataract Hooktail	Waterfalls and rapids of the Cunene and Okavango Rivers in northern Namibia.
Paragomphus elpidius (Ris, 1921)	Corkscrew Hooktail	Cunene, Okavango and Cuando River in northern Namibia and the Ikelenge Pedicle of Zambia.
Neurogomphus cocytius Cammaerts, 2004	Kokytos Siphontail	Okavango River in northern Namibia.
Libellulidae		
Parazyxomma flavicans (Martin, 1908)	Banded Duskdarter	Okavango and Cuando Rivers in northern Namibia.
Trithemis aequalis (Lieftinck, 1969)	Swamp Dropwing	Okavango and Cuando Rivers in the Namibian Caprivi.
Trithemis donaldsoni (Calvert, 1899)	Denim Dropwing	Okavango and Cunene Rivers in northern Namibia.
Trithemis hecate (Ris, 1912)	Silhouette Dropwing	Common along the Cunene, Okavango and Cuando Rivers in northern Namibia.
Trithemis morrisoni (Damm & Hadrys, 2009)	Rapids Dropwing	Okavango and Cuando Rivers in the Namibian Caprivi.
Trithetrum navasi (Lacroix, 1921)	Fiery Darter	Cunene, Okavango and Cuando Rivers in northern Namibia.

References

Clausnitzer V, Koch R, Dijkstra K-DB et al (2012) Focus on African freshwaters: hotspots of dragonfly diversity and conservation concern. Front Ecol Environ 10:129–134

Damm S, Hadrys H (2009) *Trithemis morrisoni* sp. nov. and *Trithemis palustris* sp. nov. from the Okavango and Upper Zambezi Floodplains previously hidden under *T. stictica* (Odonata: Libellulidae). Int J Odonatol 12(1):131–145

Dijkstra K-DB (2006) The *Atoconeura* problem revisited: taxonomy, phylogeny and biogeography of a dragonfly genus in the highlands of Africa (Odonata, Libellulidae). Tijdschrift voor Entomologie 149:121–144

Dijkstra K-DB (2007a) The name-bearing types of Odonata held in the Natural History Museum of Zimbabwe, with systematic notes on Afrotropical taxa. Part 1: introduction and Anisoptera. Int J Odonatol 10(1):1–29

Dijkstra K-DB (2007b) The name-bearing types of Odonata held in the Natural History Museum of Zimbabwe, with systematic notes on Afrotropical taxa. Part 2: Zygoptera and description of new species. Int J Odonatol 10(2):137–170

Dijkstra K-DB, Clausnitzer V (2014) The dragonflies and damselflies of Eastern Africa: handbook for all Odonata from Sudan to Zimbabwe. . Studies in afrotropical zoology 298. Royal Museums for Central Africa, Tervuren, 263 pp

Dijkstra K-DB, Kipping J, Mézière N (2015) Sixty new dragonfly and damselfly species from Africa (Odonata). Odonatologica 44(4):447–678

Dijkstra K-DB, Boudot J-P, Clausnitzer V et al (2011) Chapter 5. Dragonflies and damselflies of Africa (Odonata): history, diversity, distribution, and conservation. In: Darwall WRT, Smith KG, Allen DJ et al (eds) The diversity of life in African freshwaters: under water, under threat. An analysis of the status and distribution of freshwater species throughout mainland Africa. IUCN, Cambridge and Gland, 347 pp

Kipping J (2010) The dragonflies and damselflies of Botswana – an annotated checklist with notes on distribution, phenology, habitats and Red List status of the species (Insecta: Odonata). Mauritiana (Altenburg) 21:126–204

Kipping J (2012) Southern African Regional Environmental Program (SAREP) – first biodiversity field survey upper Cubango (Okavango) catchment, Angola, May 2012 – Dragonflies & Damselflies (Insecta: Odonata). Expert Report:1–108

Kipping J, Clausnitzer V, Fernandes Elizalde SRF et al (2017) The dragonflies and damselflies (Odonata) of Angola. Afr Invertebr 58(I):65–91 https://africaninvertebrates.pensoft.net/article/11382/

Kipping J, Dijkstra K-DB, Clausnitzer V et al (2009) Odonata Database of Africa (ODA). Agrion 13:20–23

Longfield C (1936) Studies on African Odonata, with synonymy and descriptions of new species and subspecies. Trans R Entomol Soc Lond 85:467–499

Longfield C (1947) The Odonata of South Angola: results of the Mission Scientifiques Suisses 1928–29, 1932–33. Arquivos do Museu Bocage 16:1–31

Longfield C (1955) The Odonata of North Angola, part 1. Publicações Culturais, Companhia de Diamantes de Angola 27:11–64

Longfield C (1959) The Odonata of North Angola, part 2. Publicações Culturais, Companhia de Diamantes de Angola 45:16–42

Mendes LF, Bivar-de-Sousa A, Figueira R et al (2013) Gazetteer of the Angolan localities known for beetles (Coleoptera) and butterflies (Lepidoptera: Papilionoidea). Boletim da Sociedade Portuguesa de Entomologia 228(VIII–14):257–292

Mills MSL, Melo M, Vaz A (2013) The Namba mountains: new hope for Afromontane forest birds in Angola. Bird Conserv Int 23:159–167

NGOWP – National Geographic Okavango Wilderness Project (2018) Initial Findings from Exploration of the Upper Catchments of the Cuito, Cuanavale and Cuando Rivers in Central and South-Eastern Angola (May 2015 to December 2016). National Geographic Okavango Wilderness Project, 352 pp

Pinhey ECG (1961a) A collection of Odonata from Dundo, Angola with the descriptions of two new species of Gomphids. Publicações Culturais, Companhia de Diamantes de Angola 56:71–78

Pinhey ECG (1961b) Some dragonflies (Odonata) from Angola and descriptions of three new species of the family Gomphidae. Publicações Culturais, Companhia de Diamantes de Angola 56:79–86

Pinhey ECG (1964) Dragonflies (Odonata) of the Angola-Congo borders of Rhodesia. Publicações Culturais, Companhia de Diamantes de Angola 63:97–130

Pinhey ECG (1965) Odonata from Luanda and the Lucala River, Angola. Revista de Biologia 5:159–164

Pinhey ECG (1966) New distributional records for African Odonata and notes on a few larvae. Arnoldia Rhodesia 2(26):1–5

Pinhey ECG (1971a) Notes on the genus *Pseudagrion* Selys (Odonata: Coenagrionidae). Arnoldia Rhodesia 5(6):1–4

Pinhey ECG (1971b) Odonata collected in Republique Centre-Africaine by R. Pujol. Arnoldia Rhodesia 5(18):1–16

Pinhey ECG (1974) A revision of the African *Agriocnemis* Selys and *Mortonagrion* Fraser (Odonata: Coenagrionidae). Occasional Papers of the National Monuments of Rhodesia B 5/4:171–278

Pinhey ECG (1975) A collection of Odonata from Angola. Arnoldia Rhodesia 7(23):1–16

Pinhey ECG (1984) A check-list of the Odonata of Zimbabwe and Zambia. Smithersia 3:1–64

Ris F (1931) Odonata aus Süd-Angola. Revue Suisse Zoologie 38(7):97–112

Suhling F, Martens A (2007) Dragonflies and damselflies of Namibia. Gamsberg Macmillan, Windhoek, 280 pp

Suhling F, Martens A (2014) Distribution maps and checklist of Namibian Odonata. Libellula Suppl 13:107–175

Tarboton W (2009) A dragonfly survey of the Humpata District. In: Huntley BJ (ed) Projecto de estudo da biodiversidade de Angola. (Biodiversity Rapid Assessment – Huíla /Namibe) Report on Pilot project. SANBI, Cape Town, 3 pp

Vick GS, Chelmick DG, Martens A (2001) In memory of Elliot Charles Gordon Pinhey (10 July 1910 – 7 May 1999). Odonatologica 30:1–11

A Checklist of Angolan Lepidoptera and Papilionoidea

Luís F. Mendes, A. Bivar-de-Sousa and Mark C. Williams

Abstract Presently, 792 species/subspecies of butterflies and skippers (Lepidoptera: Papilionoidea) are known from Angola, a country with a rich diversity of habitats, but where extensive areas remain unsurveyed and where systematic collecting programmes have not been undertaken. Only three species were known from Angola in 1820. From the beginning of the twenty-first century, many new species have been described and more than 220 faunistic novelties have been assigned. As a whole, of the 792 taxa now listed for Angola, 57 species/subspecies are endemic and almost the same number are known to be near-endemics, shared by Angola and by one or another neighbouring country. The Nymphalidae are the most diverse family. The Lycaenidae and Papilionidae have the highest levels of endemism. A revised checklist with taxonomic and ecological notes is presented and the development of knowledge of the superfamily over time in Angola is analysed.

Keywords Africa · Conservation · Ecology · Endemism · Taxonomy

---

L. F. Mendes (✉)
Museu Nacional de História Natural e da Ciência, Universidade de Lisboa, Lisboa, Portugal

CIBIO, Centro de Investigação em Biodiversidade e Recursos Genéticos, Vairão, Portugal
e-mail: luisfmendes22@gmail.com

A. Bivar-de-Sousa
Museu Nacional de História Natural e da Ciência, Universidade de Lisboa, Lisboa, Portugal

Sociedade Portuguesa de Entomologia, Lisboa, Portugal
e-mail: abivarsousa@gmail.com

M. C. Williams
Pretoria University, Pretoria, South Africa
e-mail: lepidochrysops@gmail.com

Introduction

Angola is a large country of 1,246,700 km², notable for its great diversity of physiography, climates, habitats and resultant biodiversity). The country includes seven biomes and 15 ecoregions, ranging from equatorial rainforests of the northwest (Cabinda) and along the northern border with the Democratic Republic of Congo, through the moist miombo woodlands and savannas of the central plateaus, to the dry forests and woodlands of the southeast, and to the arid shrublands and Namib Desert of the southwest. Isolated forests with Congolian affinities are found along the Angolan Escarpment, and similar remnant patches of Afromontane forests are found on some of the highest mountains such as Mount Moco and Mount Namba.

Despite the fact that at the beginning of the nineteenth century only a few species of butterflies and skippers (Insecta: Lepidoptera: Papilionoidea) were recorded from Angola, today a large number of taxa (at least 792 species and subspecies: Fig. 10.1, Table 10.1 and Appendix) are known to occur in the country. However, extensive areas are still poorly surveyed for butterflies, or have not been surveyed at all (Fig. 10.2). This applies in particular to the southern provinces of Namibe, Cunene and Cuando Cubango and the northwestern province of Zaire as well as most of southern Moxico. Furthermore, the Baixa de Cassanje (Malanje), separated from surrounding areas by steep escarpments, appears to have distinctive vegetation and may produce some interesting butterflies. Although most of the localities where but-

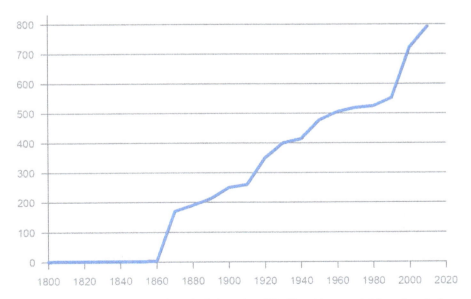

Fig. 10.1 Cumulative number of species/subspecies of Papilionoidea reported from Angola from 1801–1819 (first records) to the recent decade – 2011–2017 – according to Appendix. For practical reasons, species which first reference to the country was untraceable (marked in Appendix with a ▲) were included in the decade 2001–2010; species that are now assigned as faunistic novelties to Angola (marked in Appendix with a ◘) are included in the last decade (2011–2017)

Table 10.1 Number of species of Papilionoidea families and subfamilies known to occur in the Afrotropical Region and Angola (with % of Afrotropical species present in the country), and number of species endemic to Angola (with % of endemism shown)

Family Subfamily	Afrotropical N°	Angola N° I %	Endemism N I %
HESPERIIDAE	618	134 I 22	5 I 3.7
Coeliadinae	21	7 I 33	
Pyrginae	216	48 I 22	
Heteropterinae	27	5 I 19	
Hesperiinae	334	74 I 22	
PAPILIONIDAE	101	33 I 33	3 I 9.1
PIERIDAE	200	67 I 34	5 I 7.5
Pseudopontiinae	5	2 I 40	
Coliadinae	14	8 I 57	
Pierinae	181	57 I 32	
LYCAENIDAE	1837	210 I 11	18 I 8.6
Miletinae	119	11 I 9	
Poritiinae	658	53 I 8	
Theclinae	301	42 I 14	
Aphnaeinae	260	21 I 8	
Polyommatinae	496	83 I 17	
RIODINIDAE	15	4 I 27	0 I 0
NYMPHALIDAE	1634	344 I 21	26 I 7.6
Libytheinae	5	2 I 40	
Danainae	26	9 I 35	
Satyrinae	347	50 I 14	
Charaxinae	190	56 I 30	
Apaturinae	3	1 I 33	
Nymphalinae	73	35 I 48	
Cyrestinae	1	1 I 100	
Biblidinae	31	16 I 52	
Limenitidinae	702	97 I 14	
Heliconiinae	256	77 I 30	
TOTALS	4405	792 I 18	

terflies and skippers have been collected in Angola have been determined (Mendes et al. 2013b), some localities previously reported for a few species remain untraced despite searches by us, using the detailed maps of the *Junta de Investigações do Ultramar* (JIU 1948–1963).

The accumulation of knowledge in regard to Angolan butterflies has been constrained by several factors. The two largest Angolan entomological collections, deposited in the *Museu do Dundo* (Lunda-Norte) and in the *Instituto de Investigação Agronómica* (Huambo), have never been studied in detail. In addition little fieldwork was carried out in Angola during the post-independence period because of the protracted civil war. Finally, the vastness of the country and the difficulty in accessing many remote regions has impeded progress.

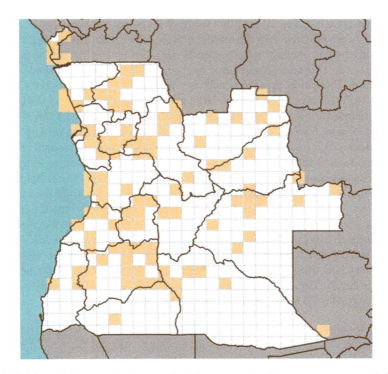

Fig. 10.2 Map of Angola showing, marked in orange, the known areas surveyed for the Papilionoidea from the beginning of their study, in the nineteenth century to the present day – each square ca. 33 × 33 km. The collecting pressure varies across the country, from 'squares' where samples were obtained only once, in passing, to others where the collectors were based for months

Until recently the Hesperiidae (skippers) were placed in the superfamily Hesperioidea, separate from the rest of the butterflies, which were placed in the superfamily Papilionoidea. However, today the skippers and butterflies are all placed in the Papilionoidea (e.g. Heikkilä et al. 2012). The classification used for butterflies in this chapter is based on Williams (2018), Espeland et al. (2018) and Dhungel and Wahlberg (2018). Six families of butterflies are represented in Angola, namely Papilionidae, Hesperiidae, Pieridae, Riodinidae, Lycaenidae and Nymphalidae.

History of Research on the Papilionoidea of Angola

The first known reference to butterflies obtained in Angola is by Latreille and Godart (1819), who reported the presence of *Colotis euippe* (Linnaeus, 1758) and described *Acraea parrhasia servona*. In the decade between 1871 and 1880, Druce (1875) reported about 90 species from Angola for the first time, a number of these being descriptions of species new to science. By the end of the nineteenth century a total of 214 butterfly taxa were known from Angola.

The first contributions to our knowledge of Angolan butterflies by Portuguese researchers were only made in the middle nineteen hundreds. These were the result of the activities in Angola of the *Centro de Zoologia* (CZ) of the *Junta de Investigações do Ultramar*, coordinated by its first director Fernando Frade. In 2014 this research institution was renamed the *Instituto de Investigação Científica Tropical* (IICT). Working from these large zoological collections, as well as from further specimens obtained in Angola by Amélia Bacelar (1948, 1956, 1958a, b, 1961) and Miguel Ladeiro (1956), considerably expanded the list of Angolan butterflies. Most of this material, obtained during colonial times, was stored in Lisbon, with corrections to the published identifications only being made recently. All of this material has now been integrated into the collections of the *Museu Nacional de História Natural e da Ciência* (MUHNAC). Albert Monard (1956) of the La Chaux-de-Fonds Swiss Museum also studied other material obtained by the CZ missions. Significant contributions in the twentieth century were also made by Weymer (1901) on the southern Angolan species, and by Evans (1937) on the Hesperiidae. All of the then known Angolan butterflies were listed by Aurivillius (in Seitz) in 1928. All of the Angolan Charaxinae were dealt with by Henning in his 1988 book on the African taxa of this family. The 339 taxa added to the faunal list during the twentieth century brought the total to 553 known butterfly taxa for Angola.

During the first 18 years of the twenty-first century, 239 further taxa were added to the total. In the first decade of the present century, most of the new information was due to several contributions by Libert (1999, 2000, 2004) on the Lycaenidae, and by Gardiner (2004). The latter author listed taxa from the southeastern Cuando Cubango province, which borders the Caprivi Strip of Namibia. Cuando Cubango and the easternmost province of Moxico are the only provinces in Angola with Zambezian fauna. To these taxa we add our own contributions (Bivar-de-Sousa and Mendes 2006, 2007, 2009a, b; Mendes and Bivar-de-Sousa 2006a, b, 2007a, b, 2009a, b, c, d). Over the last 8 years 33 species were described as new or recorded for the first time from Angola by Mendes and Bivar-de-Sousa (2012, 2017) Mendes et al. (2013a, 2017, 2018), Bivar-de-Sousa and Mendes (2014) and Bivar-de-Sousa et al. (2017), Turlin and Vingerhoedt (2013) and Pierre and Bernaud (2013). Finally, 66 further taxa are now recorded as faunistic novelties for the country (Appendix). The current total number of butterfly taxa for Angola now stands at 792.

Sources Consulted for the Checklist

In preparing this revised checklist of the Papilionoidea, the following collections of Angolan butterflies held by institutions in Portugal were examined: *Museu Nacional de História Natural e da Ciência* (MUHNAC) in Lisbon, *Museu de História Natural da Universidade do Porto* (MHNC-UP), *Liceu Nun'Álvares* in the Caldas da Saúde and the Singeverga Order of St Benedict Abbey in Areias. Major contributions to these collections were made by A Bivar-de-Sousa (Luanda district and Cuanza-Norte, Cuanza-Sul and Moxico Provinces), António Figueira, (northwestern Angola), Mário Macedo (northern Angola), Passos de Carvalho (Huambo and

Cuanza-Norte Provinces), Carneiro Mendes and Pessoa Guerreiro. The Angolan insect collection of Nozolino de Azevedo (mainly Huambo Province), maintained and made available by his widow, was also studied.

The collections in the MUHNAC in Lisbon were destroyed by a fire in March 1978. However, prior to the fire BS had studied some of the material and published his findings. In 1995 LM studied the collections, mainly of Barros Machado and Luna de Carvalho, in the Dundo Museum in Angola but there was insufficient time to do a detailed analysis. We did not inspect the entomological collections in the former *Instituto de Investigação Agronómica* de Angola, collected mainly by Passos de Carvalho, but they are apparently in good condition. No entomological collections were found by LM, in 1995 and 2013, at the *Museu de História Natural de Luanda*. Material collected from 2010 to 2014 by Ruben Capela and Carmen Van-Dúnen Santos of Agostinho Neto University, Luanda and Artur Serrano of the Faculty of Sciences, Lisbon University, was examined by us.

In addition, images of live specimens published by Lautenschläger and Neinhhuis (2014) were examined, as were several images presented by Jorge Palmeirim of Lisbon University and Pedro Vaz Pinto of the Kissama Foundation.

Taxa Excluded from the Checklist

A number of taxa have erroneously been reported to occur in Angola. This was due mainly to probable misidentifications or mislabelled specimens. Some older records are omitted because the known range of the taxon is unlikely to include Angola. A list of the omitted taxa is given below.

- Hesperiidae: *Eretis djaelaelae* (Wallengren, 1857), *Metisella metis* (Linnaeus, 1764), *Kedestes chaca* (Trimen, 1873), *Platylesches chamaeleon* (Mabille, 1891).
- Papilionidae: *Papilio menestheus* Drury, 1773, *Graphium taboranus* (Oberthür, 1886), *Graphium (Arisbe) junodi* (Trimen, 1893).
- Pieridae: *Eurema brigitta* (Stoll, 1780), *Colotis chrysonome* (Klug, 1829), *Colotis ephyia* (Klug, 1829), *Belenois theora* (Doubleday, 1846), *Mylothris rubricosta* (Mabille, 1890), *Mylothris similis* Lathy, 1906.
- Lycaenidae: *Telipna acraea* (Westwood, 1851), *Cooksonia abri* Collins & Larsen, 2008, *Mimacraea darwinia* Butler, 1872, *Liptena bassae* Bethune-Baker, 1926, *Aethiopana honorius honorius* (Fabricius, 1793), *Stempfferia uniformis* (Kirby, 1887), *Stempfferia dorothea* (Bethune-Baker, 1904), *Oxylides faunus* (Drury, 1773), *Dapidodigma hymen* (Fabricius, 1775), *Aloeides molomo* (Trimen, 1870), *Leptomyrina lara* (Linnaeus, 1764), *Deudorix livia* (Klug, 1834), *Neurellipes onias* (Hulstaert, 1924), *Zintha hintza* (Trimen, 1864).
- Riodinidae: *Afriodinia caeca semicaeca* (Riley, 1932), *Afriodinia gerontes* (Fabricius, 1781).
- Nymphalidae: *Bicyclus milyas* (Hewitson, 1864), *Ypthima congoana* Overlaet, 1955, *Charaxes jahlusa argynnides* Westwood, 1864, *Junonia touhilimasa*

Vuillot, 1892, *Neptis continuata* Holland, 1892, *Neptis strigata* Acrivirius, 1894, *Evena oberthueri* (Karsch, 1894), *Euriphene atrovirens* (Mabille, 1878), *Bebearia mardania* (Fabricius, 1793), *Euphaedra morini* Hecq, 1983, *Euphaedra xypete* (Hewitson, 1865), *Euphaedra campaspe* (Felder & Felder, 1267), *Euphaedra inanum* (Butler, 1873), *Euphaedra eupalus* (Fabricius, 1781).

A Revised Checklist of the Papilionoidea of Angola

A revised and annotated checklist of the Papilionoidea of Angola (Appendix) confirms the presence of at least 792 taxa in the country. Their presence is based mainly on verification by the authors of this chapter. Some taxa, recorded by other authors, are accepted because they were, with rare exceptions, reported by more than one author, are based on reliable literature records, or because Angola falls within their putative geographical range. In the Checklist, the first reference to their occurrence in Angola is given, followed by the sources of validation of the record and their preferred habitat(s). Occasionally more than one subspecies of a particular species occurs in the country. This is due to both the size and ecological diversity of Angola. A number of forests, especially gallery-forests, are independent of each other as are the fragmented forests of the Angolan Escarpment. In addition, the southeastern parts of Moxico and the Cuando Cubango provinces are part of the Zambesi Basin; consequently their fauna has affinities with that of eastern Africa.

As far as habitats are concerned the great majority of the Angolan Papilionoidea, as might be expected, occur in forest, both wet and dry (Appendix). However, the Hesperiidae and Pieridae appear to be almost as diverse in moist woodland (miombo) and dry woodland as they are in forest. The number of Pieridae in dry woodland and miombo is similar, while the number of species in dry woodland, arid shrubland and grassland surpasses that of wet forest. The subfamily Nymphalinae is more diverse in miombo than forest and equally diverse in savanna. The Heliconiinae (Nymphalidae) in savanna are almost as diverse as they are in wet forest.

Composition, Diversity and Endemism

All six families and all of the subfamilies (except the Lycaeninae, Leach, 1815) of Afrotropical butterflies are represented in Angola (Table 10.1).

One genus and 56 species/subspecies of Papilionoidea are endemic to Angola, many of which were described over the last few decades. The endemic genus *Mashunoides*, Mendes and Bivar-de-Sousa 2009a, b, c, d (Nymphalidae: Satyrinae) is confined to Cuando Cubango Province, in the ecotone between miombo and savanna/dry woodland mosaic. Endemism rates for Angolan butterfly families are highest for the Papilionidae and Lycaenidae and lowest for the Hesperiidae and Riodinidae (Tables 10.1 and 10.2). Examples of endemic species are illustrated in Fig. 10.3.

A Checklist of Angolan Lepidoptera and Papilionoidea

Table 10.2 Endemic butterfly species and subspecies in Angola

Family	Endemic species	Endemic subspecies
Hesperiidae	*Eagris multiplagata* *Abantis bergeri*	*Calleagris jamesoni ansorgei* *Eretis herewardi rotundimacula* *Spialia colotes colotes*
Papilionidae	*Papilio bacelarae* *Papilio chitondensis*	*Papilio macinnoni benguellae*
Pieridae	*Mylothris carvalhoi*	*Appias epaphia angolensis* *Appias phaola uigensis* *Appias sylvia ribeiroi* *Mylothris spica gabela*
Lycaenidae	*Alaena rosei* *Cooksonia nozolinoi* *Falcuna lacteata* *Deloneura barca* *Aloeides angolensis* *Zeritis krystyna* *Cupidesthes vidua* *Uranothauma nozolinoi Lepidochrysops ansorgei* *Lepidochrysops flavisquamosa* *Lepidochrysops fulvescens* *Lepidochrysops hawker* *Lepidochrysops nacrescens* *Lepidochrysops reichenowi*	*Liptena homeyeri straminea* *Falcuna libyssa angolensis* *Cigarits modestus modestus* *Leptomyrina henningi angolensis*
Nymphalidae	*Brakefieldia angolensis* *Brakefieldia ochracea* *Neita bikuarica* *Mashunoides carneiromendesi* *Charaxes figuerai* *Charaxes ehmckei* *Precis larseni* *Bebearia hassoni* *Euphaedra divoides* *Euphaedra uigensis* *Acraea bellona* *Acraea lapidorum,* *Acraea onerata*	*Amauris crawshayi angola,* *Amauris dannfelti dannfelti* *Charaxes fulvescens rubenarturi,* *Charaxes macclouni carvalhoi,* *Charaxes lucretius saldanhai* *Charaxes jahlusa angolensis* *Charaxes minor karinae* *Charaxes trajanus bambi* *Palla ussheri hassoni* *Sevenia occidentalium penricei* *Euphaedra harpalyce commineura* *Acraea violarum anchietai*

Conservation

Because butterflies are sensitive to changing environmental conditions and are taxonomically well known, they are valuable as indicators of ecological dynamics. They are also key drivers of ecological processes. In particular, adult butterflies are active pollinators of many plants and the imagos and larvae are an important source of nutrition for a diverse range of vertebrate and invertebrate predators and insect parasitoids. Their conservation importance is also due to their positive and occasionally negative economic impacts. Although humans utilise mainly moth caterpillars as a

Fig. 10.3 The holotype specimens of endemic Angolan papilionoidea: Left to right, top to bottom (*V* Ventral, *D* Dorsal): 1. *Abantis bergeri* male D (Mendes and Bivar-de-Sousa 2009a, b, c d), 2. *Eagris multiplagata* male V (Bivar-de-Sousa and Mendes 2007), 3. *Cooksonia nozolinoi* female D (Mendes and Bivar-de-Sousa 2007), 4. *Papilio bacelarae* male D (Bivar-de-Sousa and Mendes 2009a, b), 5. *Mashunoides carneiromendesi* male V (Mendes and Bivar-de-Sousa 2009a, b, c, d), 6. *Charaxes jahlusa angolensis* male D (Mendes et al. 2017), 7. *Euxanthe trajanus bambi* male D (Bivar-de-Sousa and Mendes 2006), 8. *Euphaedra (Euphaedrana) divoides* male V (Bivar-de-Sousa and Mendes 2018)

food source, the larvae of the skipper *Coeliades libeon* is much appreciated. A limited number of butterfly species are agricultural pests, including *Papilio demodocus* (young citrus orchards), *Lampides boeticus* (cultivated Leguminosae) and *Acraea acerata* (sweet-potatoes). A few species, such as *Pyrrhochalcia iphis* and *Zophopetes dysmephila*, may cause damage in coconut and oil-palm plantations.

In terms of species of conservation concern, information on the status of Angolan butterflies is very limited. Many species of Angolan butterflies are obviously abundant and widespread, both within and outside the country. Those taxa that appear to be rare and/or more localised may be genuinely rare or local but this may simply reflect a paucity of information. This makes it difficult or impossible to propose rational conservation measures at present. The urgent need for more fieldwork, particularly in regard to the endemic taxa, is thus highlighted. In the meantime habitat conservation, especially with respect to isolated forest patches, can be considered as part of a wider effort to conserve both the fauna and flora of the country.

Potential Future Discoveries and Research

Considering the number of taxa new to science described in the last few decades there are almost certainly further undiscovered butterfly taxa in Angola. Vast areas of the country remain unexplored, mainly because of inaccessibility and post-independence political instability. Not only will new taxa be found but also known taxa from bordering countries will be added to the list of Angolan butterflies during future fieldwork. This work will also improve our knowledge in regard to the distribution of the taxa in the country. Finally, almost nothing is known about the habitats, behaviours, early stages and larval host plants of Angolan butterflies, making these fertile areas for future research on the fauna. More information concerning all the endemic taxa is urgently needed in order to determine conservation priorities.

Appendix

Checklist of Papilionoidea recorded from Angola (by Family | Sub-family).
For the species' authors and references, Bivar-de-Sousa is abbreviated as BS and Mendes as M; (*) = Taxa with Angola as type locality; ◘ = Taxa now reported as new to Angola; ▲ = Previous references existing but not traced (several species when reported from Angola were assigned under names today considered synonyms, others at species level – the corresponding Angolan endemics were described later).

V = Validation of the taxon – 1: Endemic (restricted to Angola); 2: Restricted to Angola and to the neighbouring northern countries – Gabon, Congo and/or DRC; 3: Restricted to Angola and to the neighbouring eastern countries – southern DRC (former Shaba) and/or Zambia; 4: Restricted to Angola and to the southern and south-eastern neighbouring countries – Namibia and/or Botswana; 5: Species/subspecies with material studied by the authors; 6: Taxa exclusively known in the country from previously collected unstudied material – known from bibliographic references only. H – Preferred habitat: A: Humid forest – primary and secondary wet forest, gallery and riverine forest, forest edge; B: Dry forest, including dry forest and savanna mosaics; C: Brachystegia woodland (miombo) and other woodland; D: Mixed savanna with or without trees; E: Swampy areas, including the northeastern moist thicket and savanna mosaic; F: Arid shrubland and grassland; G: Rocky hillsides; H: Ubiquitous or almost ubiquitous; X: Caterpillars (almost) monophagous, the imagos range dependent on the presence of host-plants

Taxon	First reference for Angola	V	H
HESPERIIDAE | Coeliadinae			
Coeliades bixana (Evans, 1940)	Evans, 1937, as *C. bixae*	6	A
Coeliades c. chalybe (Westwood, 1852)	Evans, 1937	5	A
Coeliades libeon (Druce, 1875)	Druce, 1875 (*)	5	E,C
Coeliades f. forestan (Stoll, 1782)	Ladeiro, 1959	5	F
Coeliades hanno (Plötz, 1879)	Evans, 1937	6	A
Coeliades pisistratus (Fabricius, 1793)	Bacelar, 1948	6	C
Pyrrhochalcia iphis dejongi (Collins & Larsen, 2008)	Bacelar, 1956, as *P. iphis*	2,5	A
HESPERIIDAE | Pyrginae			
Apallaga rutilans (Mabille, 1877)	M et al., 2013a	5	A
Apallaga h. homeyeri (Plötz, 1880)	Plötz, 1880 (*)	6	A
Celaenorrhinus p. proxima (Mabille, 1877)	M & BS, 2009a, b, c, d	5	A
Tagiades flesus (Fabricius, 1781)	Evans, 1937	5	A,B
Eagris lucetia (Hewitson, 1875)	Aurivillius, 1928	5	A
Eagris decastigma fuscosa (Holland, 1893	M et al., 2013a	5	A
Eagris tigris liberti (Collins & Larsen, 2005)	Evans, 1937, as *E. tigris*	6	A
Eagris h. hereus (Druce, 1875)	Druce, 1875 (*)	6	A
Eagris t. tetrastigma (Mabille, 1891)	M et al., 2013a	5	A
Eagris multiplagata (BS & M, 2007)	BS & M, 2007 (*)	1,5	A
Ortholexis hollandi (Druce, 1909) f. *karschi* (Evans, 1937)	Evans, 1937	6	A
Calleagris hollandi (Butler, 1897)	Evans, 1937	5	C
Calleagris jamesoni ansorgei (Evans, 1951)	Weymer, 1901 (*), as *C. jamesoni*	1,5	C
Calleagris l. lacteus (Mabille, 1877)	Bacelar, 1961	6	A
Eretis lugens (Rogenhofer, 1891)	Larsen, 2005	5	D
Eretis herewardi rotundimacula (Evans, 1937)	Evans, 1937 (*)	1,5	C
Eretis melania (Mabille, 1891)	Evans, 1937	5	C,D

(continued)

A Checklist of Angolan Lepidoptera and Papilionoidea

Taxon	First reference for Angola	V	H
Sarangesa loelius (Mabille, 1877)	Bacelar, 1948	5	C
Sarangesa l. lucidella (Mabille, 1891)	M et al., 2013a	5	D
Sarangesa motozi (Wallengren, 1857)	Aurivillius, 1928	5	A,C
Sarangesa phidyle (Walker, 1870)	Evans, 1937	5	B,D
Sarangesa s. seineri (Strand, 1909)	Evans, 1937	5	D
Sarangesa p. pandaensis (Joicey & Talbot, 1921)	Evans, 1937	3,5	C
Sarangesa bouvieri (Mabille, 1877)	Evans, 1937	6	B
Sarangesa brigida sanaga (Miller, 1964)	M et al., 2013a	5	A
Sarangesa maculata (Mabille, 1891)	M et al., 2013a	5	A,B,C
Triskelionia tricerata (Mabille, 1891)	M et al., 2013a	6	A
Caprona cassualala (Bethune-Baker, 1911)	Bethune-Baker, 1911 (*)	5	D,F
Caprona pillaana (Wallengren, 1857)	M & BS, 2013a	5	D,F
Netrobalane canopus (Trimen, 1864)	M & BS, 2013a	5	D,A
Leucochitonea levubu (Wallengren, 1857)	Weymer, 1901	5	C,D
Abantis tettensis (Hopffer, 1855)	Aurivillius, 1928	5	C,D
Abantis bergeri (M & BS, 2009a, b, c, d)	M & BS, 2009a, b, c, d (*)	1,5	C
Abantis paradisea (Butler, 1870)	Weymer, 1901	6	C
Abantis zambesiaca (Westwood, 1874)	Weymer, 1901	5	D,C
Abantis contigua (Evans, 1937)	Evans, 1937	5	C
Abantis venosa (Trimen, 1889)	M et al., 2013a	5	C
Abantis vidua (Weymer, 1901)	Weymer, 1901	3	C
Spialia m. mafa (Trimen, 1870)	Weymer, 1901	6	D
Spialia spio (Linnaeus, 1764)	Weymer, 1901	5	D
Spialia delagoae (Trimen, 1898)	Larsen, 1996	5	D
Spialia c. colotes (Druce, 1875)	Druce, 1875 (*)	1,5	D,B
Spialia colotes transvaaliae (Trimen, 1889)	▲	5	D
Spialia ferax (Wallengren, 1863)	M et al., 2013a	5	C,D
Spialia dromus (Plötz, 1884)	Weymer, 1901	5	C,D
Spialia p. ploetzi (Aurivillius, 1891)	Evans, 1937, as *S. rebeli*	6	A
Spialia secessus (Trimen, 1891)	Trimen, 1891 (*)	5	F
Gomalia e. elma (Trimen, 1862)	Aurivilius, 1928	5	D
**HESPERIIDAE	Heteropterinae**			
Metisella m. midas (Butler, 1894)	Monard, 1956	5	E
Metisella a. angolana (Karsch, 1896)	Karsch, 1896 (*)	5	B
Metisella willemi (Wallengren, 1857)	M et al., 2013a, b	5	C
Metisella meninx (Trimen, 1873)	Evans, 1937	6	E
Lepella lepeletier (Latreille, 1824)	Druce, 1875	5	F
**HESPERIIDAE	Hesperiinae**			
Astictopterus abjecta (Snellen, 1872)	Snellen, 1872 (*)	6	A
Astictopterus punctulata (Butler, 1895)	M et al., 2013a	5	C
Kedestes mohozutza (Wallengren, 1857)	Monard, 1956	5	D

(continued)

Taxon	First reference for Angola	V	H
Kedestes nerva paola (Plötz, 1884)	Plötz, 1884 (*)	5	A
Kedestes brunneostriga (Plötz, 1884)	Plötz, 1884 (*)	5	C
Kedestes straeleni (Evans, 1956)	M et al., 2013a	5	C
Kedestes l. lema (Neave, 1910)	Evans, 1937	3	C
Kedestes callicles (Hewitson, 1868)	Aurivillius, 1928	5	C
Gorgyra mocquerysii (Holland, 1896)	Evans, 1937	5	A
Gorgyra diversata (Evans, 1937)	Evans, 1937	6	A
Ceratrichia nothus makomensis (Strand, 1913)	M et al., 2013a	5	A
Ceratrichia punctata (Holland, 1896)	Evans, 1937	6	A
Teniorhinus harona (Westwood, 1881)	Weymer, 1901, as *Oxypalpus ruso*	5	C
Teniorhinus ignita (Mabille, 1877)	Monard, 1956	5	B
Pardaleodes edipus (Stoll, 1781)	Bacelar, 1948	6	A
Pardaleodes i. incerta (Snellen, 1872)	Evans, 1937	5	A,D
Pardaleodes sator pusiella (Mabille, 1877)	Mabille, 1877 (*)	5	A
Pardaleodes t. tibullus (Fabricius, 1793)	M & BS, 2009	5	A
Acada biseriata (Mabille, 1893)	Evans, 1937	5	C
Parosmodes lentiginosa (Holland, 1896)	Evans, 1937	5	A
Parosmodes m. morantii (Trimen, 1873)	Weymer, 1901	5	C/D
Osmodes laronia (Hewitson, 1868)	Druce, 1875	6	A
Osmodes thora (Plötz, 1884)	Evans,1937	6	A
Acleros mackenii olaus (Plötz, 1884)	Druce, 1875, the species	5	A
Acleros nigrapex (Strand, 1913)	M et al., 2013a	5	A
Acleros ploetzi (Mabille, 1889)	M & BS, 2009	5	A
Semalea arela (Mabille, 1891)	M et al., 2013a	5	A
Semalea pulvina (Plötz, 1879)	M & BS, 2009	5	A
Semalea sextilis (Plötz, 1886)	M et al., 2013a	5	A
Hypoleucis o. ophiusa (Hewitson, 1866)	M & BS, 2009	5	A
Meza indusiata (Mabille, 1891)	Larsen, 2005	5	A,B
Meza meza (Hewitson, 1877)	Hewitson, 1877 (*)	5	A
Meza c. cybeutes (Holland, 1894)	Evans, 1937	6	A
Meza mabillei (Holland, 1893)	M et al., 2013a	5	A
Paronymus ligora (Hewitson, 1876)	Hewitson, 1876 (*)	6	A
Andronymus n. neander (Plötz, 1884)	Evans, 1937	5	A
Andronymus c. caesar (Fabricius, 1793)	Aurivillius, 1928 as *A. caesar*	5	A
Andronymus caesar philander (Hopffer, 1855)	Aurivillius, 1928 as *A. caesar*	6	A
Andronymus hero (Evans, 1937)	Evans, 1937	5	A
Andronymus helles (Evans, 1937)	Evans, 1937	5	A
Chondrolepis niveicornis (Plötz, 1882)	Plötz, 1882 (*)	5	E
Zophopetes dysmephila (Trimen, 1868)	Aurivillius, 1928, as *Z. schultzi*	6	A,D
Zophopetes cerymica (Hewitson, 1867)	M & BS, 2009	5	X
Gamia shelleyi (Sharpe, 1890)	M et al., 2013a	5	A

(continued)

A Checklist of Angolan Lepidoptera and Papilionoidea 173

Taxon	First reference for Angola	V	H
Gretna cylinda (Hewitson, 1876)	Aurivillius, 1928 (*)	5	A,C
Gretna waga (Plötz, 1886)	M et al., 2013a	5	A,C
Pteroteinon laufella (Hewitson, 1868)	Druce, 1875	5	B,C
Pteroteinon caenira (Hewitson, 1867)	M & BS, 2009	5	B
Pteroteinon concaenira (Belcastro & Larsen, 1996)	M et al., 2013a	6	B
Leona maracanda (Hewitson, 1876)	Hewitson, 1876 (*)	5	A
Caenides dacela (Hewitson, 1876)	Williams, 2007	5	A
Monza cretacea (Snellen, 1872)	Evans, 1937	5	B
Fresna nyassae (Hewitson, 1878)	Aurivillius, 1928	5	C,D
Platylesches langa (Evans, 1937)	M et al., 2013a	6	C
Platylesches moritili (Wallengren, 1857)	Trimen, 1891	5	C,D
Platylesches robustus (Neave, 1910)	M et al., 2013a	6	C
Platylesches cf. *batangae* (Holland, 1894)	▲	5	B
Brusa allardi (Berger, 1967)	M et al., 2013a	6	C
Zenonia zeno (Trimen, 1864)	Plötz, 1883, as *Hesperia coanza*	5	A,C
Pelopidas m. mathias (Fabricius, 1798)	Evans, 1937	5	C
Pelopidas thrax (Hübner, 1821)	Gardiner, 2004	5	D
Borbo fallax (Gaede, 1916)	M & BS, 2009a, b, c, d	5	D
Borbo fanta (Evans, 1937)	Evans, 1937	5	D
Borbo sirena (Evans, 1937)	M et al., 2013a	5	A,B,C
Borbo b. borbonica (Boisduval, 1833)	▲	5	D
Borbo detecta (Trimen, 1893)	Weymer, 1901	5	B
Larsenia gemella (Mabille, 1884)	Evans, 1937	5	D
Borbo micans (Holland, 1896)	M & BS, 2009	5	E
Larsenia perobscura (Druce, 1812)	M et al., 2013a	5	D
Borbo f. fatuellus (Hopffer, 1855)	M & BS, 2009	5	B,C
Larsenia holtzi (Plötz, 1883)	Plötz, 1883 (*)	5	D
Parnara monasi (Trimen, 1889)	Evans, 1937	5	E
Afrogegenes hottentota (Latreille, 1824)	Weymer, 1901	5	C,D
Afrogegenes letterstedti (Wallengren, 1857)	Evans, 1937	5	D
Gegenes pumilio gambica (Mabille, 1878)	M & BS, 2009	5	D
PAPILIONIDAE			
Papilio a. antimachus (Drury, 1782)	Carvalho, 1962	5	A
Papilio zalmoxis (Hewitson, 1864)	BS, 1983	5	A
Papilio bacelarae (BS & M., 2009)	BS & M, 2009 (*)	1,5	A
Papilio f. filaprae (Suffert, 1904)	Druce, 1875, as *P. cypraeophila*	5	A
Papilio m. mechowi (Dewitz, 1881)	Dewitz, 1881 (*)	5	A
Papilio mechowianus (Dewitz, 1885)	Aurivillius, 1928	6	A
Papilio zenobia (Fabricius, 1775)	BS & Fernandes, 1966	5	A
Papilio cynorta (Fabricius, 1793)	Druce, 1875	5	A
Papilio echerioides homeyeri (Plötz, 1880)	Plötz, 1880 (*)	5	A,C

(continued)

Taxon	First reference for Angola	V	H
Papilio chitondensis (BS & Fernandes, 1966)	BS & Fernandes, 1966 (*)	1,5	B,C
Papilio chrapkowskoides nurettini (Koçak, 1983)	Bacelar, 1956, as *P. bromius*	5	A
Papilio n. nireus (Linnaeus, 1758)	Druce, 1875	5	H
Papilio nireus lyaeus (Doubleday, 1845)	Ladeiro, 1956	5	A,D
Papilio sosia pulchra (Berger, 1950)	BS & Fernandes, 1966	5	A
Papilio mackinnoni benguellae (Jordan, 1908)	Jordan, 1908 (*)	1,5	A
Papilio d. dardanus (Brown, 1776)	Druce, 1875	5	C
Papilio phorcas congoanus (Rothschild, 1896)	BS & Fernandes, 1964	5	A
Papilio h. hesperus (Westwood, 1843)	Aurivillius, 1928	5	A
Papilio l. lormieri (Distant, 1874)	Bacelar, 1956	5	A
Papilio d. demodocus (Esper, 1798)	Weymer, 1901	5	E
Graphium a. angolanus (Goeze, 1779)	Goeze, 1779 (*)	5	F
Graphium schaffgotschi (Niepelt, 1927)	Villiers, 1979	4,5	D
Graphium ridleyanus (White, 1843)	Druce, 1875	5	A
Graphium latreillianus theorini (Aurivillius,1881)	Aurivillius, 1928	5	A
Graphium tynderaeus (Fabricius, 1793)	BS & Fernandes, 1964	5	A
Graphium a. almansor (Honrath, 1884)	Aurivillius, 1928	5	A,C
Graphium ucalegonides (Staudinger, 1884)	Smith & Vane-Wright, 2001	2	A
Graphium h. hachei (Dewitz, 1881)	Dewitz, 1881 (*)	5	A
Graphium poggianus (Honrath, 1884)	Aurivillius, 1928	3	A
Graphium u. ucalegon (Hewitson, 1865)	Bacelar, 1956	5	A
Graphium l. leonidas (Fabricius, 1793)	Druce, 1875	5	H
Graphium antheus (Cramer, 1779)	Druce, 1875	5	A,D
Graphium p. policenes (Cramer, 1775)	Druce, 1875	5	A,C
Graphium p. porthaon (Hewitson, 1865)	Gardiner, 2004	6	B,D
**PIERIDAE	Pseudopontinae**			
Pseudopontia paradoxa (Felder & Felder, 1869)	◻	5	A
Pseudopontia australis (Dixey, 1923)	Snellen, 1882, as *P. paradoxa*	3,5	A
**PIERIDAE	Coliadinae**			
Catopsilia florella (Fabricius, 1775)	Weymer, 1901	5	H
Colias e. electo (Linnaeus, 1763)	Gardiner, 2004	6	D
Colias electo hecate (Strecker, 1905)	Bacelar, 1948	5	F
Eurema b. brigitta (Stoll, 1780)	Butler, 1871	5	D,H
Eurema desjardinsi regularis (Butler, 1876)	Mabille, 1877	5	D,C
Eurema floricola leonis (Butler, 1886)	Trimen, 1891, as *E. floricola*	6	B
Eurema hapale (Mabille, 1882)	Ladeiro, 1956	5	A,B
Eurema hecabe solifera (Butler, 1875)	Butler, 1875 (*)	5	D

(Continued)

A Checklist of Angolan Lepidoptera and Papilionoidea

Taxon	First reference for Angola	V	H
Eurema senegalensis (Boisduval, 1836)	Butler, 1871	5	A
PIERIDAE | Pierinae			
Pinacopteryx e. eriphia (Godart, 1819)	Butler, 1871	5	B,D
Nepheronia a. argia (Fabricius, 1775)	Druce, 1875	5	A
Nepheronia b. buquetii (Boisduval, 1836)	Druce, 1875	5	B,D
Nepheronia p. pharis (Boisduval, 1836)	Aurivillius, 1928	5	B
Nepheronia thalassina verulanus (Ward, 1871)	Bacelar, 1958a, b	5	B
Eronia cleodora Hübner, 1823	Aurivillius, 1928	6	D
Afrodryas leda (Boisduval, 1847)	Bacelar, 1961	5	B
Teracolus a. agoye (Wallengren, 1857)	Weymer, 1901	5	D
Colotis calais williami (Henning & Henning, 1994)	Willis, 2009	6	D
Colotis antevippe gavisa (Wallengren, 1857)	Trimen, 1891	5	D
Colotis celimene pholoe (Wallengren, 1860)	Talbot, 1939	4,5	F
Colotis annae walkeri (Butler, 1884)	Butler, 1884 (*)	4,5	F
Colotis doubledayi (Hopffer, 1862)	Hopffer, 1862 (*)	5	D
Colotis e. euippe (Linnaeus, 1758)	Latreille & Godart, 1819	5	B
Colotis euippe mediata (Talbot, 1939)	Talbot, 1939	5	C,D
Colotis evagore antigone (Boisduval, 1836)	Druce, 1875	5	B
Colotis e. evenina (Wallengren, 1857)	Trimen, 1891	5	F
Colotis ione (Godart, 1819)	Bacelar, 1961	6	D
Colotis regina (Trimen, 1863)	Trimen, 1891	5	D
Colotis vesta rhodesinus (Butler, 1894)	Bacelar, 1958a, b	5	D
Teracolus e. eris (Klug, 1829)	Druce, 1875	5	D
Teracolus subfasciatus (Swainson, 1833)	Aurivillius, 1928	5	C
Belenois aurota (Fabricius, 1793)	Trimen, 1891	5	H
Belenois calypso dentigera (Butler, 1888)	Druce, 1875, as *B. calypso*	5	A
Belenois welwitschii welwitschii (Rogenhofer, 1890)	Rogenhofer, 1890 (*)	5	C
Belenois crawshayi (Butler, 1894)	Aurivillius, 1928	5	C,D
Belenois creona severina (Stoll, 1781)	Butler, 1871	5	H
Belenois g. gidica (Godart, 1819)	▲	5	D
Belenois r. rubrosignata (Weymer, 1901)	Weymer, 1901 (*)	3,5	C
Belenois s. solilucis (Butler, 1874)	Butler, 1874 (*)	5	A
Belenois sudanensis mayumbana (Berger, 1981)	▢	2,5	A
Belenois sudanensis pseudodentigera (Berger, 1981)	▢	5	C
Belenois theuszi (Dewitz, 1889)	Dewitz, 1889 (*)	5	A
PIERIDAE | Pierinae			
Belenois t. thysa (Hopffer, 1855)	▢	5	C

(continued)

Taxon	First reference for Angola	V	H
Belenois thysa meldolae (Butler, 1872)	Butler, 1872 (*)	5	A,B
Dixeia capricornus falkensteinii (Dewitz, 1879)	Dewitz, 1879 (*)	2,5	B,C
Dixeia sp.	◻	5	A
Dixeia pigea (Boisduval, 1836)	Aurivillius, 1928	5	C
Pontia h. helice (Linnaeus, 1764)	Willis, 2009	5	D
Appias epaphia angolensis (M & BS, 2006)	M & BS, 2006 (*)	1,5	A,B
Appias perlucens (Butler, 1898)	Butler, 1898 (*)	5	A
Appias phaola uigensis (M & BS, 2006)	M & BS, 2006 (*)	1,5	A
Appias s. sabina (Felder & Felder, 1865)	Druce, 1875	5	A
Appias sylvia nyassana (Butler, 1897)	Druce, 1875, as *Belenois*	6	A?
Appias sylvia ribeiroi (M & BS, 2006)	M & BS, 2006 (*)	1,5	A
Leptosia a. alcesta (Stoll, 1782)	Druce, 1875	5	A
Leptosia h. hybrida (Bernardi, 1952)	◻	5	A,C
Leptosia n. nupta (Butler, 1873)	Butler, 1873 (*)	5	A
Leptosia wigginsi pseudalcesta (Bernardi, 1965)	◻	5	A
Mylothris carvalhoi (M & BS, 2009)	M & BS, 2009 (*)	1,5	A
Mylothris mavunda (Hancock & Heath, 1985)	Koçak & Kemal, 2009	3	A?
Mylothris a. agathina (Cramer, 1779)	Trimen, 1891	5	C
Mylothris asphodelus (Butler, 1888)	Aurivillius, 1928	5	A
Mylothris elodina diva (Berger, 1954)	Berger, 1981	2,5	C
Mylothris poppea (Cramer, 1777)	Druce, 1875	5	A
Mylothris rembina (Plötz, 1880)	Talbot, 1944	6	A
Mylothris rhodope (Fabricius, 1775)	Talbot, 1944	5	D
Mylothris rueppellii rhodesiana (Riley, 1921)	Talbot, 1944	5	C
Mylothris spica gabela (Berger, 1979)	Berger, 1979 (*)	1,6	A
Mylothris sulphurea (Aurivillius, 1895)	◻	5	A
Mylothris y. cf. *yulei* (Butler, 1897)	◻	5	A
LYCAENIDAE | Miletinae			
Euliphyra mirifica (Holland, 1890)	Larsen, 2005	6	A
Aslauga m. marshalli (Butler, 1899)	Larsen, 2005	6	A
Megalopaplpus zymna (Westwood, 1851)	Ackery et al., 1995	5	A
Spalgis l. lemolea (Druce, 1890)	Ladeiro, 1956	5	A,C
Lachnocnema angolanus (Libert, 1996)	Libert, 1996 b (*)	5	A,D
Lachnocnema bamptoni (Libert, 1996)	Libert, 1996 b (*)	6	C
Lachnocnema bibulus (Fabricius, 1793)	Libert, 1996 b	5	A,C,D
Lachnocnema emperamus (Snellen, 1872)	Snellen, 1872	5	A
Lachnocnema intermedia (Libert, 1996)	Ladeiro, 1956, as *L. durbani* (*)	5	C
Lachnocnema laches (Fabricius, 1793)	Libert, 1996 a	5	A,C
Lachnocnema r. regularis (Libert, 1996)	Libert, 1996 c	6	C?

(continued)

A Checklist of Angolan Lepidoptera and Papilionoidea

Taxon	First reference for Angola	V	H
LYCAENIDAE | Poritiinae			
Alaena amazoula congoana (Aurivillius, 1914)	Aurivillius, 1914 (*)	6	G
Alaena rosei (Vane-Wright, 1980)	Vane-Wright, 1980 (*)	1,5	G
Pentila maculata pardalena (Druce, 1910)	Stempffer & Bennett, 1961	6	A
Pentila amenaida (Hewitson, 1873)	Hewitson, 1873 (*)	5	A
Pentila pauli benguellana (Stempffer & Bennett, 1961)	Stempffer & Bennett, 1961 (*)	5	A,B
Pentila t. tachyroides (Dewitz, 1879)	Dewitz, 1879 (*)	5	A
Telipna acraeoides (Grose-Smith & Kirby, 1890)	Grose-Smith & Kirby, 1890 (*)	5	A
Telipna a. albofasciata (Aurivillius, 1910)	Libert, 2005	6	A
Telipna atrinervis (Hulstaert, 1924)	◘	5	A
Telipna cuypersi (Libert, 2005)	Libert, 2005	6	A
Telipna nyanza katangae (Stempffer, 1961)	Libert, 2005	6	A
Telipna s. sanguinea (Plötz, 1880)	Aurivillius, 1928	5	A
Ornipholidotus gabonensis (Stempffer, 1947)	▲	6	A
Ornipholidotus perfragilis (Holland, 1890)	Libert, 2005	6	A
Ornipholidotus ugandae goodi (Libert, 2000)	Libert, 2005	6	A
Cooksonia nozolinoi (M & BS, 2007)	M & BS, 2007 (*)	1,5	C
Mimacraea charmian (Grose-Smith & Kirby, 1890)	Grose-Smith & Kirby, 1890 (*)	6	A
Mimacraea landbecki (Druce, 1910)	Libert, 2000 b	5	A
Mimacraea marshalli (Trimen, 1898)	Libert, 2000 b	5	A
Mimeresia debora deborula (Aurivillius, 1899)	◘	5	A
Eresiomera osheba (Holland, 1890)	▲	5	A
Citrinophila e. erastus (Hewitson, 1866)	Aurivillius, 1928	6	A
Cnodontes vansomereni (Stempffer & Bennett, 1953)	▲	5	D
Liptena evanescens (Kirby, 1887)	◘	5	A
Liptena fatima (Kirby, 1890)	◘	5	A
Liptena h. homeyeri (Dewitz, 1884)	▲	6	A
Liptena homeyeri straminea (Stempffer, Bennett & May, 1974)	Stempffer, Bennett & May, 1974 (*)	1,6	A
Liptena parva (Kirby, 1887)	◘	5	A
Liptena undularis (Hewitson, 1866)	Druce, 1875	5	A
Liptena xanthostola xantha (Grose-Smith, 1901)	Larsen, 2005	6	A
Falcuna h. hollandii (Aurivillius, 1895)	Ackery et al., 1995	6	A
Falcuna lacteata (Stempffer & Bennett, 1963)	Stempffer & Bennett, 1963 (*)	1,6	A

(continued)

Taxon	First reference for Angola	V	H
Falcuna libyssa angolensis (Stempffer & Bennett, 1963)	Stempffer & Bennett, 1963 (*)	1,5	A
Falcuna s. synesia (Hulstaert, 1924)	Stempffer & Bennett, 1963 (*)	2	A
Tetrarhanis ilala etoumbi (Stempffer, 1964)	◻	5	A
Tetrarhanis i. ilma (Hewitson, 1873)	Hewitson, 1873 (*)	6	A
Larinopoda lircaea (Hewitson, 1866)	Stempffer, 1957	6	A
Larinopoda tera (Hewitson, 1873)	Aurivillius, 1928	5	A
Hewitsonia bitjeana (Bethune-Baker, 1915)	◻	5	A
Hewitsonia k. kirbyi (Dewitz, 1879)	Dewitz, 1879 (*)	6	A
Cerautola ceraunia (Hewitson, 1873)	Larsen, 2005	6	A
Cerautola crowleyi leucographa (Libert, 1999)	Libert, 1999	6	A
Hewitola hewitsonii (Mabille, 1877)	Mabille, 1877 (*)	6	A
Cerautola miranda vidua (Talbot, 1935)	Bacelar, 1958a, b	5	A
Epitola posthumus (Fabricius, 1793)	Bacelar, 1956	5	A
Epitola urania (Kirby, 1887)	Libert, 1999	6	A
Hypophytala h. hyetta (Hewitson, 1873)	Hewitson, 1873 (*)	2	A
Stempfferia cercene (Hewitson, 1873)	Hewitson, 1873 (*)	6	A
Stempfferia cinerea (Berger, 1981)	Libert, 1999	2	A
Stempfferia michelae centralis (Libert, 1999)	Libert, 1999	5	A
Deloneura barca (Grose-Smith, 1901)	Grose-Smith, 1901 (*)	1,6	C?,D?
Deloneura cf. *subfusca* (Hawker-Smith, 1933)	◻	5	F
Epitolina dispar (Kirby, 1887)	Larsen, 2005	6	A
Epitolina melissa (Druce, 1888)	Larsen, 2005	6	A
LYCAENIDAE | Theclinae			
Myrina s. silenus (Fabricius, 1775)	Druce, 1875	5	B,D
Myrina silenus ficedula (Trimen, 1879)	Gardiner, 2004	6	D
Oxylides binza (Berger, 1981)	Druce, 1875 as *O. faunus*	2,5	A
LYCAENIDAE | Theclinae			
Oxylides feminina stempfferi (Berger, 1981)	Libert, 2004	6	A
Syrmoptera amasa (Hewitson, 1869)	Libert, 2004	6	A
Syrmoptera homeyerii (Dewitz, 1879)	Dewitz, 1879 (*)	6	A
Dapidodigma demeter nuptus (Clench, 1961)	Larsen, 2005	3,5	A
LYCAENIDAE | Aphnaeinae			
Lipaphneus a. cf. *aderna* (Plötz, 1880)	◻	5	A
Crudaria leroma (Wallengren, 1857)	Gardiner, 2004	6	D
Aloeides angolensis (Tite & Dickson, 1973)	Tite & Dickson, 1973 (*)	1,6	F
Aphnaeus erikssoni (Trimen, 1891)	Trimen, 1891 (*)	5	C

(continued)

A Checklist of Angolan Lepidoptera and Papilionoidea

Taxon	First reference for Angola	V	H
Aphnaeus orcas (Drury, 1782)	Larsen, 2005	6	A
Aphnaeus affinis (Riley, 1921)	Libert, 2013		
Erikssonia acraeina (Trimen, 1891)	Trimen, 1891 (*)	6	D
Pseudaletis a. agrippina (Druce, 1888)	▲	5	A
Cigaritis ella (Hewitson, 1865)	Gardiner, 2004	6	D
Cigaritis phanes (Trimen, 1873)	Weymer, 1901	6	F
Cigaritis homeyeri (Dewitz, 1887)	Aurivillius, 1928	5	C
Cigaritis m. modestus (Trimen, 1891)	Trimen, 1891 (*)	1,4,5	A,C
Cigaritis mozambica (Bertoloni, 1850)	▲	6	D
Cigaritis natalensis (Westwood, 1851)	Ladeiro, 1956	6	C,D
Cigaritis trimeni congolanus (Dufrane, 1954)	▲	2,5	A
Zeritis fontainei (Stempffer, 1956)	Willis, 2009	6	C,G?
Zeritis krystyna (D'Abrera, 1980)	D'Abrera, 1980 (*)	1,6	C?
Zeritis sorhagenii (Dewitz, 1879)	Dewitz, 1879 (*)	6	C?
Axiocerces a. amanga (Westwood, 1881)	Trimen, 1891	5	C
Axiocerces bambana orichalcea (Henning & Henning, 1996)	◻	5	B
Axiocerces amanga baumi (Weymer, 1901)	Weymer, 1901 (*)	,5	C
Axiocerces t. tjoanae (Wallengren, 1857)	Henning & Henning, 1996	6	B
Iolaus hemicyanus barnsi (Joicey & Talbot, 1921)	◻	2,5	A
Iolaus i. iasis (Hewitson, 1865)	Larsen, 2005	6	A
Iolaus mimosae rhodosense (Stempffer & Bennett, 1959)	Gardiner, 2004	6	D
Iolaus obscura (Aurivillius, 1923)	◻	4,5	D
Iolaus violacea (Riley, 1928)	Riley, 1928 (*)	5	C
Iolaus pallene (Wallengren, 1857)	Gardiner, 2004	6	D
Iolaus trimeni (Wallengren, 1875)	Ackery et al., 1995	5	C,D
Iolaus iturensis (Joicey & Talbot, 1921)	▲	6	C
Iolaus parasilanus mabillei (Riley, 1928)	Riley, 1928 (*)	2	A
Iolaus s. silarus (Druce, 1885)	Gardiner, 2004	5	C,D
Iolaus t. timon (Fabricius, 1787)	Ackery et al., 1995	6	A
Hemiolaus vividus (Pinhey, 1962)	Aurivillius, 1928, as *caeculus*	5	C,D
Stugeta bowkeri maria (Suffert, 1904)	Druce, 1875, as *bowkeri*	3,5	B
Stugeta bowkeri tearei (Dikson, 1980)	Gardiner, 2004	6	A,C,E
Hypolycaena a. antifaunus (Westwood, 1851)	Druce, 1875	5	A
Hypolycaena h. hatita (Hewitson, 1865)	Druce, 1875	5	A
Hypolycaena l. lebona (Hewitson, 1865)	Druce, 1875	6	A
Hypolycaena naara (Hewitson, 1873)	Hewitson, 1873 (*)	6	A
Hypolycaena nigra (Bethune-Baker, 1914)	M & BS, 2012	5	A

(continued)

Taxon	First reference for Angola	V	■
Hypolycaena p. philippus (Fabricius, 1793)	Druce, 1875	5	D
Hypolycaena buxtoni spurcus (Talbot, 1929 M & BS., 2012)	M & BS, 2012	5	F?
Pilodeudorix badhami (Carcasson, 1961)	Libert, 2004	6	E
Pilodeudorix caerulea (Druce, 1890)	Libert, 2004	5	C,D
Pilodeudorix pseudoderitas (Stempffer, 1964)	Larsen, 2005	6	A
Pilodeudorix zeloides (Butler, 1901)	Libert, 2004	6	C
Paradeudorix cobaltina (Stempffer, 1964)	Larsen, 2005	6	A
Leptomyrina henningi angolensis (M & BS, 2009)	M & BS, 2009 (*)	1,≤	E
Pilodeudorix deritas (Hewitson, 1874)	Hewitson, 1874 (*)	5	A
Pilodeudorix m. mera (Hewitson, 1873)	Hewitson, 1873 (*)	5	A
Pilodeudorix otraeda genuba (Hewitson, 1875)	▲	6	A
Hypomyrina nomenia (Hewitson, 1874)	Larsen, 2005	5	A
Deudorix antalus (Hopffer, 1855)	Bacelar, 1948	5	L,G
Deudorix caliginosa (Lathy, 1903)	▲	6	C
Deudorix dinochares (Grose-Smith, 1887)	Gardiner, 2004	5	C,D
Deudorix cf. *diocles* (Hewitson, 1869)	Libert, 2004	5	C
Deudorix lorisona coffea (Jackson, 1966)	Libert, 2004	5	A,C
Capys c. connexiva (Butler, 1896)	Henning & Henning, 1988	6	X
LYCAENIDAE | Polyommatine			
Anthene akoae (Libert, 2010)	Libert. 2010	6	?
Anthene alberta (Bethune-Baker, 1910)	Aurivillius, 1928	5	A,C
Anthene a. amarah (Guérin-Méneville, 1849)	Stempffer, 1957	5	C
Anthene lvida livida (Trimen, 1881)	Gardiner, 2004	6	C
Anthene c. crawshayi (Butler, 1899)	◻	5	C
Anthene d. definita (Butler, 1899)	Gardiner, 2004	5	A
Anthene larydas (Cramer, 1780)	Weymer, 1901	5	A
Anthene l. ligures (Hewitson, 1874)	Hewitson, 1874 (*)	6	A
Anthene liodes (Hewitson, 1874)	Aurivillius, 1909	6	A,D
Anthene l. lunulata (Trimen, 1894)	Trimen, 1894 (*)	5	D
Anthene nigropunctata (Bethune-Baker, 1910)	Gardiner, 2004	6	?
Anthene princeps (Butler, 1876)	Gardiner, 2004	5	A,D
Anthene r. rubricinctus (Holland, 1891)	Aurivillius, 1928	6	A
Anthene sylvanus (Drury, 1773)	Aurivillius, 1928	6	A
Anthene talboti (Stempffer, 1936)	Libert, 2010	6	D
Neurellipes flavomaculatus (Grose-Smith & Kirby, 1893)	Aurivillius, 1928	6	A
Neurellipes lachares (Hewitson, 1878)	Larsen, 2005	6	A
Neurellipes onias (Hulstaert, 1924)	Willis, 2009	6	?

(Continued)

A Checklist of Angolan Lepidoptera and Papilionoidea

Taxon	First reference for Angola	V	H
Neurellipes pyroptera (Aurivillius, 1895)	Libert, 2010	6	A
Neurypexina lyzanius (Hewitson, 1874)	Druce, 1875	6	A
Triclema lacides (Hewitson, 1874)	Hewitson, 1874 (*)	6	A
Triclema lucretilis (Hewitson, 1874)	Stempffer, 1957	6	A
Triclema cf. *nigeriae* (Aurivillius, 1905)	Libert, 2010	5	D
Monile g. gemmifera (Neave, 1910)	Libert, 2010	6	A
Cupidesthes vidua (Talbot, 1929)	Talbot, 1929 (*)	1,6	C?
Pseudonacaduba aethiops (Mabille, 1877)	Mabille, 1877 (*)	5	A
Pseudonacaduba s. sichela (Wallengren, 1857)	Weymer, 1901	5	D
Lampides boeticus (Linnaeus, 1767)	Butler, 1871	5	H
Uranothauma falkensteinii (Dewitz, 1879)	Dewitz, 1879 (*)	5	A?
Uranothauma antinorii cf. *felthami* (Stevenson, 1934)	BS & M, 2007	5	A
Uranothauma h. heritsia (Hewitson, 1876)	Stempffer, 1957	5	A,D
Uranothauma nozolinoi (BS & M., 2007)	BS & M, 2007 (*)	1,5	C
Uranothauma poggei (Dewitz, 1879)	Dewitz, 1879 (*)	5	A,C,D
Uranothauma c. cyara (Hewitson, 1876)	Hewitson, 1876 (*)	5	A?
Cacyreus lingeus (Stoll, 1782)	Gardiner, 2004	5	B,D
Cacyreus marshalli (Butler, 1898)	Gardiner, 2004	6	D
Cacyreus virilis (Aurivillius, 1924)	Aurivillius, 1924 (*)	6	D
Leptotes babaulti (Stempffer, 1935)	Stempffer, 1957	5	A,D
Leptotes brevidentatus (Tite, 1958)	Tite, 1958 (* – in part)	5	A,C,D
Leptotes jeanneli (Stempffer, 1935)	Stempffer, 1957	5	A,D
Leptotes p. pirithous (Linnaeus, 1767)	Snellen, 1882	5	D
Leptotes p. pulchra (Murray, 1874)	Gardiner, 2004	5	E
Tuxentius calice (Hopffer, 1855)	Gardiner, 2004	5	E
Tuxentius c. carana (Hewitson, 1876)	Hewitson, 1876 (*)	5	A
Tuxentius margaritaceus (Sharpe, 1892)	Larsen, 2005	6	A
Tuxentius m. melaena (Trimen, 1887)	Aurivillius, 1928	6	C,D
Tarucus sybaris linearis (Aurivillius, 1924)	Aurivillius, 1928	6	D
Actizera lucida (Trimen, 1883)	Stempffer, 1957	5	F
Eicochrysops eicotrochilus (Bethune-Baker, 1924)	▲	6	C
Eicochrysops hippocrates (Fabricius, 1793)	Gardiner, 2004	5	A,B,D
Eicochrysops messapus mahallakoaena (Wallengren, 1857)	Weymer, 1901, as *messapus*	6	C
Cupidopsis c. cissus (Godart, 1824)	Ladeiro, 1956	5	E
Cupidopsis j. jobates (Hopffer, 1855)	Stempffer, 1957	5	D
Euchrysops barkeri (Trimen, 1893)	Aurivillius, 1928	6	C

(continued)

Taxon	First reference for Angola	V	■
Euchrysops malathana (Boisduval, 1833)	Stempffer, 1957	5	C,D
Euchrysops osiris (Hopffer, 1855)	Druce, 1875	5	C,D
Euchrysops subpallida (Bethune-Baker, 1923)	Gardiner, 2004	6	L,G
Lepidochrysops abyssiniensis loveni (Aurivillius, 1921)			
Lepidochrysops ansorgei (Tite, 1959)	Tite, 1959 (*)	1,5	C,D
Lepidochrysops chlouages (Bethune-Baker, 1923)	▲	5	A,D
Lepidochrysops flavisquamosa (Tite, 1959)	Tite, 1959 (*)	1,6	C
Lepidochrysops fulvescens (Tite, 1961)	Tite, 1961 (*)	1,6	C
Lepidochrysops g. glauca (Trimen, 1887)	Bacelar, 1948	5	L
Lepidochrysops hawkeri (Talbot, 1929)	Talbot, 1929 (*)	1,5	C?
Lepidochrysops nacrescens (Tite, 1961)	Tite, 1961 (*)	1,6	C
Lepidochrysops reichenowi (Dewitz, 1879)	Dewitz, 1879 (*)	1,6	?
Thermoniphas distincta (Talbot, 1935)	◨	5	C
Thermoniphas p. plurilimbata (Karsch, 1895)	◨	5	A
Thermoniphas t. togara (Plötz, 1880)	▲	6	A
Oboronia guessfeldtii (Dewitz, 1879)	Larsen, 1991	5	E
Oboronia pseudopunctatus (Strand, 1912)	▲	6	A
Oboronia punctatus (Dewitz, 1879)		5	A
Actizera lucida (Trimen, 1883)	Willis, 2009	6	C
Brephidium metophis (Wallengren, 1860)	Willis, 2009	6	?
Azanus isis (Drury, 1773)	Bacelar, 1948	5	A
Azanus jesous (Guérin-Méneville, 1849)	Trimen, 1891	6	L
Azanus mirza (Plötz, 1880)	Stempffer, 1957	5	L,A
Azanus moriqua (Wallengren, 1857)	Weymer, 1901	5	L
Azanus natalensis (Trimen, 1887)	Bacelar, 1948	5	L
Azanus ubaldus (Stoll, 1782)	Bacelar, 1948	5	F
Chilades trochylus (Freyer, 1844)	Ladeiro, 1956	5	D
Zizeeria k. knysna (Trimen, 1862)	Ladeiro, 1956	5	D
Zizina otis antanossa (Mabille, 1877)	▲	5	D
Zizula hylax (Fabricius, 1775)	Gardiner, 2004	5	H
RIODINIDAE			
Afriodinia dewitzi (Aurivillius, 1899)	▲	6	A
Afriodinia intermedia (Aurivillius, 1895)	Larsen, 2005	6	A
Afriodinia r. rogersi (Druce, 1878)	Druce, 1878 (*)	5	A
Afriodinia tantalus caerulea (Riley, 1932)	Druce, 1875 as *tantalus*	6	A
NYMPHALIDAE | Libytheinae			
Libythea labdaca (Westwood, 1851)	Snellen, 1882	5	A
Libythea laius (Trimen, 1879)	Gardiner, 2004	5	A

(continued)

A Checklist of Angolan Lepidoptera and Papilionoidea

Taxon	First reference for Angola	V	H
NYMPHALIDAE | Danainae			
Danaus c. orientis (Aurivillius, 1909)	Butler, 1871	5	H
Tirumala petiverana (Doubleday, 1847)	Butler, 1866, as *Danais leonora*	5	D
Amauris n. niavius (Linnaeus, 1758)	Aurivillius, 1928	5	A,B,D
Amauris t. tartarea (Mabille, 1876)	Mabille, 1876 (*)	5	A
Amauris crawshayi angola (Bethune-Baker, 1914)	Bethune-Baker, 1914 (*)	1,6	A
Amauris h. hecate (Butler, 1866)	▣	5	A
Amauris d. dannfelti (Aurivillius, 1891)	Aurivillius, 1891 (*)	1,5	A,C
Amauris h. hyalites (Butler, 1874)	Butler, 1874 (*)	5	A
Amauris vashti (Butler, 1869)	▣	5	A
NYMPHALIDAE | Satyrinae			
Gnophodes betsimena parmeno (Doubleday, 1849)	Aurivillius, 1928	5	A
Gnophodes chelys (Fabicius, 1793)	Aurivillius, 1928	5	A
Melanitis leda (Linnaeus, 1758)	Bacelar, 1948	5	B,H
Elymnias b. bammakoo (Westwood, 1851)	Druce, 1875	5	A
Bicyclus iccius (Hewitson, 1865)	Larsen, 2005	6	A
Bicyclus sebetus (Hewitson, 1877)	Aurivillius, 1928	5	A
Bicyclus s. saussurei (Dewitz, 1879)	Dewitz, 1879 (*)	3,5	A
Bicyclus s. suffusa (Riley, 1921)	▲	3,5	C
Bicyclus taenias (Hewitson, 1877)	▣	5	A,B
Bicyclus nachtetis (Condamin, 1965)	▲	3	A
Bicyclus technatis (Hewitson, 1877)	Larsen, 2005	6	A
Bicyclus vulgaris (Butler, 1868)	Druce, 1875	5	A,D
Bicyclus moyses (Condamin & Fox, 1964)	Condamin & Fox, 1964	5	A
Bicyclus sandace (Hewitson, 1877)	Bacelar, 1958a, b	5	A,D
Bicyclus auricruda fulgida (Fox, 1963)	Aurivillius, 1928, as *auricruda*	5	A
Bicyclus collinsi (Aduse-Poku et al., 2009)	Hewitson, 1873	5	B,D
Bicyclus angulosa selousi (Trimen, 1895)	Condamin, 1963	5	C,D
Bicyclus campus (Karsch, 1893)	▲	6	A
Bicyclus a. anynana (Butler, 1879)	Gardiner, 2004	6	D
Bicyclus anynana centralis (Condamin, 1968)	Condamin, 1968	5	A
Bicyclus cottrelli (van Son, 1952)	▲	5	C
Bicyclus s. safitza (Westwood, 1850)	Butler, 1871, as *Mycalesis caffra*	5	D
Bicyclus funebris (Guérin-Méneville, 1844)	Ladeiro, 1956	5	A,B,D
Bicyclus istaris (Plötz, 1880)	▲	5	A
Bicyclus lamani (Aurivillius, 1900)	Bacelar, 1958a, b	5	A
Bicyclus golo (Aurivillius, 1893)	Monard, 1956	5	A
Bicyclus s. smithi (Aurivillius, 1899)	▲	5	A
Bicyclus vansoni (Condamin, 1965)	Condamin, 1965	5	C

(continued)

Taxon	First reference for Angola	V	H
Bicyclus buea (Strand, 1912)	Larsen, 2005	5	A
Bicyclus sanaos (Hewitson, 1866)	Druce, 1875	5	A
Hallelesis asochis congoensis (Joicey & Talbot, 1921)	Druce, 1875, as *asochis*	6	E
Brakefieldia angolensis (Kielland, 1994)	Kielland, 1994 (*)	1,6	C
Brakefieldia p. phaea (Karsch, 1894)	Kielland, 1994	5	C
Brakefieldia simonsii (Butler, 1877)	Gardiner, 2004	5	D,F
Brakefieldia centralis (Aurivillius, 1903)	Ackery et al., 1995	3	C
Brakefieldia ochracea (Lathy, 1906)	Lathy, 1906 (*)	1,5	C
Brakefieldia eliasis (Hewitson, 1866)	Druce, 1875	3,5	A,B
Mashuna upemba (Overlaet, 1955)	☐	5	E
Ypthima a. asterope (Klug, 1832)	Druce, 1875	5	D
Ypthima asterope hereroica (van Son, 1955)	Gardiner, 2004	6	D
Ypthima c. condamini (Kielland, 1982)	Larsen, 2005	6	C,D
Ypthima granulosa (Butler, 1883)	Ladeiro, 1956	6	C?,D
Ypthima recta (Overlaet, 1955)	Kielland, 1982	6	A
Ypthima doleta (Kyrby, 1880)	Aurivillius, 1928	5	A,B
Ypthima i. impura (Elwes & Edwards, 1893)	Aurivillius, 1928	5	A,D
Ypthima impura paupera (Ungemach, 1932)	Gardiner, 2004	6	A,D
Ypthima praestans (Overlaet, 1954)	▲	5	A
Ypthima pulchra (Overlaet, 1954)	▲	5	A
Ypthima diplommata (Overlaet, 1954)	Kielland, 1982	3,6	New data
Ypthimomorpha itonia (Hewitson, 1865)	Aurivillius, 1928	5	D,E
Neita bikuarica (M & BS, 2006)	M & BS, 2006 (*)	1,5	C
Neocoenyra cooksoni (Druce, 1907)	▲	6	C
Mashunoides carneiromendesi (M & BS, 2009)	M & BS, 2009 (*)	1,5	D
**NYMPHALIDAE	Charaxinae**			
Charaxes fulvescens rubenarturi (BS & M, 2017)	BS et al., 2017 (*)	1,5	A
Charaxes varanes vologeses (Mabille, 1876)	Mabille, 1876, sub *Palla* (*)	5	B,D
Charaxes candiope (Godart, 1824)	Druce, 1875	5	B,D
Charaxes cynthia kinduana (Le Cerf, 1923)	Aurivillius, 1928	5	A,D
Charaxes macclounii (Butler, 1895)	van Someren, 1970	6	A
Charaxes macclouni carvalhoi (BS, 1983)	BS, 1983 (*)	1,5	A
Charaxes protoclea protonothodes (van Someren, 1971)	Aurivillius, 1928, as *protoclea*	5	A,B
Charaxes lucretius saldanhai (BS, 1983)	BS, 1983 (*)	1,5	A

(continued)

A Checklist of Angolan Lepidoptera and Papilionoidea

Taxon	First reference for Angola	V	H
Charaxes brutus angustus (Rothschild, 1900)	Druce, 1875, as *brutus*	5	A,B,D
Charaxes brutus natalensis (Staudinger, 1885)	van Someren, 1970	6	A,B,D
Charaxes c. castor (Cramer, 1775)	Druce, 1875	5	A,B
Charaxes druceanus proximans (Joicey & Talbot, 1922)	Aurivillius, 1928	5	A,D
Charaxes eudoxus mechowi (Rothschild, 1900)	Rothschild, 1900 (*)	5	A,D
Charaxes eudoxus mitchelli (Plantrou & Howarth, 1977)	◘	5	A,D
Charaxes s. saturnus (Butler, 1866)	Druce, 1875, as *C. pelias brunnescens*	5	A,D
Charaxes p. pollux (Cramer, 1775)	Druce, 1875	5	A,C
Charaxes numenes aequatorialis (van Someren, 1972)	Aurivillius, 1928, as *numenes*	5	A
Charaxes tiridates tiridatinus (Röber, 1936)	Druce, 1875, the species	5	A
Charaxes ameliae amelina (Joicey & Talbot, 1925)	◘	5	C
Charaxes b. bohemani (Felder & Felder, 1859)	Druce, 1875	5	D
Charaxes p. pythodoris (Hewitson, 1873)	Hewitson, 1873 (*)	5	B
Charaxes smaragdalis leopoldi (Ghesquiére, 1933)	van Someren, 1964	2	A
Charaxes zingha (Stoll, 1780)	Bacelar, 1958a, b	5	A
Charaxes a. achaemenes (Felder & Felder, 1867)	van Someren, 1970	5	C,D
Charaxes e. etesipe (Godart, 1824)		5	A,E
Charaxes p. penricei (Rothschild, 1900)	Rothschild, 1900 (*)	5	C,D
Charaxes penricei dealbata (van Someren, 1966)	van Someren, 1966 (*)	2,5	A
Charaxes jahlusa angolensis (M & BS, 2017)	BS et al., 2017 (*)	1,5	A
Charaxes eupale latimargo (Joicey & Talbot, 1921)	Druce, 1875 as *eupale*	5	A
Charaxes minor karinae (Bouyer, 1999)	Bouyer, 1999 (*)	1,5	A
Charaxes anticlea proadusta (van Someren, 1971)	Aurivillius, 1928, as *anticlea*	5	A
Charaxes h. hildebrandti (Dewitz, 1879)	Dewitz, 1879 (*)	5	A
Charaxes hildebrandti katangensis (Talbot, 1928)	BS & M, 2014	5	A
Charaxes g. guderiana (Dewitz, 1879)	Dewitz, 1879 (*)	5	A,D
Charaxes brainei (van Son, 1966)	Henning, 1988	4	D
Charaxes catachrous (van Someren & Jackson, 1952)	◘	5	A

(continued)

Taxon	First reference for Angola	V	∎
Charaxes cedreatis (Hewitson, 1874)	Aurivillius, 1928	5	C
Charaxes diversiforma (van Someren & Jackson, 1957)	van Someren, 1969	5	
Charaxes etheocles silvestris (Turlin, 2011)	Druce, 1875, as *C. ephyra*	5	B
Charaxes figueirai (BS & M, 2014)	BS & M, 2014 (*)	1,5	C D
Charaxes fulgurata (Aurivillius, 1899)	Aurivillius, 1899 (*)	5	C
Charaxes howarthi (Minig, 1976)	Henning, 1988	5	C
Charaxes phaeus (Hewitson, 1877)	Gardiner, 2004	6	C D
Charaxes variata (van Someren, 1969)	◻	3,5	C
Charaxes p. paphianus (Ward, 1871)	Aurivillius, 1928	5	
Charaxes pleione congoensis (Plantrou, 1989)	van Someren 1974, as *pleione*	5	
Charaxes ehmckei (Homeyer & Dewitz, 1882)	Homeyer & Dewitz, 1882 (*)	1,5	B
Charaxes zoolina (Westwood, 1850)	Gardiner, 2004	6	
Charaxes kahldeni (Homeyer & Dewitz, 1882)	Homeyer & Dewitz, 1882 (*)	5	
Charaxes n. nichetes (Grose-Smith, 1883)	Aurivillius, 1928	5	
Charaxes nichetes pantherinus Rousseau-(Decelle, 1934)	◻	5	
Charaxes lycurgus (Fabricius, 1793)	Plantrou, 1978	5	
Charaxes zelica rougeoti (Plantrou, 1978)	Plantrou, 1978	5	
Charaxes doubledayi (Aurivillius, 1899)	Bacelar, 1956	5	
Palla decius (Cramer, 1777)	Druce, 1875	5	
Palla publius centralis (van Someren, 1975)	◻	5	
Palla ussheri hassoni (Turlin & Vingerhoedt, 2013)	Turlin & Vingerhoedt, 2013 (*)	1,5	
Palla violinitens coniger (Butler, 1896)	Aurivillius, 1928	5	
Charaxes c. crossleyi (Ward, 1871)	Aurivillius, 1928	6	
Charaxes eurinome ansellica (Butler, 1870)	Butler, 1870 (*)	5	E
Charaxes trajanus bambi BS & M, 2007	BS & M, 2007 (*)	1,5	
NYMPHALIDAE | Apaturinae			
Apaturopsis c. cleocharis (Hewitson, 1873)	Hewitson, 1873 (*)	6	
NYMPHALIDAE | Nymphalinae			
Kallimoides rumia jadyae (Fox, 1968)	Druce, 1875, as *rumia*	5	
Vanessula milca buechneri (Dewitz, 1887)	▲	6	
Antanartia d. delius (Drury, 1782)	▲	5	
Vanessa cardui (Linnaeus, 1758)	Snellen, 1882	5	C D
Precis antilope (Feisthamel, 1850)	Monard, 1956	5	C D
Precis a. archesia (Cramer, 1779)	Aurivillius, 1928	5	C D
Precis c. ceryne (Boisduval, 1847)	Druce, 1875	5	A,B,E

(continued)

A Checklist of Angolan Lepidoptera and Papilionoidea

Taxon	First reference for Angola	V	H
Precis coelestina (Dewitz, 1879)	Dewitz, 1879 (*)	5	A
Precis octavia sesamus (Trimen, 1883)	Druce, 1875, as	5	C
Precis pelarga (Fabricius, 1775)	Ladeiro, 1956	5	A,C
Precis actia (Distant, 1880)	Aurivillius, 1928	5	A,C
Precis s. sinuata (Plötz, 1880)	Bacelar, 1956	5	A,C
Precis rauana silvicola (Schultze, 1916)	◩	5	A
Precis larseni manuscript name – (M et al., 2018)	M et al., 2018 (*)	1,5	A,C
Hypolimnas a. anthedon (Doubleday, 1845)	Druce, 1875, sub *Diadema*	5	A,C
Hypolimnas misippus (Linnaeus, 1764)	Druce, 1875	5	H
Hypolimnas d. dinarcha (Hewitson, 1865)	Bacelar, 1958a, b	5	A
Hypolimnas m. monteironis (Druce, 1874)	Druce, 1874 (*)	5	A
Hypolimnas s. salmacis (Drury, 1773)	Druce, 1875	5	A
Salamis c. cacta (Fabricius, 1793)	Ackery et al., 1995	5	A
Protogoniomorpha anacardii ansorgei (Rothschild, 1904)	Rothschild, 1904 (*)	5	A,C
Protogoniomorpha parhassus (Drury, 1782)	Druce, 1875, as *Diadema salamis*	5	A,C
Protogoniomorpha t. temora (Felder & Felder, 1867)	Aurivillius, 1928	6	A
Junonia artaxia (Hewitson, 1864)	Aurivillius, 1928	5	C,D
Junonia hierta crebrene (Trimen, 1870)	Butler, 1871	5	A,C,D
Junonia n. natalica (Felder & Felder, 1860)	Gardiner, 2004	6	D
Junonia natalica angolensis (Rothschild, 1918)	Rothschild, 1918 (*)	3,5	A,C
Junonia o. oenone (Linnaeus, 1758)	Butler, 1871	5	H
Junonia orythia madagascariensis (Guenée, 1865)	Bacelar, 1948	5	A,C,D
Junonia sophia infracta (Butler, 1888)	Bacelar, 1948	5	A
Junonia stygia (Aurivillius, 1894)	Aurivillius, 1928	5	A
Junonia w. westermanni (Westwood, 1870)	Hewitson, 1873	5	A
Junonia terea elgiva (Hewitson, 1864)	Aurivillius, 1928	5	A
Junonia ansorgei (Rothschild, 1899)	◩	5	A
Junonia cymodoce lugens (Schultze, 1912)	Aurivillius, 1909, as *cymodoce*	5	A
Catacroptera c. cloanthe (Stoll, 1781)	Butler, 1871	5	A,C,D
**NYMPHALIDAE	Cyrestinae**			
Cyrestis c. camillus (Fabricius, 1781)	Ladeiro, 1956	5	A
**NYMPHALIDAE	Biblidinae**			
Biblia anvatara crameri (Aurivillius, 1894)	Bacelar, 1948	5	C,D
Byblia ilithyia (Drury, 1773)	Snellen, 1882	5	A,D

(continued)

Taxon	First reference for Angola	V	H
Mesoxantha ethosea ethoseoides (Rebel, 1914)	Aurivillius, 1928	5	A
Ariadne albifascia (Joicey & Talbot, 1921)	▲	5	A
Ariadne enotrea archeri (Carcasson, 1958)	Carcasson, 1958 (*)	5	A
Neptidiopsis ophione nucleata (Grünberg, 1911)	Aurivillius, 1928	5	A
Eurytela dryope angulata (Aurivillius, 1899)	Butler, 1871, as *dryope*	5	A,C,D
Eurytela h. hiarbas (Drury, 1782)	Druce, 1875	5	A
Sevenia boisduvali omissa (Rothschild, 1918)	▲	5	A
Sevenia occidentalium penricei (Rothschild & Jordan, 1903)	Rothschild & Jordan, 1903 (*)	1,5	A
Sevenia consors (Rothschild & Jordan, 1903)	Rothschild & Jordan, 1903 (*)	5	C
Sevenia t. trimeni (Aurivillius, 1899)	Aurivillius, 1899, as *Crenis natalensis* var. *trimeni* (*)	5	A,C
Sevenia umbrina (Karsch, 1892)	▲	5	
Sevenia amulia intermedia (Carcasson, 1961)	Aurivillius, 1928, as *amulia*	5	A,C,E
Sevenia benguelae (Chapman, 1872)	Aurivillius, 1928	5	A,C,D
Sevenia p. pechueli (Dewitz, 1879)	Dewitz, 1879 (*)	5	C
**NYMPHALIDAE	Limenitidinae**			
Harma theobene superna (Fox, 1968)	Fox, 1968	5	A
Cymothoe o. oemilius (Doumet, 1859)	Bacelar, 1958a, b	5	A
Cymothoe b. beckeri (Herrich-Schaeffer, 1858)	Druce, 1875	5	A
Cymothoe haynae fumosa (Staudinger, 1896)	Bacelar, 1956	2,5	A
Cymothoe confusa (Aurivillius, 1887)	Bacelar, 1956	5	A
Cymothoe lucasii cloetensi (Seeldrayers, 1896)	◻	5	A
Cymothoe h. harmilla (Hewitson, 1874)	Larsen, 2005	6	A
Cymothoe h. hesiodotus (Hewitson, 1869)	Aurivillius, 1928	6	A
Cymothoe h. hypatha (Hewitson, 1866)	Druce, 1875	5	A
Cymothoe lurida hesione (Weymer, 1907)	Druce, 1875, as *lurida*	5	A
Cymothoe altisidora (Hewitson, 1869)	Aurivillius, 1898	5	A
Cymothoe capella (Ward, 1871)	Bacelar, 1956	5	A
Cymothoe caenis (Drury, 1773)	Druce, 1874	5	A
Cymothoe jodutta ciceronis (Ward, 1871)	Bacelar, 1956	5	A
Cymothoe jodutta ehmckei (Dewitz, 1887)	◻	5	A
Cymothoe cf. c. coccinata (Hewitson, 1874)	◻	5	A

(continued)

A Checklist of Angolan Lepidoptera and Papilionoidea

Taxon	First reference for Angola	V	H
Cymothoe excelsa deltoides (Overlaet, 1944)	D'Abrera, 1980	3,5	A
Cymothoe s. sangaris (Godart, 1824)	Druce, 1875	5	A
Pseudoneptis bugandensis ianthe (Hemming, 1964)	Snellen, 1882, as *bugandensis*	5	A
Pseudacraea eurytus eurytus (Linnaeus, 1758)	Druce, 1875	5	A
Pseudacraea d. dolomena (Hewitson, 1865)	Aurivillius,1928	6	A
Pseudacraea b. boisduvalii (Doubleday, 1845)	Druce, 1875	5	A
Pseudacraea kuenowii gottbergi (Dewitz, 1884)	Williams, 2007, as *kuenowii*	6	A
Pseudacraea lucretia protracta (Butler, 1874)	Butler, 1874 (*)	5	A
Pseudacraea poggei (Dewitz, 1879)	Gardiner, 2004	5	A,D
Pseudacraea semire (Cramer, 1779)	Druce, 1875	5	A
Neptis saclava marpessa (Hopffer, 1855)	Druce, 1875	5	A,B,C
Neptis nemetes margueriteae (Fox, 1968)	Butler, 1871, the species	6	C,D
Neptis gratiosa (Overlaet, 1955)	▲	5	C
Neptis jordani (Neave, 1910)	Gardiner, 2004	6	D
Neptis kiriakoffi (Overlaet, 1955)	◻	5	A,C
Neptis laeta (Overlaet, 1955)	Gardiner, 2004	5	C
Neptis morosa (Overlaet, 1955)	Larsen, 2005	5	A,C
Neptis s. serena (Overlaet, 1955)	Gardiner, 2004	5	A,C,E
Neptis alta (Overlaet, 1955)	Gardiner, 2004	5	C
Neptis constantiae kaumba (Condamin, 1966)	◻	5	A
Neptis nysiades (Hewitson, 1868)	Aurivillius, 1928	5	A
Neptis nicomedes (Hewitson, 1874)	Hewitson, 1874 (*)	5	A
Neptis quintilla (Mabille, 1890)	Larsen, 2005	5	A
Neptis a. agouale (Pierre-Baltus, 1978)	◻	5	A
Neptis melicerta (Drury, 1773)	Aurivillius, 1928	5	A,D
Neptis nebrodes (Hewitson, 1874)	Hewitson, 1874 (*)	6	A
Neptis nicoteles (Hewitson, 1874)	Hewitson, 1874 (*)	6	A
Neptis e. exaleuca (Karsch, 1894)	◻	5	A
Evena crithea (Drury, 1773)	Fox, 1968	5	A
Evena angustatum (Felder & Felder, 1867)	◻	5	A
Euryphura c. chalcis (Felder & Felder, 1860)	Bacelar, 1956, as *E. fulminea*	5	A
Euryphura plautilla Hewitson, 1865	Druce, 1875	5	
Euryphura concordia (Hopffer, 1855)	Aurivillius, 1928	5	
Hamanumida daedalus (Fabricius, 1775)	Druce, 1875, as *Aterica meleagris*	5	
Pseudargynnis hegemone (Godart, 1819)	Druce, 1874, as *Aterica clorana*	5	

Taxon	First reference for Angola	V	H
Aterica g. extensa (Heron, 1909)	Druce, 1875, as *Aterica cupavia*	5	A,B,C
Cynandra opis bernardii (Lagnel, 1967)	Druce, 1875, as *Aterica afer*	5	A
Euriphene barombina (Aurivillius, 1894)	Larsen, 2005	5	A
Euriphene iris (Aurivillius, 1903)	▲	5	C
Euriphene plagiata (Aurivillius, 1897)	Larsen, 2005	6	A
Euriphene saphirina trioculata (Talbot, 1927)	▲	6	A
Euriphene t. tadema (Hewitson, 1866)	■	5	A
Euriphene gambiae gabonica (Bernardi, 1966)	Bacelar, 1958a, b	6	A
Bebearia phantasia concolor (Hecq, 1988)	Druce, 1875	6	A
Bebearia languida (Schultze, 1920)	▲	5	A
Bebearia a. absolon (Fabricius, 1793)	Bacelar, 1958a, b	5	A
Bebearia micans (Aurivillius, 1899)	■	5	A
Bebearia zonara (Butler, 1871)	■	5	A
Bebearia oxione squalida (Talbot, 1928)	Aurivillius, 1909, as *oxione*		A
Bebearia cocalia katera (van Someren, 1939)	■	5	A
Bebearia guineensis (Felder & Felder, 1867)	Holmes, 2001	6	A
Bebearia sophus aruunda (Overlaet, 1955)	Druce, 1875, as *sophus*	5	A
Bebearia plistonax (Hewitson, 1874)	Hewitson, 1874 (*)	5	F
Bebearia hassoni (Hecq, 1998)	Hecq, 1998 (*)	1	A
Euphaedra medon celestis (Hecq, 1986)	Butler, 1871, as *medon*	5	A
Euphaedra z. zaddachii (Dewitz, 1879)	Dewitz, 1879 (*)	5	A
Euphaedra cf. *sinuosa* (Hecq, 1974)	■	6	A
Euphaedra diffusa diffusa (Gaede, 1916)	■	6	A
Euphaedra ansorgei (Rothschild, 1918)	■	5	A
Euphaedra p. permixtum (Butler, 1873)	Bacelar, 1956	5	A
Euphaedra divoides (BS & M, 2018) (Manuscript name)	Staudinger, 1886, as *E. themis* var. *innocentia* (?)	1,5	A
Euphaedra adonina spectacularis (Hecq, 1997)	Bacelar, 1956	5	A
Euphaedra ceres electra (Hecq, 1983)	Butler, 1871, as *ceres*	5	A
Euphaedra fontainei (Hecq, 1977)	▲	5	A
Euphaedra v. viridicaerulea (Bartel, 1905)	■	5	A
Euphaedra preussiana robusta (Hecq, 1983)	■	5	A
Euphaedra rezia (Hewitson, 1866)	Bacelar, 1956	5	A
Euphaedra albofasciata (Berger, 1981)	■	5	A
Euphaedra disjuncta virens (Hecq, 1984)	■	5	A
Euphaedra mayumbensis (Hecq, 1984)	■	5	A

(continued)

A Checklist of Angolan Lepidoptera and Papilionoidea

Taxon	First reference for Angola	V	H
Cymothoe excelsa deltoides (Overlaet, 1944)	D'Abrera, 1980	3,5	A
Cymothoe s. sangaris (Godart, 1824)	Druce, 1875	5	A
Pseudoneptis bugandensis ianthe (Hemming, 1964)	Snellen, 1882, as *bugandensis*	5	A
Pseudacraea eurytus eurytus (Linnaeus, 1758)	Druce, 1875	5	A
Pseudacraea d. dolomena (Hewitson, 1865)	Aurivillius,1928	6	A
Pseudacraea b. boisduvalii (Doubleday, 1845)	Druce, 1875	5	A
Pseudacraea kuenowii gottbergi (Dewitz, 1884)	Williams, 2007, as *kuenowii*	6	A
Pseudacraea lucretia protracta (Butler, 1874)	Butler, 1874 (*)	5	A
Pseudacraea poggei (Dewitz, 1879)	Gardiner, 2004	5	A,D
Pseudacraea semire (Cramer, 1779)	Druce, 1875	5	A
Neptis saclava marpessa (Hopffer, 1855)	Druce, 1875	5	A,B,C
Neptis nemetes margueriteae (Fox, 1968)	Butler, 1871, the species	6	C,D
Neptis gratiosa (Overlaet, 1955)	▲	5	C
Neptis jordani (Neave, 1910)	Gardiner, 2004	6	D
Neptis kiriakoffi (Overlaet, 1955)	◻	5	A,C
Neptis laeta (Overlaet, 1955)	Gardiner, 2004	5	C
Neptis morosa (Overlaet, 1955)	Larsen, 2005	5	A,C
Neptis s. serena (Overlaet, 1955)	Gardiner, 2004	5	A,C,E
Neptis alta (Overlaet, 1955)	Gardiner, 2004	5	C
Neptis constantiae kaumba (Condamin, 1966)	◻	5	A
Neptis nysiades (Hewitson, 1868)	Aurivillius, 1928	5	A
Neptis nicomedes (Hewitson, 1874)	Hewitson, 1874 (*)	5	A
Neptis quintilla (Mabille, 1890)	Larsen, 2005	5	A
Neptis a. agouale (Pierre-Baltus, 1978)	◻	5	A
Neptis melicerta (Drury, 1773)	Aurivillius, 1928	5	A,D
Neptis nebrodes (Hewitson, 1874)	Hewitson, 1874 (*)	6	A
Neptis nicoteles (Hewitson, 1874)	Hewitson, 1874 (*)	6	A
Neptis e. exaleuca (Karsch, 1894)	◻	5	A
Evena crithea (Drury, 1773)	Fox, 1968	5	A
Evena angustatum (Felder & Felder, 1867)	◻	5	A
Euryphura c. chalcis (Felder & Felder, 1860)	Bacelar, 1956, as *E. fulminea*	5	A
Euryphura plautilla Hewitson, 1865	Druce, 1875	5	A
Euryphura concordia (Hopffer, 1855)	Aurivillius, 1928	5	C,E
Hamanumida daedalus (Fabricius, 1775)	Druce, 1875, as *Aterica meleagris*	5	C,D,F
Pseudargynnis hegemone (Godart, 1819)	Druce, 1874, as *Aterica clorana*	5	E

(continued)

Taxon	First reference for Angola	V	H
Acraea p. pseudolycia (Butler, 1874)	Butler, 1874 (*)	3,5	A,C
Acraea acara melanophanes (Le Cerf, 1927)	Pierre & Bernaud, 2013	5	D
Acraea z. zetes (Linnaeus, 1758)	Druce, 1875	5	C,D
Acraea admatha (Hewitson, 1865)	Pierre, 1979	6	A
Acraea endoscota (Le Doux, 1928)	Larsen, 2005	6	A
Acrae l. leucographa (Ribbe, 1889)	Larsen, 2005	6	A
Acraea q. quirina (Fabricius, 1781)	▲	5	A
Acraea camaena (Drury, 1773)	Larsen, 2005	6	E
Acraea n. neobule (Doubleday, 1843)	Druce, 1875	5	A,C,D
Acraea eugenia ochreata (Grünberg, 1910)	Larsen, 2005	5	A
Acraea brainei (Henning, 1986)	Ackery et al., 1995	6	?
Acraea e. epaea (Cramer, 1779)	Aurivillius, 1928	5	A
Acraea formosa (Butler, 1874)	Butler, 1874 (*)	5	A,B
Acraea l. leopoldina (Aurivillius, 1895)	▲	3	A
Acraea p. poggei (Dewitz, 1879)	Dewitz, 1879 (*)	6	C
Acraea alcinoe camerunica (Aurivillius, 1893)	▲	6	A
Acraea umbra macarioides (Aurivillius, 1893)	Aurivillius, 1928	5	A,B
Acraea consanguinea intermedia (Aurivillius, 1899)	Le Doux, 1937	2,5	A
Acraea excisa (Butler, 1874)	Bacelar, 1948	5	A
Acraea pseuderyta (Godman & Salvin, 1890)	Aurivillius, 1928	5	A
Acraea vestalis congoensis (Le Doux, 1937)	◻	5	A
Telchinia p. perenna (Doubleday, 1847)	Aurivillius, 1928	5	A
Telchinia p. penelope (Staudinger, 1896)	◻	5	A
Telchinia o. oreas (Sharpe, 1891)	Lathy, 1906	6	C
Telchinia circeis (Drury, 1782)	Larsen, 2005	6	A
Telchinia parrhasia servona (Godart, 1819)	Godart, 1819 (*)	6	A
Telchinia peneleos pelasgia (Grose-Smith, 1900)	Larsen, 2005	5	A,C
Telchinia p. pharsalus (Ward, 1871)	Aurivillius, 1928	5	A,E
Telchinia encedana (Pierre, 1976)	Pierre, 1976	6	E
Telchinia e. encedon (Linnaeus, 1758)	Druce, 1875	5	A,C,D
Telchinia alciope (Hewitson, 1852)	Bacelar, 1956	5	A
Telchinia a. aurivillii (Staudinger, 1896)	▲	6	A
Telchinia esebria (Hewitson, 1861)	Butler, 1874	5	B,C
Telchinia j. jodutta (Fabricius, 1793)	Druce, 1875	5	A
Telchinia lycoa (Godart, 1819)	Druce, 1875	5	A,C
Telchinia serena (Fabricius, 1775)	Butler, 1871	5	A,C,D

(continued)

A Checklist of Angolan Lepidoptera and Papilionoidea 193

Taxon	First reference for Angola	V	H
Telchinia v. ventura (Hewitson, 1877)	Monard, 1956	6	E
Telchinia acerata (Hewitson, 1874)	Snellen, 1882	5	A,C
Telchinia oberthueri (Butler, 1895)	Bacelar, 1948	6	A
Telchinia sotikensis karschi (Aurivillius, 1899)	Aurivillius, 1928	5	A
Telchinia b. bonasia (Fabricius, 1775)	Druce, 1875	5	A,C,D
Telchinia uvui balina (Karsch, 1892)	Larsen, 2005	6	A
Telchinia o. orestia (Hewitson, 1874)	Snellen, 1882	5	A
Telchinia p. pentapolis (Ward, 1871)	▲	6	A
Telchinia induna imduna (Trimen, 1895)		5	C
Telchinia r. rahira (Boisduval, 1833)	Druce, 1875	5	C,D,E
Telchnia mirifica (Lathy, 1906)	Lathy, 1906 (*)	3	C
Lachnoptera anticlia (Hübner, 1819)	Bacelar, 1958a, b, as *L. iole*	6	A
Phalanta e. eurytis (Doubleday, 1847)	Bacelar, 1956	5	A
Phalanta phalantha aethiopica (Rothschild & Jordan, 1903)	Druce, 1875	5	C,D

References

Ackery PR, Smith CR, Vane Wright RI (1995) Carcasson's African Butterflies. An annotated catalogue of the Papilionoidea and Hesperioidea of the Afrotropical Region. CSIRO, Victoria, i–xii + 803 pp

Aduse-Poku K, Vingerhoedt E, Wahlberg N (2009) Out-of-Africa again: A phylogenetic hypothesis of the genus Charaxes (Lepidoptera: Nymphalidae) based on five genes region. Mol Phylogenet Evol 53:463–478

Aurivillius C In Seitz A (1928) Les Macrolepidopteres du Globe. Les Macrolepidoptères de la Faune Ethiopienne 13(4):615 pp + 80 pl. E. Le Moult, Paris

Bacelar A (1948) Lepidópteros de África principalmente das colónias portuguesas (colecção do Museu Bocage). Arquivos do Museu Bocage 19:165–207

Bacelar A (1956) Lepidópteros (Rhopalocera) do Buco Zau, enclave de Cabinda, Angola. Anais da Junta de Investigações do Ultramar 11(3):175–197

Bacelar A (1958a) Alguns Lepidópteros (Rhopalocera) do enclave de Cabinda. Revista portuguesa de Zoologia e Biologia Geral 1(2/3):197–217

Bacelar A (1958b) Alguns Lepidópteros (Rhopalocera) da África Ocidental portuguesa. Revista portuguesa de Zoologia e Biologia Geral 1(4):311–330

Bacelar A (1961) Lepidópteros do Bié (Rhopalocera) da colecção do Colégio de São Bento, em Luso (Angola). Memórias da Junta de Investigações do Ultramar (2)23:61–81

Berger LA (1979) Espèces peu connues et descriptions de nouvelles sous-espèces de Mylothris (Lepidoptera Pieridae). Revue de Zoologie Africaine 93(1):1–9

Bethune-Baker GT (1914) Notes on the taxonomic value of genital armature in Lepidoptera. Trans Entomol Soc London 1914:314–337

Bivar-de-Sousa A (1983) Contribuição para o conhecimento dos lepidópteros de Angola (3ª nota). Dados sobre a ocorrência do género Charaxes (Lep. Nymphalidae) em Angola (1ª parte). Actas do I Congresso Ibérico de Entomologia 1:107–119

Bivar-de-Sousa A, Fernandes JA (1964) Contribuição para o conhecimento dos Lepidópteros de Angola. Boletim da Sociedade Portuguesa de Ciências Naturais (2) 10(25):104–115

Bivar-de-Sousa A, Mendes LF (2006) On the genus Euxanthe Hübner, 1819 in Angola, with description of a new subspecies (Lepidoptera, Nymphalidae, Charaxinae). Nouvelle Revue d'Entomologie 23(4):369–376

Bivar-de-Sousa A, Mendes LF (2007) New data on the Uranothauma from Angola, with description of a new species (Lepidoptera: Lycaenidae: Polyommatinae). Boletin Sociedad Entomológica Aragonesa 41:73–76

Bivar-de-Sousa A, Mendes LF (2009a) On a new species of the genus Princeps Hübner, (1807) from Cabinda (Angola) (Lepidoptera: Papilionidae). SHILAP, Revista de Lepidopterologia 37(147):313–318

Bivar-de-Sousa A, Mendes LF (2009b) New data on the Angolan Charaxes of the "etheocles group" with description of a new species (Lepidoptera, Nymphalidae, Charaxinae). Boletim da Sociedade portuguesa de Entomologia 8(15)(229):293–309

Bivar-de-Sousa A, Mendes LF (2014) New data on the Angolan Charaxes of the "etheocles group" with description of a new species (Lepidoptera, Nymphalidae, Charaxinae). Boletim da Sociedade portuguesa de Entomologia 229(8–15):293–309

Bivar-de-Sousa A, Mendes LF, Vasconcelos S (2017) Description of one new species and one new subspecies of Nymphalidae from Angola (Lepidoptera: Papilionoidea). SHILAP, Revista de Lepidopterologia 45(178):227–236

Bivar-de-Sousa A, Mendes LF (2018, in press) The "themis group" of Euphaedra (Euphaedrana) in Angola. Revision and description of one new species (Lepidoptera: Nymphalidae: Limenetidinae). Boletim da Sociedade portuguesa de Entomologia

Bouyer T (1999) Note sur les Charaxes du «groupe eupale» avec description d'une nouvelle sous-espèce (Lepidoptera Nymphalidae). Entomologia Africana 4(1):37–40

Butler AG (1872) On a small collection of butterflies from Angola. Proc Zool Soc London 1871:721–725

Carvalho EL (1962) Alguns Papilionídeos da Lunda (Lepidoptera). Publicações Culturais da Companhia dos Diamantes de Angola 60:163–170

Condamin M (1966) Mise au point sur les Neptis au facies d' «14avann» (Lepidoptera: Nymphalidae). Bulletin de l'Institut Fondamental de l'Afrique Noire 28(A)(3):1008–1029

D'Abrera B (1980) Butterflies of the afrotropical region. Lansdowe Ed., Melbourne, pp i–x + 1–593

Dhungel B, Wahlberg N (2018) Molecular systematics of the subfamily Limenitidinae (Lepidoptera: Nymphalidae). PeerJ 6:e4311

Druce H (1875) A list of the collections of diurnal Lepidoptera made by J. J. Monteiro in Angola with description of some new species. Proc Zool Soc London 27:406–417

Eltringham H (1912) A monograph of the African species of the genus Acraea Fab., with a supplement of those of the Oriental Region. Trans Entomol Soc London 1912(1):1–374

Espeland M, Breinholt J, Willmott KR et al (2018) A comprehensive and dated phylogenomic analysis of butterflies. Curr Biol 28:1–9

Evans WH (1937) Catalog of African Hesperiidae (Indicating the classification and nomenclature adopted in the British Museum). British Museum, London, 212pp + 30 pl

Gardiner A (2004) Chapter 10. Butterflies of the four corners area. In: Timberlake JR, Chillers SL (eds) Biodiversity of the Four Corners Area: Technical Review, vol 15. Occasional Publications on Biodiversity, Bulawayo, pp 381–397

Hecq J (1997) Euphaedra. Lambillionea & Hecq, Tervuren & Mont-Sur-Marchienne, 121 pp + 48 pl

Heikkilä M, Kaila L, Mutanen M et al (2012) Cretaceous origin and repeated Tertiary diversification of the redefined butterflies. Proc R Soc Lond B 279:1093–1099

Henning SF (1988) The Charaxes Butterflies of Africa. Aloe Books, Johannesburg, 457 pp

Holmes CWN (2001) A reaprisal of the Bebearia mardania complex (Lepidoptera, Nymphalidae). Trop Zool 14:31–62

Homeyer A, Dewitz H (1882) Drei neue westafrikanische Charaxes. Berliner Entomologisches Zeitschrift 26(II):381–383

JIU (1948–1963) Cartas do levantamento aerofotogramétrico de Angola (escala 1: 250,000) Junta de Investigações do Ultramar: folhas:8–471

Kielland J (1982) Revision of the genus Ypthima in the Ethiopian Region excluding Madagascar (Lepidoptera, Satyridae). Tijdschrift voor Entomologie 125(5):99–154

,vision of the genus *Henotesia* (excluding Madagascar and other Indian
J (1994)pidoptera Satyridae). Lambillionea 94(2):235–273
islandpidópteros de Angola (estudo de uma colecção oferecida ao Museu Zoológico
JM (1 ais da Junta de Investigações do Ultramar 11(3):151–172
Coimputterflies of West Africa. Text volume: 1–595. Apollo Books, Stenstrup
TB urt JB (1819) Encyclopédie Méthodologique. Histoire Naturelle (Zoologie), 9,
Paris, i–iv + 1–328
7, Neinhuis C (eds) (2014) Riquezas Naturais do Uíge – Uma breve Introdução
do atual. A utilização, a Ameaça e a Preservação da Biodiversidade. Technische
Dresden, Dresden, 125 pp
9) Revision des Epitola (ls). *Révision des genres* Epitola *Westwood,* Hypophytala
t Stempfferia Jackson et description de trois nouveaux genres (Lepidoptera
dae). ABRI & Lambillionea, Nairobi/Tervuren, pp 1–219, pl. I–XVI
000) Révision du genre Mimacraea Butler, avec description de quatre nouvelles espèces
x nouvelles sous-espèces (Lepidoptera, Lycaenidae). Lambillionea & ABRI, Tervuren/
bi, pp 1–73
A (2004) Revision du genre *Oxylides* Hübner (Lepidoptera, Lycaenidae). Lambillionea
(2):143–158
des LF, Bivar-de-Sousa A (2006a) A new species of *Neita* van Son (Nymphalidae, Satyrinae)
from southern Angola. Boletin Sociedad Entomológica Aragonesa 39:95–96
Mendes LF, Bivar-de-Sousa A (2006b) Notes and descriptions of Afrotropical *Appias* butterflies
(Lepidoptera: Pieridae). Boletin Sociedad Entomológica Aragonesa 39:151–160
Mendes LF, Bivar-de-Sousa A (2007a) New species of *Cooksonia* Druce, 1905 from Angola
(Lepidoptera: Lycaenidae, Lipteninae). SHILAP, Revista de Lepidopterologia 35(138):265–268
Mendes LF, Bivar-de-Sousa A (2007b) On the genus Eagris Guenée, 1863 in Angola (Lepidoptera:
Hesperiidae). SHILAP, Revista de Lepidopterologia 35(139):311–316
Mendes LF, Bivar-de-Sousa A (2009a) Description of a new species of *Mylothris* from northern
Angola (Lepidoptera Pieridae). Bolletino della Societá Entomológica Italiana 141(1):55–58
Mendes LF, Bivar-de-Sousa A (2009b) New account on the butterflies of Angola. The genus
Leptomyrina (Lepidoptera Lycaenidae). Bolletino della Societá Entomológica Italiana
141(2):109–112
Mendes LF, Bivar-de-Sousa A (2009c) On a new south-eastern Angolan Satyrine butterfly belong-
ing to a new genus (Lepidoptera, Nymphalidae). Entomologia Africana 14(2):5–8
Mendes LF, Bivar-de-Sousa A (2009d) The genus *Abantis* Hopffer, 1855 in Angola and description
of a new species (Lepidoptera: Hesperiidae, Pyrginae). SHILAP, Revista de Lepidopterologia
37(147):313–318
Mendes LF, Bivar-de-Sousa A (2012) Notes on the species of *Hypolycaena* (Lepidoptera,
Lycaenidae, Theclinae) known to occur in Angola. Boletin Sociedad Entomológica Aragonesa
50:193–197
Mendes LF, Bivar-de-Sousa A, Figueira R (2013a) Butterflies of Angola / Borboletas diurnas de
Angola. Lepidoptera. Papilionoidea, I. Hesperiidae, Papilionidae. IICT and CIBIO, Lisboa and
Porto, 288 pp
Mendes LF, Bivar-de-Sousa A, Figueira R et al (2013b) Gazetteer of the Angolan localities known
for beetles (Coleoptera) and butterflies (Lepidoptera: Papilionoidea). Boletim da Sociedade
Portuguesa de Entomologia 8(14/228):257–290
Mendes LF, Bivar-de-Sousa A, Vasconcelos S et al (2017) Description of two new subspecies
and notes on *Charaxes* Ochenheimer, 1816 of Angola (Lepidoptera: Nymphalidae). SHILAP,
Revista de Lepidopterologia 45(178):299–315
Mendes LF, Bivar-de-Sousa A, Vasconcelos S (2018, in press) On the butterflies of genus *Precis*
Hübner, 1819 from Angola and description of a new species (Lepidoptera: Nymphalidae:
Nymphalinae). SHILAP Revista de Lepidopterologia
Monard A (1956) Compendium Entomologicum Angolae – 1. Insecta. VI – Ord. Lepidoptera.
Anais da Junta de Investigações do Ultramar 11(3):119–128
NHM – Natural History Museum (2004). Wallowtails. Available at: http://internet.nhm.ac.uk/cgi-
bin/perth/wallowtails/list.dsml

Pierre J (1988) Les Acraea du super-group «egina»: Révision et phylog Nymphalidae). Annales de la Société Entomologique de France 24(3):263–Lepidoptera :

Pierre J, Bernaud D (2013) Butterflies of the world. NymphalidaeXXIII. Acraea Goecke & Evers, Keltern, 39:1–8, pl. 1–28

Rothschild W, Jordan K (1903) Lepidoptera collected by Oscar Neumann in nor cra . Novitates Zoologicae 10(3):491–542

Smith CR, Vane-Wright RI (2001) A review of the Afrotropical species of the gen ica (Lepidoptera: Rhopalocera: Papilionidae). Bull Nat His Mus London (Entomol) 70 2

Stempffer H (1957) Les Lépidoptères de l'Afrique noire française. Lycaenidés l Africaines 14(3):1–228

Stempffer H, Bennett NH (1963) A new genus of Lipteninae (Lepidoptera: Lycaenidae). E Mus (Nat His) (Entomol) 13:171–194

Talbot G (1944) A preliminary revision of the genus *Mylothris* Hübn. (1819) (Lep. Rhop Pierid. T Roy Ent Soc London 94(2):155–185

Tite GE (1959) New species and notes on the genus *Lepidochrysops* (Lepidoptera, Ly aeridae) The Entomologist 92:158–163

Tite GE (1961) New species of the genus *Lepidochrysops* (Lepidoptera, Lycaenicae) The Entomologist 94:21–25

Tite GE, Dickson CGC (1973) The genus *Aloeides* and allied genera. Bull Brit Mus (Nat His) (Entomol) 29:227–280

Trimen R (1891) On butterflies collected in tropical south-western Africa by Mr. A. W. Eriks n. Proc Zool Soc 1891:59–107

Turlin B, Vingerhoedt E (2013) Butterflies of the World, supl. 23. Les Charaxinae de la fa ne Afrotropicale. Les genres Palla et Euxanthe. Nymphalidae: Charaxinae: Pallini et Euxanth ri. Goecke & Evers, Keltern

Weymer G (1901) Beitrag zur Lepidopterofauna von Angola. Entomologischen Zeitsch iften. Stuttgart 15(17):61–64, 65–67, 69–70

Weymer G (1908) Einige neuer Lepidopteren des Deutschen Entom. National-Museums, gesam melt von Dr. F. Cr. Wellman in Benguella. Deutsche Entomologische Zeitschrift 1908:507–413

Willis CK (2009) Amist the butterflies of southwestern Angola. Metamorphosis 20(3):74–88

Williams MC (2018) Afrotropical butterflies. Available at: http://www.lepsocafrica.org/?p=publ cations&s=atb

Part IV
Diversity and Distribution of the Vertebrate Taxa

11

Angolan Freshwater Fishes: A Biogeographic Model

Paul H. Skelton

Abstract The discovery and exploration of Angolan freshwater fishes was largely effected by foreign scientists on expeditions organised by European and North American parties. Current knowledge of Angolan freshwater fishes is briefly described according to the main drainage systems that include Cabinda, Lower Congo, Angolan Coastal region including the Cuanza, the southern Congo tributaries, the Zambezi, Okavango, Cunene and Cuvelai drainages. A biogeographic model to explain the freshwater fish fauna of Angola is presented. The need for the conservation of Angolan freshwater fishes will rise with rapidly increasing pressures on aquatic ecosystems from urbanisation, dams for power, agriculture and human needs, habitat destruction from mining and deforestation, pollution, the introduction of alien species and overfishing.

Keywords Africa · Cuanza · Cunene · Cuvelai · Okavango · Southern Congo · Zambezi

Historical Review

Despite Poll's work (1967) over a very limited area, Angola remains a poorly known region in which there remains much to be discovered (Lévêque and Paugy 2017a: 93)

The quotation above sums up the current state of knowledge for the freshwater fishes of Angola. Poll (1967) is a landmark publication that reviews the historical literature and records the known species and their distribution within the major river basins of the country at that time. No other account of Angolan fishes as a whole has been published. The current situation of a poorly known region is due to a number of factors including the historical neglect of scientific exploration by the colonial

---

P. H. Skelton (✉)
South African Institute for Aquatic Biodiversity (SAIAB), Grahamstown, South Africa

Wild Bird Trust, National Geographic Okavango Wilderness Project, Hogsback, South Africa
e-mail: P.skelton@saiab.ac.za

authorities, widely dispersed collections in international institutions from various expeditions, the relative inaccessibility to scientists and collectors of the inland rivers and biologically rich areas, and the difficulties of aquatic exploration relative to terrestrial fauna. The fact that there is no national Angolan depository for wet collections such as fishes fostered by local expertise is a further hindrance to discovery. This aspect is fundamental to effective and sustained scientific productivity in any endeavor such as ichthyology (Skelton and Swartz 2011). This accentuates the situation for Angolan freshwater fishes when it is recognised that Poll's (1967) account rested largely on the collection in the Diamond Company of Angola (DIAMANG) museum in Dundo, which to a large extent is a product of the industrial diamond mining activity in the mainly local drainages.

There are four distinct phases of scientific discovery of Angolan freshwater fishes, Phase 1 – early explorations in the second half of the nineteenth century; Phase 2 – scientific expeditions in the twentieth century until World War II; Phase 3 – post WWII to Angolan independence in 1975; and, Phase 4 – post independence investigations.

Although several of Castelnau's (1861) Lake Ngami fishes occur in the Angolan reaches of the Okavango River system, the discovery and scientific description of Angolan freshwater fishes was initiated by Steindachner (1866) describing a collec⁻ tion derived from the Atlantic coastal rivers. Steindachner's species include some iconic species such as his *Kneria angolensis, Clarias angolensis* and *Enteromius kessleri* that help define the Angolan Atlantic coastal fauna. Guimarães (1884) working with specimens in the Lisbon Museum (subsequently lost in the fire of 1978) (Saldanha 1978) submitted by the Portuguese explorer José Alberto de Oliveira Anchieta provided detailed descriptions and illustrations of three species taken from the Cunene and the Curoca Rivers from 1873–1884, viz. *Schilbe stein⁻ dachneri, Mormyrus anchietae* and *Enteromius mattozi.*

The second phase of discovery (early twentieth century) is marked by a series of expedition reports that include freshwater fishes. Boulenger's (1909–1916) cata⁻ logue of fishes in the British Museum (Natural History) provided the basis of Angolan freshwater fish fauna. Again the fauna included collections such as Woosnam's Okavango collection described by Boulenger (1911) that includes spe⁻ cies which occur in Angolan reaches. Boulenger (1910) described a collection by Ansorge from the Cuanza and Bengo Rivers that set the scene for considering the uniqueness of the fauna of these rivers of the Atlantic coast. Other notable expedi⁻ tions that included descriptions of freshwater fishes are the Vernay Angola Expedition of 1925 (Nichols and Boulton 1927), the Gray African Expedition of 1929 (Fowler 1930), the Vernay-Lang Kalahari Expedition of 1930 (Fowler 1935), the Swiss Scientific Mission to Angola 1928–1929 and 1932–1933 (Pellegrin 1936), and Karl Jordan's Expedition to South West Africa and Angola of 1933–1934 (Trewavas 1936). All these expeditions realised new species but were somewhat limited in geographical scope to the southern Atlantic coastal rivers and to the upland western reaches of the Cubango (Okavango) tributaries, the Oshana-Etosha system and the plateau reaches of the Cuanza. This restriction emanates from the access realised by the Benguella Railway constructed from 1903 to 1928 from Lobito port to Huambo and beyond (Ball 2015).

The third phase of scientific exploration of the freshwater fishes of Angola after the WWII up until independence in 1975 is significant in that studies into ecological aspects as well as the beginnings of a synthesis of the Angolan fauna occurred. Ladiges and Voelker (1961) studied the fish fauna of the Longa River in the Angolan watershed highlands. In addition to providing an ecological description and zonation of the river they described a few new species – *Kneria maydelli* from the Cunene, *Enteromius* (as *Barbus*) *roussellei* and *Chiloglanis sardinhai*. Ladiges (1964) followed up this article with an account of the zoogeography and ecology of Angolan freshwater fishes based on a present/absent list of fishes in the Angolan Coastal region, the Cunene, the Okavango Basin, and the Zambezi. Trewavas (1973) recorded the cichlids of the Cuanza and Bengo Rivers that exposed the independent derivation of the cichlid fauna of the Cuanza River in terms of the inland and coastal reaches. An unpublished collection by Graham Bell-Cross from the Okavango and the Cunene basins was deposited in the NHM in 1965, and this together with collections made by Mike Penrith from the State Museum in Windhoek, Namibia, provided specimens essential for Greenwood's (1984) revision of serranochromine species. Mike Penrith's collections from the Cunene and Okavango in the early 1970s led to a few descriptions of new species by Penrith (1970) and Penrith (1973).

A major milestone account of Angolan freshwater fishes was Max Poll's (1967) *Contribution à la Faune Ichthyologique de l'Angola* – based largely on the extensive collections made by Barros Machado and others and lavishly illustrated with excellent fish drawings as well as a gallery of photographs drawn from the Dundo Museum in Lunda-Norte. Poll (1967) summarised the history of freshwater ichthyology and provided a full checklist of 264 species in 18 families and 54 genera as then recorded from the inland waters of the country (excluding the Cabinda enclave). A faunistic and zoogeographical account considered five ichthyological regions (see below). Acknowledging a clearly incomplete inventory Poll listed the diversity of his regions as follows: The Congo tributaries with 121 species are richest and most diverse with characteristic families and genera known from the Congo Basin. Next in diversity was the western Atlantic coastal region with 109 species, followed by the Zambezi (62 species) but Poll pointed out that Bell-Cross had then recently recorded 77 species from the upper Zambezi, also of tropical diversity but distinct in character from the Congolean rivers. The Okavango-Cubango (57 species) reflected its close connections to the Zambezi as well as to the Cunene (55 species) in the west. The Cunene presented a mixed fauna of both the Zambezian elements as well as Atlantic coastal nature.

Poll (1967) pointed out and summarised a few notable ichthyological characteristics of the Angolan fauna – there was no pronounced endemic character to the fauna as a whole. The occurrence of lungfish (*Protopterus*) in Angola is known only from records in Congo tributaries and from Cabinda, but Poll mentions that he was shown a photograph by Ladiges of *Protopterus annectens brieni* from the Cubango region (see this recorded in Ladiges 1964: 265). Such occurrence of lungfish in the Cubango or Okavango system has not yet been confirmed in spite of extensive collecting in that drainage. Polypterids are restricted to Congo tributaries as are freshwater clupeids (however marine or estuarine species also occur in coastal Atlantic rivers). The presence of kneriids is a distinct feature of the fauna especially of the escarpment reaches of rivers of the coastal region. Mormyrid diversity (36 species) is relatively high, most especially in the southern Congo rivers. Characins (17 species) are less diverse but there is an equivalent representation of Citharinids (16 species). The largest family represented in the country is the Cyprinidae (79 species) and this is especially notable for the Atlantic coastal drainages (43 species) that is even richer than the Congo tributaries (27 species). However the chedrins (*Raiamas* and *Opsaridium* and *Engraulicypris*) are poorly represented – two species in Congo tributaries, one in the Zambezian region and one in the Atlantic coastal drainages. Of the catfishes, the claroteids (10 species) are a presence as are the clariids (17 species) of which the majority (11 species) are represented across different provinces. Other catfish families present include schilbeids (eight species) mochokids (15 species), amphiliids (six species) and one malapterurid. Cyprinodonts are relatively few (eight species) but show a particular relationship across the Cassai-Zambezi watershed. Cichlids (31 species) are well represented but not as well as the cyprinids. They are however more endemic in nature, in particular the Atlantic coastal fauna (19 species with eight endemic). The anabantids (three species) and mastacembelids (three species) are poorly represented.

The last phase of ichthyological exploration informing on the fishes of Angola, since independence in 1975, includes several taxonomic or systematic articles (e.g. Greenwood 1984, Musilová et al. 2013); published river faunal accounts (Skelton et al. 1985; Hay et al. 1997) and several unrestricted informal fish survey reports emanating from specific projects (Bills et al. 2012, 2013; Skelton et al. 2016). These surveys have exposed several new species to the fauna and together with phylogenetic studies on a wide range of lineages that include Angolan representatives, have led to a vastly improved understanding of the distributional nuances that give explanation to improved biogeographical insights.

Freshwater Drainages and Ecoregions of Angola

The drainages of Angola include southern source tributaries of the Congo, western source tributaries of the Zambezi, coastal rivers to the Atlantic from the Chiloango in Cabinda to the Cunene in the south, and the endorheic Etosha and Okavango Basin drainages in the south (Fig. 11.1). The watershed between the Congo system and the coastal Atlantic and Zambezian rivers is a major ichthyofaunal divide of considerable biogeographic significance (Poll 1967, Jubb 1967, Skelton 1994, Snoeks et al. 2011, Paugy et al. 2017).

The freshwater fishes of Angola fall within four major African ichthyological provinces (Fig. 11.1) – Lower Guinea, Congolese, Angolan (coastal) or Cuanza and Zambezian (Roberts 1975, Snoeks et al. 2011, Lévêque and Paugy 2017b).

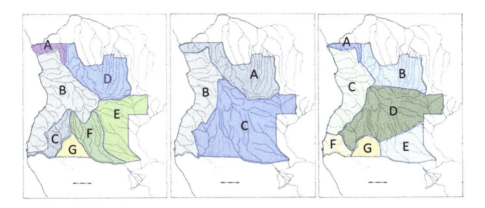

Fig. 11.1 *Left*: **Main drainage basins of Angola**. A: Lower Congo, B: Cuanzan or Atlantic Coastal, C: Cunene, D: Southwest Congo E: (west) Upper Zambezi, F: Okavango, G: Cuvelai. Chiloango in Cabinda not illustrated. *Center*: **Ichthyological provinces in Angola**, modified from Lévêque and Paugy (2017b) to include upper Cuanza and upper Cunene in the Zambezian Province. A: Congolian, B: Angolan or Cuanzan, C: Zambezian. Cabinda and the Chiloango River in the Lower Guinea Province not illustrated. *Right*: **Aquatic ecoregions in Angola**, modified, after Thieme et al. (2005). A – Lower Congo, B – Cassai, C – Cuanza, D – Zambezian headwaters, E – Okavango and Upper Zambezian floodplains, F – Namib coastal, G – Etosha. Southern West Coastal Equatorial (Cabinda) not illustrated

Previously Poll (1967) considered the freshwater fishes of Angola to be from five ichthyological regions drawn along watershed lines – Congo tributaries, Zambezi region, Angolan (western) coastal drainages excluding the Cunene, the Cubango-Okavango, and the Cunene. Thieme et al. (2005) defined ecoregions as "a large area containing a distinct assemblage of natural communities and species, whose boundaries approximate the original extent of natural communities before major land use change", and often reach across watershed lines. The Thieme et al. (2005) aquatic ecoregion map for Africa included Angolan inland waters within eight aquatic ecoregions as follows (Fig. 11.1): Floodplains, Swamps and Lakes – Region 12 Okavango Floodplains, Region 16 Upper Zambezi Floodplains; Moist Forest Rivers – Region 22 Lower Congo, Region 29 Southern West Coastal Equatorial; Savannah Dry Forest – Region 63 Cuanza, Region 76 Upper Zambezian headwaters; Xeric Systems – Region 82 Etosha, Region 88 Namib Coastal. Both the Ichthyological Provinces and the Ecoregions are convenient categories to consider the freshwater fishes of Angola.

Cabinda

Pellegrin (1928) recorded 28 species from the Chiloango River from Republic of Congo (formerly 'French' Congo). The freshwater fishes of Lower Guinea, Central West Africa that includes Cabinda were considered in detail through the two volumes edited by Stiassny et al. (2007). This rich Central West African fauna includes 555 species in 147 genera and 38 families of which 78 species, 52 genera and 25 families have been recorded from the Chiloango (also Shiloango) River in Cabinda (Appendix 1). This Atlantic coastal river is clearly boosted by the large number of marine and estuarine species that enter freshwaters sporadically or regularly (Whitfield 2007). A number of species from here such as *Enteromius holotaenia, Enteromius musumbi, Aplocheilichthys spilauchen* and *Oreochromis angolensis*, and estuarine species of marine origin also occur in the lower reaches of Angolan Atlantic coastal rivers, some to as far south as the Cunene River (Penrith 1982, Hay et al. 1997). Fowler (1930) recorded a number of species in the Academy of Natural Sciences of Philadelphia collection taken from the Chiloango region as it was known at the time. The fauna from the Chiloango in Angolan territory is likely underrepresented in most groups due to lack of sampling.

Lower Congo

There are no records available of fishes collected in Angolan waters of the Lower Congo mainstream or of the southern bank tributaries. The largest of these tributaries is the Inkisi River of which the fish fauna is known from the studies of Wamuini Lunkayilakio et al. (2010) supplemented by the description of new species described in association with that work (Wamuini Lunkayilakio and Vreven 2008, 2010). Based on these studies it is likely that most of the species in the DRC from the reaches above the Sanga Falls at least are likely to occur in Angola as well. The nature of the likely fauna of this neglected area of Angola as far as fish exploration is concerned (Appendix 2) indicates that the species are essentially of Congolian or Lower Guinean affinity with a few endemic species indicative of the isolation of fauna in the river reaches above the Sanga Falls. The widespread presence of *Oreochromis niloticus* is attributed to introduction for aquaculture (Wamuini Lunkayilakio et al. 2010).

Cuanza and Atlantic Coastal Rivers

Poll (1967) listed 110 species in 32 genera and 15 families from the Atlantic coastal region that includes the Cuanza River, which is revised (Appendix 3) to 105 species in 45 genera and 17 families in the light of more recent surveys in the Cuanza. There are very few species recorded from the Angolan Coastal rivers other than the Cuanza, and in areas north of the Bengo to the mouth of the Congo records from Angola are practically non-existent. Devaere et al. (2007) record *Channallabes apus* as being described from this region. Fowler (1930) noted species of the Cuanza and the Bengo rivers received from the British Museum on exchange, in many instances as described by Boulenger (1910) or as recorded in Boulenger

(1909–1916). Trewavas (1936) recorded and described seven species from a headwater stream of the Cuvo River arising on Mount Moco including the only *Amphilius* species (*Amphilius lentiginosus*) described from the region. A second undescribed *Amphilius* species has been recorded in the Cuanza (South African Institute for Aquatic Biodiversity – SAIAB – collection). Both these species differ in key morphological characteristics from the *Amphilius* of the Zambezian region that indicate their faunal connections lie primarily with the Lower Guinean or Congolean regions. Trewavas (1936) also described species from the Longa (*Enteromius breviceps*), the Catumbela (*Enteromius dorsolineatus, E evansi*), and the Balombo (*Enteromius dorsolineatus*). Pellegrin (1936) described the fishes collected by two Swiss expeditions (1928–1929 and 1932) made under the direction of Monard from the *Musée d'Histoire Naturelle de la Chaux-de-Fonds* included two species, *Enteromius kessleri, Clarias dumerilii*, that were drawn from the Cueve, with the bulk of the collections coming from the Cunene, the Cuvelai and the Cubango. Ladiges and Voelker (1961) described *Kneria maydelli* from the Cunene, and *Enteromius rousellei* and *Chiloglanis sardinhai* from the Longa. Poll (1967) described *Kneria sjolandersi* and *Chiloglanis angolensis* from the Bero, to the north of the Cunene. Trewavas (1973) recorded *Oreochromis angolensis* and *Tilapia cabrae* from the Bengo. Bills et al. (2012) made a small collection from the upper reaches of the Cueve River that included species of the following genera – *Petrocephalus, Enteromius, Labeobarbus, Micralestes, Amphilius, Chiloglanis, Clarias, Pharyngochromis, Thoracochromis, Tilapia, Coptodon,* and *Mastacembelus*. The list is typically 'Zambezian' and the species positively identified are closely linked to the upper Cuanzan and Cubango fauna. The indication from these references is therefore that the Angolan Coastal fauna is a mix of Lower Guinean (along the coastal plain) and Zambezian (above the escarpment) with some Congolean elements in the upper Cuanza/Lucala (see below).

The 'Cuanzan or Angolan Coastal' ichthyofaunal region is drawn primarily on what is known of the fishes of the Cuanza River as described by Boulenger (1910) and in Boulenger's (1909–1916) catalogue of fishes in the British Museum (Natural History) now the Natural History Museum (NHM). Fowler's (1930) account of the fishes of the Gray African Expedition in 1929 included records from the Bengo and the lower Cuanza, but also a collection of species from Chouzo on the upper reaches of the Cutato-Cuanza tributary, that provided a first strong indication that the fauna of these reaches is 'Zambezian' in character and different to those from the coastal reaches as reported by Boulenger (1910) and others. This association was later reiterated by Trewavas (1973) when considering the cichlid species of the Cuanza and Bengo rivers and has been firmly supported by the extensive surveys conducted by SAIAB and INIP (*Instituto Nacional de Investigação Pesqueira*) between 2005 and 2010. The current assessment records at least 102 species, some of which are undescribed (Appendix 3). The collections indicate that the river basin is even more heterogenous in fish faunal characteristics than simply 'lower' and 'upper' and the different zones distinguishable include (1) the lower reaches from the escarpment base to the sea, (2) the escarpment reaches, (3) the upper Cuanza and (4) the Lucala tributary, itself probably sub-zoned into the middle and upper reaches separated by the Calandula Falls (formerly 'Duque de Bragança' Falls).

Two ecophysiological components derive the fishes of the lower Cuanza: a diverse Tropical West African or Lower Guinean brackish water or marine component, and secondly the primary and secondary freshwater fishes. The known Tropical West African brackish water fishes from the system are generally widespread species and do not include endemics. Some species such as the Bull Shark (*Carcharhinus leucas*) and the Atlantic Tarpon (*Megalops atlanticus*) are well known as gamefish from this river. Two clupeid species include the recognised freshwater species (*Pellonula vorax* and *Odaxothrissa ansorgii*) and probably other brackish water forms. One haemulid (*Pomadasys* sp.) and one polynemid threadfin, possibly *Polydactylus quadrifilis* as known from Central West Africa (Snoeks and Vreven 2007), have been recorded (SAIAB records). Mullets (Mugilidae), as yet unidentified at species level, are present as are the sleepers (Eleotridae) and gobies (*Awaous* and *Periopthalmus*). Two pipefish have been positively identified: *Enneacampus ansorgii* and *Microphis brachyurus aculeatus*. The tonguefish *Cynoglossus senegalensis* was collected in the downstream reaches.

The freshwater species of this lower zone are mostly widespread species that also occur in coastal reaches of rivers to the north, well into the adjacent Lower Guinean Province and beyond, and many probably also to the south. An example of this is *Parailia occidentalis* that has a range through to the Senegal River in West Africa (de Vos 1995). The species that occur are found generally throughout the region to the escarpment, with a few ascending into middle Cuanza sections. Other characteristic species in this zone include mormyrids of the genera *Hippopotamyrus, Marcusenius* and *Petrocephalus*, the alestid *Alestes ansorgii*, cyprinids of the genus *Labeo*, two *Enteromius* species (*E. holotaenia* and *E. musumbi*), and several distinctive claroteid catfishes (two *Chrysichthys* species *C. acutirostris* and *C. ansorgii*), as well as *Schilbe bocagii*, and the widespread *Clarias gariepinus*. The *Chrysichthys* species confirm the West Africa coastal affinities of the assemblage as the genus is not known from the upper reaches nor from the upper Zambezian floodplain fauna. The cichlid fauna, as detailed systematically by Trewavas (1973) is in part also restricted to the zone – *Oreochomis angolensis, Hemichromis angolensis* and *Tilapia* cf. *cabrae*. The range of the procatopodid lampeye *Aplocheilichthys spilauchen* previously known from the Senegal River to the Bengo River has been extended to the Cuanza. The absence of the anabantid genus *Ctenopoma* from this zone is remarkable.

The Escarpment Zone of the Cuanza is characterised by a stepwise series of rapids, cascades and falls interspersed by rocky pools and runs. The fish fauna of this important zone for hydropower development is rich but relatively poorly known or described. The SAIAB-INIP collections are extensive and indicate that few species from the coastal zone penetrate high up into the zone. This is most probably partly an artefact of the Cambambe Dam near the base, in existence for several decades, which has likely affected the natural penetration of many species. The major freshwater families are represented; the smaller cyprinids, various catfish families, and cichlids are particularly well represented. The generic composition includes: *Hippopotamyrus, Petrocephalus, Marcusenius, Parakneria, Enteromius, Labeobarbus, Labeo, Raiamas, Brycinus, Rhabdalestes, Hepsetus, Schilbe, Chrysichthys, Clarias, Clariallabes, Parauchenoglanis, Chiloglanis, Synodontis, Micropanchax, Hemichromis, Pharyngochromis, Pseudocrenilabrus, Serranochromis, Tilapia, Oreochromis*, and *Mastacembelus*. Only a single *Labeobarbus* species was recorded in this zone during the survey and it also occurs in the Lucala tributary. Boulenger (1910) recorded two *Labeobarbus* species from the Cuanza at Dondo – *L. rocadasi* and *L gulielmi*. A unique morphotype of *Labeobarbus* with an extremely pointed tiny mouth, collected during the Capanda pre-impoundment surveys is present in the Luanda Museum (pers. obs., Fig. 11.2) and is likely to be an undescribed species.

The Upper Cuanza extends from a waterfall on the mainstream above the Capanda dam to the watershed and consists largely of relatively low-gradient floodplain rivers on Kalahari sand formations, similar to that of the upper reaches of the Zambezi and Okavango systems in Angola. Characteristic genera from this zone are: *Hippopotamyrus, Petrocephalus, Marcusenius, Parakneria, Enteromius, Labeobarbus, Labeo, Brycinus, Rhabdalestes, Hepsetus, Schilbe, Chrysichthys, Doumea, Clarias, Clariallabes, Parauchenoglanis, Chiloglanis, Synodontis, Micropanchax, Hemichromis, Pharyngochromis, Pseudocrenilabrus, Serranochromis, Tilapia, Oreochromis*, and *Mastacembelus*. Fishes from Chouzo in the upper Cuanza described by Fowler (1930), include species such as *Marcusenius angolensis, Hepsetus cuvieri, Labeo rocadasi, Enteromius evansi* (type locality), *Enteromius lujae* (identity of this species is still debated but the same species occurs in the Okavango headwaters), *Clarias gariepinus, Clarias theodorae* (as *C. fouloni*), *Clarias ngamensis* (as *Dinotopterus prentissgrayi*), *Ctenopoma machadoi* (type locality), *Serranochromis macrocephalus* (as *Tilapia acuticeps*, see Trewavas 1973). Norman (1923) described *Synodontis laessoei*, synonymised with *Synodonti nigromaculatus* by Poll (1971), as the only species of this genus in the Cuanza, contrast to the specious lineage in the Okavango-Zambezi region (Day et al. 20(Pinton et al. 2013). Few species characteristic of the upper Cuanza are found bey(the zone within the basin. This agrees with the notion that the fauna in this zor historically and biogeographically an integral part of the 'Zambezian' f

Fig. 11.2 An extraordinary undescribed *Labeobarbus* species from the Cu Luanda Museum, 2005. (Photo PH Skelton)

(Trewavas 1973). Ladiges (1964) and Poll (1967) showed this to be general for the fauna as a whole, and specific studies on species like *Hepsetus cuvieri* (Zengeya et al. 2011) and cichlids like *Serranochromis* and *Tilapia sparrmanii* (Musilová et al. 2013) confirm this relationship. Recent surveys across the watershed between the Cuanza and the Okavango indicate that a number of other species like *Parakneria fortuita*, and several *Enteromius* species like *E mocoensis*, *E evansi*, *E breviceps*, *E brevidorsalis* occur in streams on either side and have helped to define the Upper Zambezi headwaters ecoregion that embraces this trans-system conformance.

An early indication that the Lucala River, a major tributary that joins the system in the lower reaches, is exceptional for its fishes was the fauna collected by Ansorge using a wide range of methods including explosives (Boulenger 1910). It is however only in the escarpment and upper reaches that such exception occurs. An assemblage of large fishes of the genus *Labeobarbus* in particular is outstanding, and Boulenger (1910) described 12 species now in *Labeobarbus* (Vreven et al. 2016), all of which remain valid at this time. In addition to these species, unpublished barcode studies conducted by SAIAB on the fauna indicates that several lineages in the ystem are restricted to the Lucala, including an *Alestes*, *Pharyngochromis*, rranochromis, *Tilapia*, two *Enteromius species*, a *Parakneria*, *Hippopotamyrus*, a undescribed *Congoglanis*.

significance of the use of explosives in assembling Ansorge's collection
d by Boulenger (1910) is that it included a number of large mainstream spe-
otherwise are extremely difficult to collect. The assemblage of large
us described in the paper has defined the Cuanza Basin since that time.
faunal characteristics of the Lucala include species of the following
otamyrus, Petrocephalus, Kneria, Alestes, Enteromius, Labeobarbus,
Amphilius, Congoglanis, Schilbe, Clarias, Chiloglanis, Synodontis,
Pharyngochromis, Serranochromis, Tilapia, and Mastacembelus.
pper reaches is poorly known. Only a single collection made by
rom the Lucala above the Calandula Falls. This limited sample
ge the full character of the zone, but does indicate a degree of
dle Lucala zone, and differing through the absence of major
arbus species so characteristic of the latter. The physical
hes suggests there is a zonal distinction in the ecological
al elements. The known fauna includes species of the
otamyrus, Petrocephalus, Parakneria, 'Barbus',
glanis, Clarias, Micropanchax, Pharyngochromis,
be stated at this point except that an investigation
iven the unique nature of the Middle Lucala.

atershed with tributaries of the Congo-Cuango
ing reasons for its unique character. A high
is therefore evident and with further taxo-
d enhanced.

Cunene

Poll (1967), from the ichthyological perspective, treated the Cunene River system as a separate entity to the Atlantic coastal region, whereas it has been regarded as part of the Zambezian Province by Roberts (1975), part of the 'Angola' ichthyofaunal province by Lévêque and Paugy (2017a, b), and divided as part of the Namib aquatic ecoregion and part upper Zambezian headwaters ecoregions by Thieme et al. (2005). The reason for these varied treatments is that the river system is geo-eco-historically complex. Thus it has a dual geomorphological origin (the upper reaches being a natural part of the Kalahari Basin that has been captured by an Atlantic coastal river) and environmentally the lower reaches sit within the 'xeric' Namib region and the inland upper reaches within the savanna dry forest environs.

The fishes of the Cunene River are relatively well documented, starting with *Schilbe steindachneri* (a synonym of *S. intermedius)* and *Mormyrus anchietae* (a synonym of *M. lacerda*) described by Guimarães (1884), and summarised in the most recent checklist by Hay et al. (1997). Excluding the more strictly marine families there are 82 species recorded from the Cunene (Appendix 3). Hay et al. (1997) also record the broad distribution of species within the system according to three sections, the upper reaches down to Ruacana Falls, a middle section down to Epupa Falls and the lower river from below Epupa Falls to the mouth. Of 65 species recorded above Ruacana Falls 13 are restricted to that section. At least one species, *Marcusenius deserti*, is restricted to the lower reaches close to the coast (Kramer et al. 2016). Apart from the several marine species recorded in the extreme lower reaches by Penrith (1970) and Hay et al. (1997) that reflect a southernmost extension of the tropical (Lower Guinean) fauna, the general composition is clearly Zambezian in character. There are few representatives indicative of the Angolan (Cuanzan) Province, e.g. *Enteromius mattozi* (described by Guimarães (1884) from the Curoca River to the north of the Cunene). Pellegrin's (1921) *Enteromius* (formerly *Barbus) rohani*, probably a synonym of *E. mattozi*, was likely taken from the Caculovar River, a tributary of the Cunene, and not from the Lomba (neither the Zambezi as Pellegrin claimed, nor the Longa coastal Atlantic as suggested by Poll 1967). *Enteromius argenteus* is another minnow that has been reported from the Cunene but whose identity is unconfirmed – and is likely to be juveniles of *E. mattozi* (Skelton Unpublished Data).

There are also several isolated endemics from the system such as *Marcusenius deserti, Marcusenius magnoculis, Marcusenius multisquamatus, Hippopotamyrus longilateralis, Engraulicypris howesi, Zaireichthys cuneneensis, Orthochromis machadoi, Thoracochromis albolabrus, Thoracochromis buysi*, that suggests a degree of isolation probably reflecting older biogeographic connections. The absence of certain conspicuous families or genera such as *Parakneria, Labeobarbus, Opsaridium, Hydrocynus, Parauchenoglanis, Amphilius, Hemichromis,* and *Mastacembelus* is also noteworthy and perhaps indicative of a lack of more recent connections with the Zambezian and Cuanza systems.

Cassai and Southern Congo Rivers

Collections from the Lulua River, a tributary of the Cassai in Congo by Fowler (1930) whilst not strictly in Angola, probably pertain to Angola as well. Thus although not the only source, Poll (1967) is the current practical published source for the fishes of the southern Congo river tributaries in Angola. There are three main tributaries draining the region, from the east the Cassai including the Luangwe, the Cuilu (or Kwilu) and the Cuango. Poll (1967: 18–23) plotted the records of the fishes of each of these in his distribution table, recording 108, 28 and 37 species respectively and in the addendum supplemented the Cassai with three species and the Cuango with 24 species. The figure for the southern catchments of the Congo in Angola is now estimated at around 162 species (Appendix 3). The Cuilo and Cuango faunas are most evidently far from well explored. The Cassai fish fauna is better represented but still poorly explored, and includes species both typical of the Congo (e.g. *Polypterus ornatipinnis, Channallabes apus,* several mormyrid species, *Bryconaethiops microstoma, Alestes grandisquamis, Distichodus fasciolatus, Distichodus lusosso, Mastacembelus congicus*), and many species found also in the Upper Zambezi or the Okavango (e.g. *Hydrocynus vittatus, Hepsetus cuvieri, Pollimyrus castelnaui, Enteromius brevidorsalis, Parauchenoglanis ngamensis, Clarias stappersii, Clarias theodorae, Schilbe yangambianus, Micropanchax katangae, Oreochromis andersonii, Coptodon rendalli, Tilapia sparrmanii, Tilapia ruweti, Hemichromis elongatus, Serranochromis microcephalus, Serranochromis robustus jallae, Pseudocrenilabrus philander, Ctenopoma multispine, Microctenopoma intermedium*). The presence of *Dundocharax bidentatus* in the Cassai and the rare Zambezian endemic not yet found in Angola, *Neolebias lozii* are further good indicators of geographical connection. The strong Cassai-Zambezian faunal association is attributed to the clear evidence of hydrological pattern that the upper Cassai was formerly part of the Upper Zambezi system (Bell-Cross 1965).

Zambezian-Cuando-Cubango Headwaters and Floodplains

There is sufficient direct connection between the Zambezi, the Cuando and the Okavango river basins and similarity of the fish faunas in each to consider these under a single heading.

The Zambezi headwaters in Angola drain the Kalahari sand formation over an extensive divide with the Cassai to form major floodplains known as the Bulozi Floodplains. There are a number of lakes associated with the drainage including the largest freshwater lake in Angola, Lake Dilolo. The Okavango drainage is divided into two branches, the Cuito-Cuanavale in the east and the Cubango in the west. The Cuito-Cuanavale drains Kalahari sand formations giving rise to extensive low-gradient seepage bog and floodplain rivers in slump valleys extending into miombo savanna woodlands in the upper reaches. There are several lakes in these headwaters.

The Cubango branch arises as several relatively steep gradient rocky rivers in the Angolan highlands on the Bié plateaux before descending to the low-gradient reaches along the Namibian border to join with the Cuito before crossing to Botswana and forming the mostly endorheic Okavango Delta. The watershed of the system is shared with the Cuando, the Zambezi (mainly the Lungwe-Bungo), the Cueve-Cuanza, and the Cuanza as well as the Cunene and Cuvelai oshanas in the west.

The fishes of the Upper Zambezi are well studied and documented (e.g. Jackson 1961, Jubb 1961, 1967, Balon 1974, Bell-Cross and Minshull 1988, Tweddle 2010) with numbers now estimated at around 100–120 species (Appendix 3; Tweddle et al. 2004), possibly with as many as 20–25 undescribed. However published records from the Angolan territory are sparse, and limited in the published literature to Poll's (1967) 41 species (against his checklist of 62) taken mostly from two localities close to the watershed (Lagoa Calundo and the Longa-Luena tributary). Recent collections from the source reaches of Zambezian tributaries in Angola made by the National Geographic Okavango Wilderness Project (NGOWP 2018) are still being assessed but include 39 species from 12 families that have been included in the checklist of fishes from this region (Appendix 3). One notable new record is *Enteromius chiumbeensis* described by Poll (1967) from the Chiumbe River a tributary of the Cassai, reinforcing the close connections between these adjacent trans-watershed systems.

The upper Zambezian fish fauna is distinctive in several respects, most notably for the relatively speciose endemic *Synodontis* catfishes and the serranochromine cichlids (Trewavas 1964, Bell-Cross 1975, Greenwood 1993, Day et al. 2009; Pinton et al. 2013). To a large extent, in Angola, the fauna is ecologically tuned to the extensive seepage and floodplain drainages within a band of miombo savanna woodland on Kalahari sand deposits. Overall the known Angolan Upper Zambezi fish fauna is similar to that of the better-studied (in Angola) Okavango Basin fishes (often with the same or closely related species e.g., mormyrids of the genera *Hippopotamyrus, Marcusenius, Petrocephalus, Pollimyrus* – Kramer et al. 2003, 2004, 2012, 2014, and *Zaireichthys* species – Eccles et al. 2011). Whilst there are a few endemics, only one, *Paramormyrops jacksoni* Poll 1967 is restricted to Angola. The isolated *Neolebias lozii* is known only from the Barotse floodplains in Zambia.

Fishes of the Cuando-Linyanti-Chobe system have not been reported on from the Angolan section of that Zambezi tributary but van der Waal and Skelton (1984) provided a checklist of fishes in the Cuando River in Namibian waters. The 56 species recorded were all also found in the Zambezi system in Namibia. The Pallid Sand Catlet, *Zaireichthys pallidus* Eccles et al. (2011) is described from the Cuando but is not restricted to that system. Kramer et al. (2014) described a new species of *Pollimyrus* from the Cuando, a species possibly endemic to that tributary. Recent collections by the National Geographic Okavango Wilderness Project (NGOWP/SAIAB) from the upper reaches of the Cuando in Angola further inform the list of species (Appendix 3).

The fishes of the Okavango Basin have been studied and reported on in the literature for over 150 years since Castelnau (1861) described 14 species from Lake

Ngami, including the iconic Tigerfish (*Hydrocynus vittatus*) the Southern African Pike (*Hepsetus cuvieri*), the large Blunttooth Catfish (*Clarius ngamensis*) and the Three Spot Bream (*Oreochromis andersonii*). Fifty years later Boulenger (1911) reported on a collection from the Okavango-Lake Ngami made by RB Woosnam and described six new species including one named for Castelnau – *Pollimyrus castelnaui*. These fishes were all included in Gilchrist and Thompson (1913, 1917) and Boulenger (1909–1916). Fowler (1935) described a collection made from the Delta by the Vernay-Lang Expedition of 1930. Pellegrin (1936) described fishes collected by two Swiss expeditions of 1929 and 1933 from the Cunene, the Cuvelai and the Cubango. Barnard (1948) described in detail a collection from Rundu, Namibia. The results of all these efforts were summarised in checklists published by Poll (1967), Jubb (1967), Jubb and Gaigher (1971) and Skelton et al. (1985). More recently surveys of Angolan Okavango Basin rivers have been made (Bills et al. 2012, 2013, Skelton et al. 2016) that have reached little-explored areas, encountered additional species and provide for a more complete assessment of the fishes and their intra-basin distributions.

The additional species recently discovered include new species of *Clariallabes,* several serranochromine cichlids, and a dwarf climbing perch (*Microctenopoma* sp). Recent distribution records extend the range of several species from the Congo tributaries or in the case of *Clypeobarbus bellcrossi* from Zambezi headwaters in Zambia to the Okavango. Congolean species such as *Marcusenius moorii* (Günther) and *Enteromius chicapaensis* (Poll), and *Nannocharax lineostriatus* (Poll), and several *Micropanchax* as *M. luluae, M. nigrolateralis, M. lineolateralis.* The known range of a number of species from the Atlantic Coastal and Cuanza systems has been extended to the Okavango, e.g. *Enteromius breviceps, E. brevidorsalis, E. evansi, E. mocoensis, E. greenwoodi.* A new understanding of the complex distribution of the twin species *Enteromius trimaculatus* and *E. poechii* has also been reached – the former being found in the Cunene and the extreme upper reaches of the Cubango in place of the latter which is widespread in the downstream floodplain reaches of the Okavango and Upper Zambezi system.

Cuvelai

The Cuvelai drainage lies in a triangle between the Okavango in the east and the Cunene in the west and the streams known as 'iishanas' are intermittent, only flowing during periods of sustained rainfall into the endorheic Etosha Pan in Namibia (van der Waal 1991, Hipondoka et al. 2018). The 1929 and 1932–1933 Swiss expeditions to Angola collected the following species from Mupa (Pellegrin 1936): *Marcusenius altidorsalis* (?), *Mormyrus lacerda, Enteromius paludinosus, Tilapia sparrmanii,* and *Pseudocrenilabrus philander.* Seventeen species, all conforming to Cunene fauna, have been confirmed from the western iishanas of the system by Hipondoka et al. (2018), and connections with the Cunene substantiated through remote sensing techniques. Four widespread pioneering species are consistently present in collections, viz., *Clarias gariepinus, Enteromius paludinosus, Oreochromis andersonii* and *Pseudocrenilabrus philander* and several others are common – *Clarias ngamensis, Schilbe intermedius* and *Enteromius trimaculatus*.

Biogeography

The biogeography of Angolan freshwater fishes is closely tied to the geomorphology and the geomorphological history of the territory. In brief, Angola consists of a narrow coastal plain, a distinct escarpment and an interior plateau that is being eroded most rapidly from the Congo Basin. The coastal plain consists of a series of rivers flowing from the escarpment or – in the case of the Congo in the north, the Cuanza in the middle and the Cunene in the south – where the escarpment has been penetrated, from the interior plateau or the Congo Basin. The fish fauna of the coastal plain is primarily a southern extension of the tropical coastal fauna of West Africa and Central West Africa. River connections along this narrow strip are either via sea-level fluctuations or via river captures between watersheds, either as adjacent systems or via extended reaches through captured inland drainages that are not determined by the coastal gradients and processes. According to Lévêque and Paugy (2017a,b) the primary direction of dispersal of the coastal west African fauna was northwards from the Congo. Present day ocean currents off Angola are counter clockwise (http://oceancurrents.rsmas.miami.edu) and it is possibly only inshore counter currents that might have facilitated faunal dispersal southwards from the Congo, especially after the capture and penetration of the Congo Basin by the Lower Congo in the late Cretaceous (Flügel et al. 2015). Such would certainly explain much of the marine derived elements of the region. Given favourable currents it is likely also that the considerable volumes of freshwater entering the sea from the Congo at various times would facilitate even freshwater fishes down the coast and might explain the presence of such species as *Enteromius musumbi, Physailia occidentalis, Chysichthys spp, Oreochromis angolensis* and *Aplocheilichthys spilauchen* in the Angolan region. An alternative and complementary explanation for some freshwater faunal elements such as *Marcusenius deserti* and *Raiamas ansorgii* of the Angolan Coastal reaches is that it is primarily derived via the Cuanzan and Cunene gateways through capture of portions of the Kalahari Basin drainage. It is not only the Cuanza and the Cunene that have breached the escarpment but also the Cuvo and the Longa and possibly others, as is evident in the list of freshwater fishes reported from these lesser rivers (see above).

The evolution of the extensive Kalahari Basin is certainly key to understanding the majority of the freshwater fish fauna of Angola. Haddon and McCarthy (2005), Key et al. (2015), Moore and Larkin (2001), and Moore et al. (2012) sketch the evolution of the Kalahari Basin and its drainage since the breakup of Gondwanaland and the isolation of Africa in the late Cretaceous. Following rifting, the continental margins were probably elevated and this formed an escarpment that separated the narrow coastal plain from the elevated Kalahari sedimentary basin that was drained primarily by the palaeo-upper Zambezi, the predominant system in the Angolan region (Fig. 11.3). The western portion of the system flowed from the escarpment highlands of the extreme northwest of the basin, now part of the Cuanza, generally southeast through the Limpopo valley to the Indian Ocean. The eastern parts of the upper Zambezi reached northeastwards to as far as pre-rift East African plateaus and

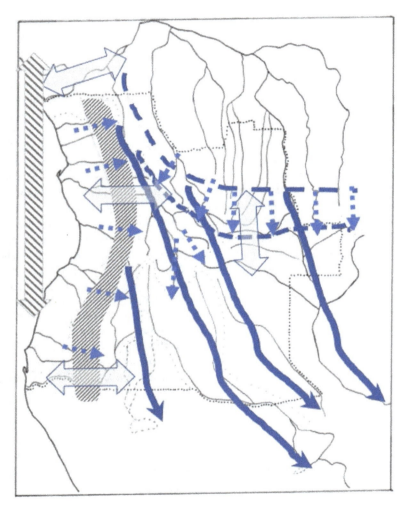

Fig. 11.3 A diagrammatic model for the post-Cretaceous biogeography of Angolan freshwater fishes. Angolan border – fine dotted line; present day drainage – thin lines; present day inter basin watersheds – open dotted lines; paleo drainage lines – thick extended arrows; paleo and present escarpment retreat – dashed arrows; paleo and present south and southwestern Congo Basin watershed – thick dashed lines; Angolan escarpment – right slanted hash; gateway drainage captures – large open bi-directional arrows. Coastal dispersal of fishes – large left-slanted bi-directions arrow. The model is based on geomorphological interpretations by Flügel et al. (2015), Haddon and McCarthy (2005), Moore and Larkin (2001), Moore et al. (2012), and others

included the proto-Luangwa and the proto-Chambeshi-Kafue-upper Zambezi as well as the Okavango. These drained into an interior basin to form, at times, a mega palaeo lake – Palaeolake Magadigadi (Burroughs et al. 2009, Moore and Larkin 2001, Moore et al. 2012, Podgorski et al. 2013). The proto-Cunene consisted of an upper portion draining endorheically to the Etosha basin. The most significant events in the history of the Kalahari Basin were firstly the downwarping and back-tilting of drainage coupled with upwarping along the southern margins that severed the initial Indian Ocean outlet via the Limpopo; the tapping of the Congo Basin by the lower Congo River that advanced the erosion and southern retreat of the northern watershed of the basin, especially in the northeast (Luapula-Chambeshi) and, in the Angola area, the Cassai-Zambezi. The dismemberment and tapping of drainage portions from the Kalahari Basin to coastal outlets including the Cuanza, the Cunene, and the Zambezi also affects the biogeographical history significantly (Moore and Larkin 2001, Moore et al. 2012, Key et al. 2015).

The most profound biogeographic significance to emerge from this geomorpho logical narrative is that the Kalahari Basin has been an evolutionary basin for fishes over a long period of time. The evidence is exemplified in the serranochromine cichlid radiation and the clade of *Synodontis* catfish and the radiation of several mormyrid genera that characterise the Zambezian fauna (Bell-Cross 1975, Greenwood 1984, Kramer et al. 2003, 2004, Day et al. 2009, Kramer and Swartz 2010, Kramer et al. 2012, Schwarzer et al. 2012, Pinton et al. 2013, Kramer and Wink 2013). The strong identity of the upper Zambezian fauna further exemplifies this notion. That the fauna has been supplemented with species from neighbouring ichthyological provinces, especially the Congo, is also evident in species or genera with internally restricted distributions such as *Hepsetus cuvieri, Hydrocynus vittatus, Parauchenoglanis ngamensis, Mastacembelus, Hemichromis elongatus, Amphilius* and others. The broader distributions of some species into basins like the east coast rivers (e.g. *Enteromius bifrenatus, Microctenopoma intermedium, Clarias theodorae, Brycinus lateralis*) gives biological credence to the former east coast linkage and subsequent drainage dismemberment on the proto-upper Zambezi (Skelton 1994, 2001).

There are other emerging details of biogeographical interest to Angola that will in time lead to a detailed accounting of the origins and development of the freshwater fishes. Thus the presence of doumeine catfishes in the Cuanza, southwest of the Congo, indicates clearly insemination from the Congo. The flock of *Labeobarbus* species in the Lucala-Cuanza probably also indicates a Congolian insemination. However the assumption that all traffic was from the Congo is not necessarily correct and *Neolebias bidentatus* in the Cassai, for example, as with other 'Zambezian' elements in that system, more likely reflects a Zambezian (i.e. Kalahari) insemination to the Congo. This, in essence, is the basis of the 'Upper Zambezi headwater' freshwater ecoregion (Fig. 11.1: basin C).

Conservation

Angola is an emerging African economy with a rapidly growing human population and increasing demand on freshwater resources. The rapid population growth and expansion of urban areas in places such as Luanda but also in the more rural districts (Mendelsohn and Weber 2015) is placing an ever increasing stress on the environment, especially that of the rivers for which such urban growth centres are dependent on for water and power. Although many Angolan rivers are relatively unregulated there are dams on several systems such as the three major hydroelectric dams on the Cuanza. A further four hydroelectric dams are planned for the escarpment section of this system alone. In the case of certain transboundary rivers like the Okavango, the threat of increased river regulation is of serious concern to the integrity of the Okavango Delta in Botswana, a World Heritage and Ramsar site (King and Chonguic 2016).

Diamond mining activities along the southern Congo tributaries have had environmental impacts of unknown severity as practically no public investigations or information is available.

With human populations, urbanisation and development comes pollution and other direct threats to aquatic life such as fishing and the introduction of invasive alien species. Few alien fishes have been recorded from Angola, but two species that have been introduced are *Oreochromis mossambicus* (SAIAB, in the Cuanza) and *Oreochromis niloticus* in Cabinda and, as recently confirmed, in the upper Cubango. The threats these particular species pose as aliens is well documented (e.g. Wise et al. 2007, Zengeya et al. 2013, Bbole et al. 2014). This is the first record of an alien species with high impact potential in the Okavango system and the threat posed is transboundary in nature. Potential transboundary threats from outside Angola include that of alien crayfish from the Zambezi (Nunes et al. 2016).

Indigenous fishery practices in Angola include a range of gear ranging from simple traps to elaborate fishing fences and walls (Poll 1967, Mendelsohn and Weber 2015). In places such traditional practices are still in evidence (Fig. 11.4 *top*), but elsewhere traditional practices are being replaced by modern gear such as monofilament gillnets and mosquito-net seines (Fig. 11.4 *bottom*) that are excessively destructive and unsustainable (Tweddle et al. 2015).

The current IUCN redlist assessments for Angolan freshwater fishes (Appendix 3) reflects the relatively weak knowledge of the species – a third of the known species are either not assessed or are Data Deficient (DD). One species (*Oreochromis lepidurus*) is listed as Endangered (see Moelants 2010), three are Vulnerable (1%) and 185 (65%) are Least Concern. The endangered species is a Lower Congo endemic found mainly in the DRC and is primarily threatened by oil pollution derived from boats. The Vulnerable species are also cichlids of the genus *Oreochromis* – *O. andersonii* (see Marshall and Tweddle 2007) and *O. macrochir* (see Marshall and Tweddle 2007), both are threatened through hybridisation from the alien invasive species *Oreochromis niloticus*. The latter species has recently been confirmed as present in Angola, within the Okavango catchment and its impact

Fig. 11.4 *Top* – traditional fishing fence on the Cacuchi River, 2012 (Photo PH Skelton). *Bottom* – Drying fish caught with monofilament gillnets on the Cuito River, 2015. (Photo G Neef)

on the native *Oreochromis* is now an imminent threat. Given the situation of rapidly escalating changes to the natural aquatic environment in Angola it is likely that the IUCN redlist score for the country will rise rapidly.

Acknowledgements I am supported in my research by the Director and staff of SAIAB, in particular Roger Bills and members of the collections division, administration staff, and by Maditaba Meltaf in the library for the provision of literature. Steve Boyes and John Hilton of the Wild Bird Trust have provided me with excellent opportunity to study fishes in Angola since 2015. I have been supported in the field and laboratory by Adjani Costa, Roger Bills, Ben van der Waal, Götz Neef and others of the National Geographic Okavango Wilderness Project. SAIAB engagement with Angolan fishes was initiated in 2005 in partnership with INIP (Instituto Nacional de Investigação Pesqueira). Ernst R Swartz (SAIAB) and D Neto (INIP) were instrumental in opening the channels of new knowledge on Angolan freshwater fishes.

Appendices

Appendix 1

Freshwater and brackish water fishes of Shiloango River, Cabinda, as recorded by Stiassny et al. (2007)

Species	Author & Date
Clupeidae	
Pellonula vorax	Günther, 1868
Mormyridae	
Isichthys henryi	Gill, 1863
Marcusenius moorii	Günther, 1863
Paramormyrops kingsleyae	(Günther, 1863)
Brienomyrus brachyistius	(Gill, 1862)
Hepsetidae	
Hepsetus lineatus	(Pellegrin, 1926)
Alestidae	
Brycinus longipinnis	(Günther, 1864)
Brycinus macrolepidotus	Valenciennes, 1850
Brycinus kingsleyae	(Günther, 1896)
Nannopetersius ansorgii	(Boulenger, 1910)
Distichodontidae	
Distichodus notospilus	Günther, 1867
Eugnathichthys macroterolepis	Boulenger, 1899
Nannaethiops unitaeniatus	Günther, 1872
Nannocharax parvus	Pellegrin, 1906
Neolebias ansorgii	Boulenger, 1912
Neolebias spilotaenia	Boulenger, 1912

(continued)

Angolan Freshwater Fishes: A Biogeographic Model

Species	Author & Date
Cyprinidae	
Enteromius carens	(Boulenger, 1912)
Enteromius jae	(Boulenger, 1903)
Enteromius guirali	(Thominot, 1886)
Enteromius callipterus	(Boulenger, 1907)
Enteromius camptacanthus	(Bleeker, 1863)
Enteromius rubrostigma	(Poll & Lambert, 1964)
Enteromius holotaenia	(Boulenger, 1904)
Labeobarbus aspius	(Boulenger, 1912)
Labeobarbus cardozoi	(Boulenger, 1912)
Labeobarbus roylii	(Boulenger, 1912)
Labeobarbus batesii	(Boulenger, 1903)
Labeobarbus sandersi	(Boulenger, 1912)
Labeo batesii	Boulenger, 1911
Labeo lukulae	Boulenger, 1902
Opsaridium ubangiense	(Pellegrin, 1901)
Ariidae	
Arius latiscutatus	Günther, 1864
Claroteidae	
Anaspidoglanis macrostoma	(Pellegrin, 1909)
Parauchenoglanis altipinnis	(Boulenger, 1911)
Chrysichthys auratus	(Geoffroy Saint-Hilaire, 1809)
Chrysichthys nigrodigittatus	(Lacépède, 1803)
Schilbeidae	
Parailia occidentalis	(Pellegrin, 1901)
Pareutropius debauwi	(Boulenger, 1900)
Clariidae	
Clarias angolensis	Steindachner, 1866
Clarias gabonensis	Günther, 1867
Malapteruridae	
Malapterurus beninensis	Murray, 1855
Procatopodidae	
Aplocheilichthys spilauchen	(Duméril, 1861)
Plataplochilus loemensis	(Pellegrin, 1924)
Nothobranchiidae	
Epiplatys singa	(Boulenger, 1899)
Aphyosemion escherischi	(Ahl, 1924)
Anabantidae	
Ctenopoma nigropannosum	Reichenow, 1875
Microctenopoma ansorgii	(Boulenger, 1912)
Microctenopoma nanum	(Günther, 1896)
Microctenopoma congicum	(Boulenger, 1887)

(continued)

Species	Author & Date
Cichlidae	
Pelvicachromis subocellatus	(Günther, 1872)
Chilochromis duponti	Boulenger, 1902
Coptodon tholloni	(Sauvage, 1884)
Pelmatolapia cabrae	(Boulenger, 1899)
Coptodon guineensis	(Günther, 1862)
Oreochromis schwebischi	(Sauvage, 1884)
Sarotherodon nigripinnis	(Guichenot, 1861)
Lutjanidae	
Lutjanus dentatus	(Duméril, 1861)
Monodactylidae	
Monodactylus sebae	(Cuvier, 1829)
Polynemidae	
Polydactylus quadrifilis	(Cuvier, 1829)
Mugilidae	
Mugil bananensis	(Pellegrin, 1927)
Neochelon falcipinnis	(Valenciennes, 1836)
Chelon dumerili	(Steindachner, 1870)
Eleotridae	
Eleotris daganensis	Steindachner, 1870
Eleotris senegalensis	Steindachner, 1870
Eleotris vittata	Duméril, 1861
Bostrychus africanus	(Steindachner, 1879)
Dormitator lebretonis	(Steindachner, 1870)
Gobiidae	
Periopthalmus barbarus	(Linnaeus, 1766)
Gobionellus occidentalis	(Boulenger, 1909)
Bathygobius soporator	(Valenciennes, 1837)
Bathygobius casamancus	(Rochebrune, 1880)
Nematogobius maindroni	(Sauvage, 1880)
Microdesmidae	
Microdesmus aethiopicus	(Chabanaud, 1927)
Mastacembelidae	
Mastacembelus shiloangoensis	(Vreven, 2004)
Mastacembelus niger	Sauvage, 1879
Syngnathidae	
Enneacampus ansorgii	(Boulenger, 1910)
Microphis aculeatus	(Kaup, 1856)
Cynoglossidae	
Cynoglossus senegalensis	(Kaup, 1858)
Citharichthys stampflii	(Steindachner, 1894)

Appendix 2

Freshwater fishes of the Inkisi River DRC, from above the Sangha waterfalls, after Wamuini Lunkayilakio et al. (2010)

Species	Author & Date
Mormyridae	
Hippopotamyrus cf. ansorgii	(Boulenger, 1905)
Paramormyrops cf. kingsleyae	(Günther, 1896)
Paramormyrops cf. sphekodes	(Sauvage, 1879)
Cyprinidae	
Enteromius miolepis	(Boulenger, 1902)
Enteromius unitaeniatus	(Günther, 1867)
Enteromius vandersti	(Poll, 1945)
Garra congoensis	Poll, 1959
Labeo macrostomus	Boulenger, 1898
Labeobarbus sp. nov.	
Labeobarbus boulengeri	Vreven, Musschoot, Snoeks & Schliewen, 2016
Labeobarbus robertsi	(Banister, 1984)
Raiamas kheeli	Stiassny, Schelly & Schliewen, 2006
Alestidae	
Nannopetersius mutambuei	Wamuini Lunkayilakio & Vreven, 2008
Claroteidae	
Parauchenoglanis balayi	(Sauvage, 1879)
Clariidae	
Clarias angolensis	Steindachner, 1866
Clarias buthupogon	Sauvage, 1879
Clarias camerunensis	Lönnberg, 1895
Clarias gariepinus	(Burchell, 1822)
Clarias gabonensis	Günther, 1867
Schilbeidae	
Schilbe zairensis	de Vos, 1995
Cichlidae	
Haplochromis snoeksi	Wamuini Lunkayilakio & Vreven, 2010
Hemichromis elongatus	(Guichenot, 1861)
Oreochromis niloticus	(Linnaeus, 1758)
Sarotherodon galilaeus	(Linnaeus, 1758)
Coptodon tholloni	(Sauvage, 1884)
Anabantidae	
Ctenopoma nigropannosum	Reichenow, 1875
Chanidae	
Parachanna obscura	(Günther, 1861)

Appendix 3

Freshwater fishes of the (A) Cuanza (Atlantic coastal), (C) southern Congo, (Z) Upper Zambezian, (O) Okavango, and (K) Cunene basins in Angola, after Poll (1967) with updated adjustments for taxonomy and known records by the author

Species	Author & Date	A	C	Z	O	K	I[a]
Protopteridae							
Protopterus aethiopicus	Heckel, 1851		x				DD
Protopterus dolloi	Boulenger, 1900		x				LC
Polypteridae							
Polypterus ornatipinnis	Boulenger, 1902		x				LC
Clupeidae							
Pellonula vorax	Günther, 1868	x	x				LC
Odaxothrissa ansorgii	Boulenger, 1910	x	x				LC
Kneriidae							
Kneria angolensis	Steindachner, 1866	x	x	?			LC
Kneria maydelli	Ladiges & Voelker, 1961					x	LC
Kneria polli	Trewavas, 1936	x	x				LC
Kneria sjolandersi	Poll, 1967	x					DD
Kneria ansorgii	(Boulenger, 1910)	x	x				DD
Parakneria marmorata	(Norman, 1923)	x					DD
Parakneria vilhenae	Poll, 1965		x				DD
Parakneria fortuita	Penrith, 1973	x		x	x		DD
Mormyridae							
Mormyrops attenuatus	Boulenger, 1898		x				LC
Mormyrops anguilloides	(Linnaeus, 1758)		x				LC
Petrocephalus okavagoensis	Kramer et al., 2012			x	x		NE
Petrocephalus magnitrunci	Kramer et al., 2012				x		NE
Petrocephalus magnoculis	Kramer et al., 2012					x	NE
Petrocephalus longicapitis	Kramer et al., 2012			x	x		NE
Petrocephalus christyi	Boulenger, 1920		x				NE
Petrocephalus cunganus	Boulenger, 1910	x					DD
Petrocephalus micropthalmus	Pellegrin, 1909		x				LC
Petrocephalus simus	Sauvage, 1879	x	x	?			LC
Hippopotamyrus ansorgii	(Boulenger, 1905)	x		x	x		LC
Hippopotamyrus longilateralis	Kramer & Swartz, 2010					x	NE
Pollimyrus brevis	(Boulenger, 1913)		x				LC
Pollimyrus castelnaui	(Boulenger, 1911)			x	x	x	LC
Pollimyrus cuandoensis	Kramer, van der Bank & Wink, 2013			x			NE
Pollimyrus marianne	Kramer et al., 2003			x			NE
Cyphomyrus cubangoensis	(Pellegrin, 1936)			x	x		NE

(continued)

Angolan Freshwater Fishes: A Biogeographic Model

Species	Author & Date	A	C	Z	O	K	I^a
Cyphomyrus psittacus	(Boulenger, 1897)		x				LC
Paramormyrops jacksoni	(Poll, 1967)			x			DD
Marcusenius altisambesi	Kramer et al., 2007		x	x	x		LC
Hippopotamyrus pappenheimi	(Boulenger, 1910)	x					LC
Heteromormyrus pauciradiatus	(Steindachner, 1866)	x					DD
Pollimyrus tumifrons	(Boulenger, 1902)		x				NE
Marcusenius desertus	Kramer, vanderBank & Wink, 2016					x	NE
Marcusenius multisquamatus	Kramer & Wink, 2013					x	NE
Marcusenius angolensis	(Boulenger, 1905)	x	x	x	x	x	LC
Marcusenius cuangoanus	(Poll, 1967)		x				VU
Marcusenius dundoensis	(Poll, 1967)		x				DD
Marcusenius moorii	(Günther, 1867)		x				LC
Marcusenius stanleyanus	(Boulenger, 1897)		x				LC
Campylomormyrus alces	(Boulenger, 1920)		x				LC
Campylomormyrus cassaicus	(Poll, 1967)		x				DD
Campylomormyrus elephas	(Boulenger, 1898)		x				LC
Campylomormyrus numenius	(Boulenger, 1898)		x				LC
Campylomormyrus luapulaensis	(David & Poll, 1937)		x				DD
Campylomormyrus rhynchophorus	(Boulenger, 1898)		x				LC
Campylomormyrus tshokwe	(Poll, 1967)		x				LC
Gnathonemus barbatus	Poll, 1967		x				DD
Gnathonemus petersii	(Günther, 1862)		x				LC
Mormyrus caballus	Boulenger, 1898		x				NE
Mormyrus lacerda	Castelnau, 1861	x		x	x	x	LC
Mormyrus rume	Valenciennes, 1847		x				NE
Cyprinidae							
Garra dembeensis	(Rüppell, 1835)		x				LC
Clypeobarbus bellcrossi	(Jubb, 1965)			x	x		DD
Coptostomabarbus wittei	David & Poll, 1937			x	x		LC
Enteromius afrovernayi	(Nichols & Boulton, 1927)		x	x	x	x	LC
Enteromius amphigramma	(Boulenger, 1903)	x					
Enteromius ansorgii	(Boulenger, 1904)		x				LC
Enteromius argenteus	(Günther, 1868)	x				x	LC
Enteromius barotseensis	(Pellegrin, 1920)			x	x	x	LC
Enteromius barnardi	(Jubb, 1965)			x	x	x	LC
Enteromius bifrenatus	(Fowler, 1935)			x	x	x	LC
Enteromius breviceps	(Trewavas, 1936)	x			x	x	LC
Enteromius brevidorsalis	(Boulenger, 1915)	x	x	x	x	x	LC
Enteromius brevilateralis	(Poll, 1967)	x	x				DD
Enteromius caudosignatus	(Poll, 1967)		x				DD

(continued)

Species	Author & Date	A	C	Z	O	K	I[a]
Enteromius chicapaensis	(Poll, 1967)		x		x		LC
Enteromius chiumbeensis	(Pellegrin, 1936)		x		x		LC
Enteromius dorsolineatus	(Trewavas, 1936)	x				x	LC
Enteromius eutaenia	(Boulenger, 1904)	x	x	x	x	x	DD
Enteromius evansi	(Fowler, 1930)	x			x		LC
Enteromius fasciolatus	(Günther, 1868)	x	x	x	x	x	LC
Enteromius greenwoodi	(Poll, 1967)	x			x		DD
Enteromius haasianus	(David, 1936)	x	x	x	x		LC
Enteromius holotaenia	(Boulenger, 1904)	x	x				LC
Enteromius kerstenii	(Peters, 1868)			x	x	x	LC
Enteromius kessleri	(Steindachneri, 1866)	x	x	x	x		LC
Enteromius lineomaculatus	(Boulenger, 1903)		x	x	x	x	LC
Enteromius lujae	(Boulenger, 1913)	x	x	x	x	x	DD
Enteromius machadoi	(Poll, 1967)		x				DD
Enteromius mattozi	(Guimarães, 1884)	x	x			x	LC
Enteromius mediosquamatus	(Poll, 1967)		x				DD
Enteromius miolepis	(Boulenger, 1902)		x	x	x		LC
Enteromius mocoensis	(Trewavas, 1936)	x			x		DD
Enteromius multilineatus	(Worthington, 1933)		x	x	x	x	LC
Enteromius musumbi	(Boulenger, 1910)	x					LC
Enteromius paludinosus	(Peters, 1852)	x	x	x	x	x	LC
Enteromius petchkovski	(Poll, 1967)		x				DD
Enteromius poechii	(Steindachneri, 1911)	?	x	x		x	LC
Enteromius radiatus	(Peters, 1853)	x	x	x	x	x	LC
Enteromius rousellei	(Ladiges & Voelker, 1961)		x				DD
Enteromius thamalakanensis	(Fowler, 1935)			x	x	x	LC
Enteromius trimaculatus	(Peters, 1852)	x	x		x	x	LC
Enteromius unitaeniatus	(Günther, 1867)	x	x	x	x	x	LC
Enteromius cf viviparus	(Weber, 1897)			x	x	x	NE
Enteromius wellmani	(Boulenger, 1911)	x					DD
Labeobarbus caudovittatus	(Boulenger, 1902)		x				LC
Labeobarbus codringtonii	(Boulenger, 1908)		x	x	x		LC
Labeobarbus ensis	(Boulenger, 1910)	x					LC
Labeobarbus gulielmi	(Boulenger, 1910)	x					DD
Labeobarbus girardi	(Boulenger, 1910)	x					DD
Labeobarbus jubbi	(Poll, 1967)		x				DD
Labeobarbus lucius	(Boulenger, 1910)	x					DD
Labeobarbus marequensis (Cassai)	(Smith, 1841)		x				LC
Labeobarbus nanningsi	de Beaufort, 1933	x	x				DD
Labeobarbus rhinophorus	(Boulenger, 1910)	x					DD
Labeobarbus rocadasi	(Boulenger, 1910)	x					DD
Labeobarbus rosae	(Boulenger, 1910)	x					DD

(continued)

Angolan Freshwater Fishes: A Biogeographic Model

Species	Author & Date	A	C	Z	O	K	Iᵃ
Labeobarbus ansorgii	(Boulenger, 1906)	x					LC
Labeobarbus ensifer	(Boulenger, 1910)	x					LC
Labeobarbus boulengeri	Vreven et al., 2016	x					NE
Labeobarbus macrolepidotus	(Pellegrin, 1928)		x				LC
Labeobarbus steindachneri	(Boulenger, 1910)	x					LC
Labeobarbus stenostomata	(Boulenger, 1910)	x					DD
Labeobarbus varicostoma	(Boulenger, 1910)	x					DD
Labeo annectens	Boulenger, 1903	x	x				LC
Labeo ansorgii	Boulenger, 1907	x	x			x	LC
Labeo chariensis	Pellegrin, 1904		x				LC
Labeo cylindricus	Peters, 1852		x	x	x		LC
Labeo greeni	Boulenger, 1902		x		?		LC
Labeo lineatus	Boulenger, 1898		x				LC
Labeo longipinnis	Boulenger, 1898		x				LC
Labeo macrostoma	Boulenger, 1898		x				LC
Labeo parvus	Boulenger, 1902	x	x				LC
Labeo rocadasi	Boulenger, 1907	x					LC
Labeo ruddi	Boulenger, 1907					x	LC
Labeo velifer	Boulenger, 1898		x				NE
Labeo weeksii	Boulenger, 1909		x				LC
Engraulicypris howesi	Ridden, Bills & Villet, 2016					x	NE
Opsaridium zambezense	(Peters, 1852)		x	x	x		LC
Raiamas ansorgii	(Boulenger, 1910)	x					DD
Raiamas christyi	(Boulenger, 1920)		x				LC
Hepsetidae							
Hepsetus cuvieri	(Castelnau, 1861)	x	x	x	x	x	NE
Alestidae							
Bryconaethiops microstoma	Günther, 1873		x				LC
Alestes macropthalmus	Günther, 1867		x				LC
Brycinus kingsleyae	(Günther, 1896)		x				LC
Brycinus grandisquamis	(Boulenger, 1899)		x				LC
Brycinus humilis	(Boulenger, 1905)	x	x				DD
Brycinus imberi	(Peters, 1852)	?	x				LC
Brycinus lateralis	(Boulenger, 1900)		x	x	x	x	LC
Micralestes acutidens	(Peters, 1852)		x	x	x		L
Micralestes argyrotaenia	Trewavas, 1936					x	J
Micralestes humilis	Boulenger, 1899		x				
Nannopetersius ansorgii	(Boulenger, 1910)	x					
Rhabdalestes maunensis	(Fowler, 1935)			x	x	x	
Hydrocynus vittatus	Castelnau, 1861		x	x	x		
Distichodontidae							
Distichodus fasciolatus	Boulenger, 1898		x				
Distichodus lusosso	Schilthuis, 1891		x				
Distichodus maculatus	Boulenger, 1898		x				

Species	Author & Date	A	C	Z	O	K	I[a]
Distichodus notospilus	Günther, 1867		x				LC
Distichodus sexfasciatus	Boulenger, 1897		x				LC
Nannocharax macropterus	Pellegrin, 1926		x	x	x		LC
Nannocharax procatopus	Boulenger, 1920		x				LC
Nannocharax angolensis	(Poll, 1967)		x				LC
Nannocharax lineostriatus	(Poll, 1967)		x	x	x		DD
Nannocharax machadoi	(Poll, 1967)			x	x	x	LC
Nannocharax multifasciatus	Boulenger, 1923			x	x	x	DD
Dundocharax bidentatus	Poll, 1967		x				DD
Claroteidae							
Chrysichthys ansorgii	Boulenger, 1910	x					LC
Chrysichthys bocagii	Boulenger, 1910	x					LC
Chrysichthys cranchii	(Leach, 1818)		x				LC
Chrysichthys delhezi	Boulenger, 1899		x				LC
Chrysichthys macropterus	Boulenger, 1920		x				DD
Chrysichthys nigrodigitatus	(Lacepède, 1803)	x					LC
Parauchenoglanis ngamensis	(Boulenger, 1911)		x	x	x		LC
Amphiliidae							
Zaireichthys dorae	(Poll, 1967)		x				DD
Zaireichthys flavomaculatus	(Pellegrin, 1926)		x				DD
Zaireichthys pallidus	Eccles, Tweddle & Skelton, 2011			x	x		NE

Angolan Freshwater Fishes: A Biogeographic Model

Species	Author & Date	A	C	Z	O	K	Iᵃ
Clarias dumerilii	Steindachner, 1866	x	x			x	LC
Clarias platycephalus	Boulenger, 1902		x				NE
Clarias gariepinus	(Burchell, 1822)	x	x	x	x	x	LC
Clarias ngamensis	Castelnau, 1861	x	x	x	x	x	LC
Clarias nigromarmoratus	Poll, 1967		x				LC
Clarias stappersii	Boulenger, 1915	x	x	x	x	x	LC
Clarias liocephalus	Boulenger, 1898		x	x	x	x	LC
Clarias theodorae	Weber, 1897		x	x	x	x	LC
Clariallabes heterocephalus	Poll, 1967		x				LC
Clariallabes variabilis	Pellegrin, 1926		x				LC
Clariallabes platyprosopos	Jubb, 1965				x	x	LC
Clariallabes sp						x	NE
Platyclarias machadoi	Poll, 1977		x				DD
Schilbeidae							
Parailia occidentalis	(Pellegrin, 1901)	x					LC
Schilbe intermedium	Rüppell, 1832		x	x	x	x	LC
Schilbe angolensis	(De Vos, 1984)	x					DD
Schilbe ansorgii	(Boulenger, 1910)	x					LC
Schilbe bocagii	(Guimarães, 1884)	x					LC
Schilbe grenfelli	(Boulenger, 1900)		x				LC
Schilbe yangambianus	(Poll, 1954)		x	x			LC
Mochokidae							
Synodontis laessoei	Norman, 1923	x					DD
Synodontis leopardinus	Pellegrin, 1914			x	x	x	LC
Synodontis longirostris	Boulenger, 1902		x				LC
Synodontis macrostigma	Boulenger, 1911			x	x	x	LC
Synodontis macrostoma	Skelton & White, 1990			x	x	x	LC
Synodontis nigromaculatus	Boulenger, 1905		x	x	x	x	LC
Synodontis ornatipinnis	Boulenger, 1899	x	x				LC
Synodontis thamalakanensis	Fowler, 1935			x	x	x	LC
Synodontis woosnami	Boulenger, 1911			x	x	x	LC
Synodontis vanderwaali	Skelton & White, 1990			x	x	x	LC
Chiloglanis angolensis	Poll, 1967	x				x	DD
Chiloglanis fasciatus	Pellegrin, 1936			x	x		LC
Chiloglanis lukugae	Poll, 1944		x				LC
Chiloglanis micropogon	Poll, 1952		x				NE
Chiloglanis sardinhai	Ladiges & Voelker, 1961	x					LC
Euchilichthys astatodon	(Pellegrin, 1928)		x				LC
Euchilichthys royauxi	Boulenger, 1902		x				LC
Atopochilus macrocephalus	Boulenger, 1906		x				DD
Chiloglanis sp. (dark)				x	x		NE
Chiloglanis sp. (gold)				x	x		NE

(continued)

Species	Author & Date	A	C	Z	O	K	I[a]
Procatopodidae							
Aplocheilichthys spilauchen	(Duméril, 1861)	x					LC
Micropanchax hutereaui	(Boulenger, 1913)		x	x	x		LC
Micropanchax johnstonii	(Günther, 1894)		x	x	x	x	LC
Micropanchax katangae	(Boulenger, 1912)		x	x	x	x	LC
Micropanchax luluae	(Fowler, 1930)		x		x		NE
Micropanchax macrurus	(Boulenger, 1904)	x	x			x	LC
Micropanchax mediolateralis	(Poll, 1967)		x		x		LC
Micropanchax myaposae	(Boulenger, 1908)	x					LC
Micropanchax nigrolateralis	(Poll, 1967)		x		x		DD
Micropanchax 'pigmy'				x	x		NE
Cichlidae							
Hemichromis elongatus	(Guichenot, 1861)	x	x	x	x		LC
Hemichromis angolensis	Steindachner, 1865	x					NE
Pharyngochromis acuticeps	(Steindachner, 1866)	x		x	x	x	LC
Pseudocrenilabrus philander	(Weber, 1897)	x	x	x	x	x	LC
Oreochromis andersonii	(Castelnau, 1861)			x	x	x	VU
Oreochromis macrochir	(Boulenger, 1912)			x	x	x	VU
Oreochromis angolensis	(Trewavas, 1973)	x					LC
Coptodon rendalli	(Boulenger, 1897)	x	x	x	x	x	LC
Pelmatolapia cabrae	(Boulenger, 1899)	x	x				LC
Oreochromis lepidurus	(Boulenger, 1899)	x	x				EN
Oreochromis schwebischi	(Sauvage, 1884)	x	x				LC
Tilapia sparrmanii	Smith, 1840	x	x	x	x	x	LC
Tilapia ruweti	(Poll & Thys van den Audenaerde, 1965)	x	x	x		x	LC
Serranochromis altus	Winemiller & Kelso-Winemiller, 1991			x	x		LC
Serranochromis angusticeps	(Boulenger, 1907)		x	x	x	x	LC
Serranochromis longimanus	(Boulenger, 1911)			x	x		LC
Serranochromis macrocephalus	(Boulenger, 1899)	x	x	x	x	x	LC
Serranochromis robustus jallae	(Boulenger, 1864)		x	x	x	x	LC
Serranochromis thumbergi	(Castelnau, 1861)		?	x	x	x	LC
Sargochromis greenwoodi	(Bell-Cross, 1975)			x	x		LC
Sargochromis carlottae	(Boulenger, 1905)			x	x		LC
Sargochromis giardi	(Pellegrin, 1903)			x	x	x	LC
Sargochromis coulteri	(Bell-Cross, 1975)				x		LC
Sargochromis codringtonii	(Boulenger, 1908)			x	x	x	LC
Thoracochromis lucullae	(Boulenger, 1913)	x					LC
Orthochromis machadoi	(Poll, 1967)					x	LC
Sargochromis thysi	(Poll, 1967)			x			DD
Chetia welwitschi	(Boulenger, 1898)	x				x	DD

(continued)

Species	Author & Date	A	C	Z	O	K	Iᵃ
Chetia gracilis	(Greenwood, 1984)				x		LC
Thoracochromis albolabrus	(Trewavas & Thys vd Audenaerde, 1969)					x	LC
Thoracochromis buysi	(Penrith, 1970)					x	LC
Anabantidae							
Ctenopoma machadoi	(Fowler, 1930)	x					LC
Ctenopoma multispine	Peters, 1844		x	x	x	x	LC
Microctenopoma intermedium	(Pellegrin, 1920)		x	x	x		LC
Microctenopoma sp.		x			x		NE
Mastacembelidae							
Mastacembelus ansorgii	Boulenger, 1905	x					DD
Mastacembelus niger	Sauvage, 1879		x				LC
Mastacembelus congicus	Boulenger, 1896		x				LC
Mastacembelus frenatus	Boulenger, 1901			x	x		LC
Mastacembelus sp.		x					NE
Eleotridae							
Eleotris vittata	Duméril, 1861					x	LC
Dormitator lebretonis	(Steindachner, 1870)					x	NE
Gobiidae							
Awaous lateristriga	(Duméril, 1861)					x	NE
Nematogobius maindroni	(Sauvage, 1880)					x	NE
Ctenogobius lepturus	(Pfaff, 1933)					x	NE
Periophthalmus barbarus	(Linnaeus, 1766)	x					LC
Syngnathidae							
Enneacampus ansorgii	(Boulenger, 1910)	x					LC
Enneacampus kaupi	(Bleeker, 1863)	x					LC
	TOTALS	**104**	**161**	**93**	**103**	**82**	

IUCN status (I) as recorded by Darwell et al. (2011) and IUCN (2018). The table is for tentative indications of distribution and IUCN status

DD data deficient, *EN* endangered, *LC* least concern, *NE* not evaluated, *VU* vulnerable

ᵃIUCN Red List Categories Codes

References

Ball P (2015) Benguela – more than just a current. The Heritage Portal, p 13. http://www.the-heritageportal.co.za/article/Benguela-more-just-current

Balon EK (1974) Fishes from the edge of Victoria Falls, Africa: demise of a physical barrier for downstream invasions. Copeia 1974(3):643–660

Barnard KH (1948) Report on a collection of fishes from the Okavango River, with notes on Zambesi fishes. Ann S Afr Mus 36:407–458

Bbole I, Katongo C, Deines AM et al (2014) Hybridization between non-indigenous *Oreochromis niloticus* and native *Oreochromis* species in the lower Kafue River and its potential impacts on fishery. J Ecol Nat Environ 6(6):215–225

Bell-Cross G (1965) Movement of fish across the Congo-Zambezi watershed in the Mwinilunga district of Northern Rhodesia. Proceedings of the Central African Scientific and Medical Congress, Lusaka, 1963, pp 415–424

Bell-Cross G (1975) A revision of certain *Haplochromis* species (Pisces: Cichlidae) of Central Africa. Occas Pap Natl Mus Monuments Rhod Ser B 5(7):405–464

Bell-Cross G, Minshull JL (1988) The fishes of Zimbabwe. National Museums and Monuments of Zimbabwe, Harare

Bills IR, Skelton PH, Almeida F (2012) A survey of the fishes of the upper Okavango system in Angola. SAIAB Investigational Report 73, 61 pp

Bills IR, Mazungula N, Almeida F (2013) A survey of the fishes of upper Okavango River system in Angola. SAIAB Investigational Report 74, 21 pp

Boulenger GA (1909–1916) Catalogue of the fresh-water of Africa in the British Museum (Natural History), Vol 1 (1909) Vol 2 (1910), Vol 3 (1915), Vol 4 (1916). Trustees of the British Museum, London

Boulenger GA (1910) LXI.–on a large collection of fishes made by Dr. W. J. Ansorge in the Quanza and Bengo Rivers, Angola. Ann Mag Nat Hist 6(36):537–561

Boulenger GA (1911) V. on a collection of fishes from the Lake Ngami Basin, Bechuanaland. Trans Zool Soc London 18(5):399–418, pls XXXVIII-XLIII

Burrough SL, Thomas DSG, Bailey RM (2009) Mega-lake in the Kalahari: a late Pleistocene record of the Palaeolake Magadigadi system. Quat Sci Rev 28:1392–1411

Castelnau F (1861) Mémoire sur les Poissons de l'Afrique Australe. J-B Baillière et Fils, Paris, p 78

Day JJ, Bills R, Friel JP (2009) Lacustrine radiation in African *Synodontis* catfish. J Evol Biol 22:805–817

De Vos LDG (1995) A systematic revision of the African Schilbeidae (Teleostei, Siluriformes). With an annotated bibliography. Annalen Zoologische Wetenschappen 271:1–450

Devaere S, Adriaens D, Verraes W (2007) *Channallabes sanghaensis* sp.n. a new anguilliform catfish from the Congo River basin, with some comments on other anguilliform clariids (Teleostei, Siluriformes). Belg J Zool 137:17–26

Eccles DH, Tweddle D, Skelton PH (2011) Eight new species in the dwarf catfish genus *Zaireichthys* (Siluriformes: Amphiliidae). Smithiana Bull 13:3–28

Flügel TJ, Eckardt FD, Cotterill FPD (2015) Chapter 15: the present day drainage patterns of the Congo river system and their Neogene evolution. In: de Wit MJ et al (eds) Geology and resource potential of the Congo basin, Regional geology reviews. Springer, Berlin/Heidelberg, pp 315–337

Fowler HW (1930) The fresh-water fishes obtained by the gray African expedition – 1929. With notes on other species in the academy collection. Proc Acad Natl Sci Phila 82:27–83

Fowler HW (1935) Scientific results of the Vernay-Lang Kalahari Expedition, March to September, 1930. The freshwater fishes. Ann Transv Mus 16(2):251–293

Gilchrist JDF, Thompson WW (1913) The freshwater fishes of South Africa. Ann S Afr Mus 11(5):321–463

Gilchrist JDF, Thompson WW (1917) The freshwater fishes of South Africa (continued). Ann S Afr Mus 11(6):465–575

Greenwood PH (1984) The haplochromine species (Teleostei, Cichlidae) of the Cunene and certain other Angolan rivers. Bull Brit Mus (Nat Hist) 47(4):187–239

Greenwood PH (1993) A review of the serranochromine cichlid fish genera *Pharyngochromis, Sargochromis, Serranochromis and Chetia* (Teleostei, Labroidei). Bull Brit Mus (Nat Hist) 59:33–44

Guimarães ARP (1884) 1. Diagnoses de trois nouveaux poisons d'Angola. J Sci Math Phys Lisboa 37:1–10

Haddon IG, McCarthy TS (2005) The Mesozoic–Cenozoic interior sag basins of Central Africa: the late-cretaceous–Cenozoic Kalahari and Okavango basins. J Afr Earth Sci 43:316–333

Hay CJ, van Zyl BJ, van der Bank FH et al (1997) A survey of the fishes of the Kunene River, Namibia. Modoqua 19:129–141

Hipondoka MHT, van der Waal BCW, Ndeutapo MH, Hango L (2018) Sources of fish in the ephemeral western iishana region of the Cuvelai–Etosha Basin in Angola and Namibia. Afr J Aquat Sci 43(3):199–214.https://doi.org/10.2989/16085914.2018.1506310

Jackson PBN (1961) The fishes of northern Rhodesia: a checklist of indigenous species. Department of Game and Fisheries, Lusaka

Jubb RA (1961) An illustrated guide to the freshwater fishes of the Zambezi River, Lake Kariba, Pungwe, Sabi, Lundi and Limpopo Rivers. Stuart Manning, Bulawayo

Jubb RA (1967) The freshwater fishes of southern Africa. AA Balkema, Cape Town

Jubb RA, Gaigher IG (1971) Checklist of the fishes of Botswana. Arnoldia, Rhodesia 5(97):1–22

Key RM, Cotterill FPD, Moore AE (2015) The Zambezi river: an archive of tectonic events linked to the amalgamation and disruption of Gondwana and subsequent evolution of the African plate. S Afr J Geol 118:425–438

King J, Chonguic E (2016) Integrated management of the Cubango-Okavango River basin. Ecohydrol Hydrobiol 16:263–271

Kramer B, Swartz ER (2010) A new species of slender Stonebasher within the *Hippopotamyrus ansorgii* complex from the Cunene River in southern Africa (Teleostei: Mormyriformes). J Nat Hist 44(35–36):2213–2242

Kramer B, Wink M (2013) East–west differentiation in the *Marcusenius macrolepidotus* species complex in southern Africa: the description of a new species for the lower Cunene River, Namibia (Teleostei: Mormyridae). J Nat Hist 47(35–36):2327–2362

Kramer B, van der Bank FH, Flint N et al (2003) Evidence for parapatric speciation in the Mormyrid fish, *Pollimyrus castelnaui* (Boulenger, 1911), from the Okavango–upper Zambezi River systems: *P. marianne* sp. nov., defined by electric organ discharges, morphology and genetics. Environ Biol Fish 77:47–70

Kramer B, van der Bank FH, Wink M (2004) The *Hippopotamyrus ansorgii* species complex in the upper Zambezi River system with a description of a new species, *H. szaboi* (Mormyridae). Zool Scr 33:1–18

Kramer B, Bills IR, Skelton PH et al (2012) A critical revision of the churchill snoutfish, genus *Petrocephalus* Marcusen, 1854 (Actinopterygii: Teleostei: Mormyridae), from southern and eastern Africa, with the recognition of *Petrocephalus tanensis*, and the description of five new species. J Nat Hist 46:2179–2258

Kramer B, van der Bank H, Wink M (2014) Marked differentiation in a new species of dwarf stonebasher, *Pollimyrus cuandoensis* sp. nov. (Mormyridae: Teleostei), from a contact zone with two sibling species of the Okavango and Zambezi rivers. J Nat Hist 48(7–8):429–463

Kramer B, van der Bank FH, Wink M (2016) *Marcusenius desertus* sp. nov. (Teleostei: Mormyridae), a mormyrid fish from the Namib desert. Afr J Aquat Sci 41(1):1–18

Ladiges W (1964) Beiträge zur zoogeographie und Oekologie der süßwasserfische Angolas. Die Mitteilungen aus dem Hamburgischen Zoologischen Museum und Institut 61:221–272

Ladiges W, Voelker J (1961) Untersuchungen über die Fishfauna in Gebirgsgewässern des Wasserscheidenhochlands in Angola. Die Mitteilungen aus dem Hamburgischen Zoologischen Museum und Institut 59:117–140

Lévêque C, Paugy D (2017a) General characteristics of ichthyological fauna. In: Paugy D, Lévêque C, Otero O (eds.) The inland water fishes of Africa, diversity, ecology and human use. IRD Éditions, Paris, & Royal Museum for Central Africa, Tervuren, pp 83–96

Lévêque C, Paugy D (2017b) Geographical distribution and Affinities of African freshwater fishes. In: Paugy D, Lévêque C, Otero O (eds) The inland water fishes of Africa, diversity, ecology and human use. IRD Éditions. France, & Royal Museum for Central Africa, Belgium, pp 97–114

Marshall BE, Tweddle D (2007) *Oreochromis macrochir*. The IUCN Red List of Threatened Species 2007: e.T63336A12659168

Mendelsohn J, Weber B (2015) An atlas and profile of Moxico, Angola. RAISON, Windhoek

Moelants T 2010. *Oreochromis lepidurus*. The IUCN Red List of Threatened Species2010: e.T182875A7991695

Moore AE, Larkin PA (2001) Drainage evolution in south-Central Africa since the break-up of Gondwana. S Afr J Geol 104:47–68

Moore AE, Cotterill FPD, Eckardt FD (2012) The evolution and ages of Makgadikgadi palaeolakes: Consilient evidence from Kalahari drainage evolution. S Afr J Geol 115:385–413

Musilová Z, Kalous L, Petrtýl M et al (2013) Cichlid fishes in the Angolan headwaters region:

molecular evidence of the ichthyofaunal contact between the Cuanza and Okavango-Zambezi systems. PLoS One 8(5):e65047

NGOWP – National Geographic Okavango Wilderness Project (2018) Initial findings from exploration of the upper catchments of the Cuito, Cuanavale and Cuando Rivers in Central and South-Eastern Angola (May 2015 to December 2016). National Geographic Okavango Wilderness Project, 352 pp

Nichols JT, Boulton R (1927) Three new minnows of the genus *Barbus*, and a new characin from the Vernay Angola expedition. Am Mus Novit 264:1–8

Norman JR (1923) A new cyprinoid fish from Tanganyika territory, and two new fishes from Angola. Ann Mag Nat Hist 12(72):694–696

Nunes AL, Douthwaite RJ, Tyser B et al (2016) Invasive crayfish threaten Okavango Delta. Front Ecol Environ 14(5):237–238

Paugy D, Lévèque C, Otero O (eds) (2017) The inland water fishes of Africa, IRD Éditions. Institut de Recherche pour de Developpement/RMCA Royal Museum for Central Africa, Paris/ Tervuren

Pellegrin J (1921) Description d'un Barbeau nouveau de l'Angola. Bull Soc Zool Fr 46:118–120

Pellegrin J (1928) Poissons du Chiloango et du Congo receuillis par l'expédition du Dr Schouteden (1920–1922). Annales du Musée Royal du Congo Belge, Zoologie Série 1 3(1):1–50

Pellegrin J (1936) Contribution à l'ichthyologie de l'Angola. Arquivos do Museu Bocage 7:45–62

Penrith M-L (1970) Report on a small collection of fishes from the Kunene River mouth. Cimbebasia Series A 1:165–176

Penrith MJ (1973) A new species of *Parakneria* from Angola (Pisces: Kneriidae). Cimbebasia Series A 11:131–135

Penrith MJ (1982) Additions to the checklist of southern African freshwater fishes and a gazetteer of south-western Angolan collecting localities. J Limnol Soc South Afr 8(2):71–75

Pinton A, Agnèse J-F, Paugy D, Otero O (2013) A large-scale phylogeny of *Synodontis* (Mochokidae, Siluriformes) reveals the influence of geological events on continental diversity during the Cenozoic. Mol Phylogenet Evol 66:1027–1040

Podgorski JE, Green AG, Kgotlhang L et al (2013) Paleo-megalake and paleo-megafan in southern Africa. Geology 11:1155–1158

Poll M (1967) Contribution à la Faune Ichthyologique de l'Angola. Publicações Culturais 75 75. Companhia dos Diamentes de Angola (DIAMANG), Lisbon, 381 pp

Poll M (1971) Révision des *Synodontis* Africains (Famille Mochocidae). Annales Musée Royal de l'Afrique Centrale Serie IN-8 Sciences Zoologiques No. 191. Musée Royal de l'Afrique Centrale, Tervuren, 497 pp

Roberts TC (1975) Geographical distribution of African freshwater fishes. Zool J Linnean Soc 57(4):249–319

Saldanha L (1978) Museu Bocage. Copeia 1978(4):739–740

Schwarzer J, Swartz ER, Vreven E et al (2012) Repeated trans-watershed hybridization among haplochromine cichlids (Cichlidae) was triggered by Neogene landscape evolution. Proc R Soc London, Ser B 279:4389–4398

Skelton PH (1994) Diversity and distribution of freshwater fishes in East and Southern Africa. Annales Musée Royal de l'Afrique Centrale, Sciences Zoologiques 275:95–131

Skelton PH (2001) A complete guide to the freshwater fishes of Southern Africa. Struik, Cape Town

Skelton PH, Swartz ER (2011) Walking the tightrope: trends in African freshwater systematic ichthyology. J Fish Biol 79:1413–1435

Skelton PH, Bruton MN, Merron GS et al (1985) The fishes of the Okavango drainage system in Angola, South West Africa and Botswana: taxonomy and distribution. Ichthyol. Bull. JLB Smith Inst Ichthyol 50:1–21

Skelton PH, Neef G, Costa A (2016) Into the wilderness expedition 2015: the fishes. SAIAB Investigational Report No 75, 49 pp

Snoeks J, Vreven EJ (2007) Chapter 38: Polynemidae, 445-449 in: Stiassny, MLJ, Teugels GG, Hopkins CD (eds) The fresh and brackish water fishes of lower Guinea, west-Central Africa. Collection Faune et Flore tropicales 42, vol 2. Institut de recherché pour le développement,

Paris, France/Muséum national d'histoire naturelle, Paris, France/Musée royal de l'Afrique Centrale, Tervuren

Snoeks J, Harrison IJ, Stiassny MLJ (2011) Chapter 3: The status and distribution of freshwater fishes. In: Darwall WRT, Smith KG, Allen DJ, Holland RA, Harrison IJ, Brooks EGE (eds) The diversity of life in African freshwaters: under water, under threat. An analysis of the status and distribution of freshwater species throughout mainland Africa. IUCN, Cambridge/Gland, pp 42–73

Steindachner F (1866) Ichthyologische Mittheilungen. (IX.) [With subtitles I-VI.]. Verh Zool Bot Ges Wien 16:761–796

Stiassny MLJ, Teugels GG, Hopkins CD (eds) (2007) The fresh and brackish water fishes of Lower Guinea, West-Central Africa. Collection Faune et Flore Tropicales 42, Volume 1 and 2. IRD & Muséum National d'Histoire Naturelle, Paris & Musée Royal de l'Afrique Centrale, Tervuren

Thieme ML, Abell R, Stiassny ML et al (eds) (2005) Freshwater ecoregions of Africa and Madagascar, a conservation assessment. Island Press, Washington

Trewavas E (1936) Dr. Karl Jordan's expedition to south-West Africa and Angola: the fresh-water fishes. Novitates Zoologicae 40:63–74

Trewavas E (1964) A revision of the genus *Serranochromis* Regan (Pisces, Cichlidae). Annales Musée Royal de l'Afrique Centrale Serie IN-8 Sciences Zoologiques No. 125, Musée Royal de l'Afrique Centrale, Tervuren, 58 pp

Trewavas E (1973) A new species of cichlid fishes of rivers Quanza and Bengo, Angola, with a list of the known Cichlidae of these rivers and a note on *Pseudocrenilabrus natalensis* fowler. Bull Brit Mus (Nat Hist) 25(1):28–37

Tweddle D (2010) Overview of the Zambezi River system: its history, fish fauna, fisheries, and conservation. Aquat Ecosyst Health Manage 13(3):224–240

Tweddle D, Skelton, PH, van der Waal et al (2004) Aquatic biodiversity survey "four corners" transboundary natural resources management area. SAIAB Investigational Report No 71 202 pp

Tweddle D, Cowx IG, Peel RA et al (2015) Challenges in fisheries management in the Zambezi, one of the great rivers of Africa. Fish Manag Ecol 22:99–111

Van der Waal BCW (1991) A survey of the fisheries in Kavango, Namibia. Modoqua 17(2):113–122

Van der Waal BCW, Skelton PH (1984) Checklist of fishes of Caprivi. Modoqua 13(4):303–321

Vreven EJ, Musschoot T, Snoeks J et al (2016) The African hexaploid Torini (Cypriniformes: Cyprinidae): review of a tumultuous history. Zool J Linnean Soc 177(2):231–305

Wamuini Lunkayilakio S, Vreven E (2010) 'Haplochromis' snoeksi, a new species from the Inkisi River basin, lower Congo (Perciformes: Cichlidae). Ichthyol Explor Freshwaters 21(3):279–287

Wamuini Lunkayilakio SW, Vreven E (2008) *Nannopetersius mutambuei* (Characiformes: Alestidae), a new species from the Inkisi River basin, Democratic Republic of Congo. Ichthyol Explor Freshwaters 19:367–376

Wamuini Lunkayilakio S, Vreven E, Vandewalle P et al (2010) Contribution à la connaissance de l'ichtyofaune de l'Inkisi au Bas-Congo (RD du Congo). Cybium 34(1):83–91

Whitfield AK (2007) Estuary associated fish species. In: Stiassny MLJ, Teugels GG, Hopkins CD (eds) The fresh and brackish water fishes of Lower Guinea, West-Central Africa. Collection Faune et Flore Tropicales 42, vol 1. IRD & Muséum National d'Histoire Naturelle, Paris & Musée Royal de l'Afrique Centrale, Tervuren, pp 46-56

Wise RM, van Wilgen BW, Hill MP et al (2007) The economic impact and appropriate management of selected invasive alien species on the African continent. Final report for GISP. CSIR report number CSIR/RBSD/ER/2007/0044/C

Zengeya TA, Decru E, Vreven EJ (2011) Revalidation of *Hepsetus cuvieri* (Castelnau, 1861) (Characiformes: Hepsetidae) from the Quanza, Zambezi and southern part of the Congo ichthyofaunal provinces. J Nat Hist 45:1723–1744

Zengeya TA, Robertson MP, Booth AJ et al (2013) Qualitative ecological risk assessment of the invasive Nile tilapia, *Oreochromis niloticus* in a sub-tropical African river system (Limpopo river, South Africa). Aquat Conserv Mar Freshwat Ecosyst 23:51–64

12

Research on Angolan Amphibians: Past and Present

Ninda Baptista, Werner Conradie, Pedro Vaz Pinto and William R. Branch

Abstract Angolan amphibians have been studied since the mid-nineteenth century by explorers and scientists from all over the western world, and collections have been deposited in around 20 museums and institutions in Europe, Northern America, and Africa. A significant interruption of this study occurred during Angola's liberation struggle and civil war for nearly four decades and, as a consequence, knowledge about the country's biodiversity became outdated with critical gaps. Since 2009, a new era in Angolan biodiversity studies started as expeditions scattered in southwest-

---

N. Baptista (✉)
Instituto Superior de Ciências da Educação da Huíla, Rua Sarmento Rodrigues, Lubango, Angola

National Geographic Okavango Wilderness Project, Wild Bird Trust, Parktown, Gauteng, South Africa

CIBIO-InBIO, Centro de Investigação em Biodiversidade e Recursos Genéticos, Laboratório Associado, Universidade do Porto, Vairão, Portugal
e-mail: nindabaptista@gmail.com

W. Conradie
National Geographic Okavango Wilderness Project, Wild Bird Trust, Hogsback, South Africa

School of Natural Resource Management, Nelson Mandela University, George, South Africa

Port Elizabeth Museum (Bayworld), Humewood, South Africa
e-mail: werner@bayworld.co.za

P. Vaz Pinto
Fundação Kissama, Luanda, Angola

CIBIO-InBIO, Centro de Investigação em Biodiversidade e Recursos Genéticos, Universidade do Porto, Campus de Vairão, Vairão, Portugal
e-mail: pedrovazpinto@gmail.com

W. R. Branch (deceased)
National Geographic Okavango Wilderness Project, Wild Bird Trust, Parktown, Gauteng, South Africa

Department of Zoology, Nelson Mandela University, Port Elizabeth, South Africa

ern, northeastern, southeastern, and northwestern Angola lead to exciting discoveries, including new records for the country, descriptions of new species, range extensions and taxonomical updates. Currently 111 amphibian species are listed for the country (of which 21 are endemic), but this number is an underestimate and the various unresolved taxonomical issues challenge the study of every other aspect of this group. The Angolan amphibian fauna remains one of the most poorly known in Africa and much still has to be done in order to understand its diversity, evolution and conservation needs. An overview of existing knowledge of Angolan amphibians is presented, including an updated checklist for the country, comments on problematic groups, endemic species, biogeography, recent findings, and priority research topics.

Keywords Angolan escarpment · Checklist · Endemism · Herpetology · Research priorities · Taxonomy

Introduction

Amphibians are a fascinatingly diverse group that plays crucial ecological roles (Beard et al. 2002; Davic and Welsh 2004; Regester et al. 2006) and are useful as indicators of ecosystem health (Waddle 2006), thus the relevance of their study surpasses herpetological curiosity. Despite the fact that the rate of description of amphibian species in the world is continuously increasing, current taxonomic research is still insufficient to properly inform conservation planning (Köhler et al. 2005; Brito 2010).

Like other groups presented in this book, Angolan amphibians are among the most poorly known in Africa (Conradie et al. 2016). To study this group it is necessary to deal with historical as well as scientific issues including: many species are known from holotypes collected more than a century ago and which may have been subsequently lost; collection localities had old colonial names, some no longer used and others confused with homonyms; a considerable amount of early literature is written in diverse languages (Portuguese, French, German, English and even Latin) and is not easily accessible; and many names used for Angolan taxa have been lost in synonymies and their current status remains problematic. Overviews of the history and evolution of the southern African amphibian taxonomy exist, mentioning Angolan taxa briefly (Poynton 1964; Channing 1999; Du Preez and Carruthers 2009, 2017). This chapter focuses on Angola, and the compiled information is intended to serve as a baseline that facilitates the study of this group. It consists of an essentially chronological summary of the studies of Angolan amphibians since the very first to the most recent findings, presents a checklist of species, and identifies some of the most evident challenges and exciting research priorities. Given the complicated status of many names available for Angolan taxa, species considered as valid in this review follow Frost (2018). An Atlas of historical and bibliographic records of Angolan herpetofauna has been released subsequent to the compilation of information for this chapter (Marques et al. 2018).

Early Beginnings

The European exploration and settlement in Africa resulted in the discovery of strange and wondrous animals. As these were sent in increasing numbers to European centres of learning and study, they stimulated the departure of expeditions to explore the Angolan flora and fauna by Portugal and by other nations. The exotic collections obtained by these explorers were then shipped to their home countries, and so, in the nineteenth century, the study of amphibians from Angola started in Europe. This was the case for the rest of southern Africa, the only exception being South Africa, which in the early 1800s already had Andrew Smith, a British explorer and researcher, based in the Cape (Channing 1999; Branch and Bauer 2005).

In 1866, José Vicente Barbosa du Bocage made the first list of amphibians and reptiles from Angola based on assorted specimens deposited in the Natural History Museum of Lisbon (Bocage 1866a, b). It documented only 19 amphibian species, eight of which were new to science and which Bocage (1866b) described *(Hyperolius cinnamomeoventris, H. tristis, H. fuscigula, H. quinquevittatus, H. steindachneri, Rana (=Ptychadena) subpunctata, Rana (=Amietia) angolensis, Bufo funereus (=Sclerophrys funerea)*. The material came from two expeditions, one by José de Anchieta in 1864 to Cabinda, and the other from Duque de Bragança (now Calandula) by Pinheiro Bayão.

During this period, Europeans were exploring Angola, either on their own initiative, or on behalf of various institutions that promoted scientific expeditions to Angola. Publications from this era consist essentially of descriptions of new species and new distribution records for known species. The renowned Austrian explorer and botanist Friedrich Martin Josef Welwitsch (1806–1872) explored Angola for the Portuguese government, arriving in 1853 and undertaking almost a decade of strenuous exploring and collecting. After his return to Europe his collections were donated to the British Museum, later shared with Portugal, and the Angolan amphibians were reported on by Günther (1865), who described new species of reed frog *(Hyperolius nasutus, H. parallelus)*.

Collections from the Austrian frigate Novara were deposited in the Natural History Museum of Vienna and studied by Steindachner (1867), who described *Ptychadena porosissima* and *Hyperolius bocagei* from no precise locality. Anchieta persisted in his extensive exploration of Angola, and Bocage (1867, 1873, 1879a, b, 1882, 1893, 1897b) examined his specimens, as well as the herpetological collections of Capello & Ivens (Bocage 1879a, b), describing *Hylambates (=Leptopelis) anchietae, Hylambates (=Leptopelis) cynnamomeus,* and *Rappia (=Hyperolius) benguellensis* among other species currently not valid. The German explorers von Homeyer, who collected in Pungo-Andongo, and von Mechow, who collected in Malanje and Cuango, had their specimens deposited in the Zoological Museum of Berlin and studied by Peters (1877, 1882), who described *Bufo buchneri* from Cabinda. Boulenger (1882) studied the material from the British Museum and described *Tomopterna tuberculosa,* and Rochebrune (1885) described four new *Hyperolius* species from Cabinda (*H. lucani, H. maestus, H. protchei, H. rhizophilus*).

Bocage (1895a) compiled the extant information about the herpetology of Angola and Congo, using all the above-mentioned references, except for Rochebrune's (1885). A total of 40 species of amphibians were listed for Angola. Even today, more than a century after its release, this work is still a valuable reference on Angolan herpetology. After this, Bocage published several other findings (Bocage 1895b, 1896a, b, 1897a, b), mostly from Anchieta's new collections, with new locality records for many frogs, and the description of a new pygmy toad, *Bufo (=Poyntonophrynus) dombensis*.

From 1898 to 1906, José Júlio Bethencourt Ferreira studied Angolan material collected by Anchieta, Francisco Newton and Pereira do Nascimento (Ferreira 1897, 1900, 1904, 1906), and described new species (*Rappia (=Afrixalus) osorioi, Arthroleptis carquejai, Rappia (=Hyperolius) nobrei*) and some species and varieties that have been subsequently synonymised.

From 1903 to 1905, William John Ansorge collected considerable material in northern, central and southwestern Angola. The collected amphibians are deposited in the British Museum, and were studied by Boulenger (1905, 1907a, b). *Arthroleptis (=Phrynobatrachus) parvulus, Arthroleptis xenochirus, Rana (=Ptychadena) ansorgii, Rana (=Tomopterna) cryptotis*, and *Rana (=Ptychadena) bunoderma* were all described from this material.

A number of expeditions in Angola included herpetological surveys, and had their reptiles studied, but the amphibians were not reported. Examples of this are the Rohan-Chabot Mission (1912–1914), which explored the south of Angola and had its specimens deposited in the Paris Natural History Museum, and the Vernay Angola Expedition (1925), from which the large collection is housed in the American Museum of Natural History.

Analysing material from the Berlin Zoological Museum, Ahl (1925) described *Hylarthroleptis (=Phrynobatrachus) brevipalmatus* from Angola, and several species of reed frogs, two of which are endemic to Angola (*Hyperolius bicolor, Hyperolius gularis)* and others which have later been synonymised into larger species complexes such as *Hyperolius parallelus* complex (*H. angolensis, H. huillensis, H. microstictus), Hyperolius marmoratus* complex (*H. decoratus, H. marungaensis)*, and *Hyperolius platyceps* complex (*H. angolanus*).

In 1930–1931, the Pulitzer-Angola Expedition surveyed southwestern and central areas of the country. Over 400 specimens of amphibians were collected and deposited in the Carnegie Museum, in the United States of America. These were studied by Karl Patterson Schmidt (1936), who reported on 17 species. Although no new species were described, some were synonymised and others revived from synonymy leading the author to highlight the importance of understanding the Angolan fauna for clarifying African amphibian taxonomy.

During two trips to central and southern Angola (1928–1929 and 1932–1933) Albert Monard made important collections of amphibians and reptiles, as well as other groups. The herpetological material was deposited in the La Chaux-de-Fonds Museum, Switzerland. Monard (1937) provided an updated compilation of Angolan amphibians with a revision of the existing literature (including Ahl, Bocage, Boulenger and Schmidt's publications), as well as his own findings. Five new species of frog were described: *Hyperolius cinereus, Cassiniopsis (=Kassina) kuvan-*

gensis, Rana (=Ptychadena) keilingi, Hyperolius erythromelanus, Rana (=Ptychadena) buneli, the last two now considered synonyms of *H. paralleus* and *Ptychadena bunoderma*, respectively. In total, 80 species of amphibians were mentioned, meaning that in the four decades since Bocage's (1895a) first synthesis the known frog species for Angola had doubled.

In 1933–1934, Karl Jordan's expedition to South West Africa (now Namibia) and Angola surveyed localities on the Angolan escarpment (Congulo and Quirimbo) and afromontane forest (Mount Moco) (Jordan 1936). This material is deposited in the British Museum and the herpetofauna studied by Parker (1936). One new species of treefrog (*Leptopelis jordani*) and a new subspecies of white-lipped frog (*Rana (=Amnirana) albolabris acustirostris*) were described from this expedition. As the name *acustirostris* was preoccupied, Mertens (1938b) proposed the replacement name *Rana (=Amnirana) albolabris parkeriana*, which was later elevated to a full species by Perret (1977). Both these species remain known only from their type localities, and are escarpment-endemics.

In the 1930s W Schack visited Angola and made a collection of amphibians which were deposited in the *Natur-Museum Senckenberg*, Frankfurt, and studied by Mertens (1938a), who recorded only eight species, none of which was new.

In 1952–1954, within the scope of the Hamburg Museum expeditions, GA von Maydell made significant herpetological collections from north to south of Angola. The reptiles were studied by Walter Hellmich (1957a), but the amphibians have never been studied until recently (Ceríaco et al. 2014b). Hellmich made a trip to the Angolan region of Entre-Rios, and reported on new localities for frog species (Hellmich 1957b), also commenting on the Angolan biogeography.

From 1957 to 1959, the Portuguese Mission of Apiarian Studies of the Overseas collected amphibians especially in central and eastern Angola (Luando and Cameia), which were deposited in the Zoology Center of the Institute of Tropical Scientific Research, in Lisbon. These were studied only decades later, by Clara Ruas (1996, 2002).

Raymond F Laurent worked extensively on the herpetofauna of the Congo Basin. He studied material from Museu do Dundo, Lunda-Norte, including the extensive collection made in southwestern Angola by the Museum Director, António Barros Machado. During this period, he recorded several new frogs for Angola (Laurent 1950, 1954, 1964), and described four new species (*Ptychadena grandisonae, P. guibei, P. perplicata* and *Hyperolius vilhenai*).

In 1971 and 1974, Wulf Haacke, from the then Transvaal Museum, South Africa, made two trips to Angola to search mainly for geckos, but incidentally collecting amphibians that were later studied by John Poynton (Poynton and Haacke 1993).

Until the 1970s, zoological expeditions surveyed mostly southwestern and central parts of the country, which were more easily accessible than the inland plateau and the moist forests of the north. Herpetological knowledge about the northeastern region was greatly improved by Laurent's studies. The most poorly studied areas of Angola remained the northwest (the region of Zaire and Uíge provinces, and northern Malange, Bengo, and Cuanza-Norte provinces), followed by the southeastern 'lands at the end of the world', a commonly used expression that refers to the very remote and extensive regions of Moxico and Cuando Cubango provinces.

Recent History and Increase of Information

For almost three decades, in the period between Angola's independence and the end of the civil war (1975–2002), the country's instability precluded virtually all field surveys. Every amphibian publication dating from this period involved taxonomic revisions based on existing literature and museum collections, e.g. Perret's (1976) revision of the amphibians, particularly types, deposited in the Lisbon Museum of Natural History. This has become an extremely valuable work given the subsequent loss of these important specimens following the fire that destroyed the museum in 1978.

A key for the identification of Angolan amphibians mainly based on literature revision, including all the species listed for Angola at the time, was published (Cei 1977). With dichotomous keys, drawings, and insights on the Angolan amphibian biogeography, it was intended to make Angolan amphibian identification more accessible to the general public and particularly to students. Poynton (1964) published a faunal study of the southern African amphibians, which referred to Angolan material. This was later updated from 1985 to 1991, when Poynton & Broadley published *Amphibia Zambesiaca*, a series of papers (Poynton and Broadley 1985a, b, 1987, 1988, 1991) that addressed in detail all the amphibian families occurring in the Zambezi drainage region, including many that extend into Angola. The publication of a toponymic index of the zoological collections made in Angola (Crawford-Cabral and Mesquitela 1989) was a valuable contribution to the study of vertebrates of the country. It provided an overview of the zoological collections performed in Angola and studies related to these expeditions, including a section of type localities and the list of described vertebrates per locality, which lists amphibian species, subspecies and varieties.

In 1993, Poynton & Haacke described the first new Angolan amphibian species in decades: *Bufo (= Pontynophrynus) grandisonae*, based on Haacke's expeditions of the 1970s. In 1996, the re-examination of Monard's collection of amphibians from 1928, revealed an 'enigmatic' ranid originally identified as *Aubria subsigillata* that could not be assigned to any known genus (Perret 1996), but which was later assigned to *Aubria masako* (Channing 2001) following features described by Ohler (1996). A comprehensive revision of the Angolan amphibians and mapping of each species' distribution based on museum and literature records was made by Ruas (1996), providing taxonomic comments on some species, but not addressing the Hyperoliidae family (then including the current Leptopelinae subfamily). Ruas (2002) described in detail the contents of the amphibian collection deposited in the Zoology Center of the Institute of Tropical Scientific Research in Lisbon, again excluding the Hyperoliidae and Leptopelinae, which are still to be examined. Channing (1999) discussed aspects of Angolan amphibian taxonomy within a southern African historical perspective. Blanc and Frétey (2000) analysed the biogeography, species richness and endemism of the central African and Angolan amphibians, based on the number of species per country. They highlighted the discrepancy in species richness among genera in Angola, with *Bufo* (currently *Mertensophryne, Sclerophrys* and *Poyntonophrynus*), *Hyperolius* and *Ptychadena* being the most specious genera, which totalled 42 species, almost half of the species known for the country at the time (86).

Only in 2009 did Angolan-international collaboration lead to a new era of field surveys, initiated with an expedition to Huíla and Namibe provinces in southwestern Angola. This trip, organised by Brian Huntley, can be considered as a historical landmark for research on Angolan biodiversity. Numerous groups were surveyed (plants, invertebrates, mammals, birds, reptiles and amphibians). A new escarpment-endemic reed frog, *Hyperolius chelaensis,* was described from Serra da Chela (Conradie et al. 2012), and the colourful ashy reed frog, *Hyperolius cinereus* Monard 1937 was rediscovered (Conradie et al. 2013). Later in the same year, Alan Channing and Pedro Vaz Pinto surveyed Cangandala National Park and made a trip to Calandula, revisiting this important type locality of several amphibian species, and rediscovered *Hyperolius steindachneri* Bocage, 1866 in Angola (Channing and Vaz Pinto Unpublished Data). The material obtained from these trips was important for a number of taxonomic revisions. The Angolan river frog *Amietia angolensis,* previously thought to be widespread in Africa, was found to occur only in Angola (Channing and Baptista 2013; Channing et al. 2016), reed frogs of the *Hyperolius nasutus* complex (Channing et al. 2013) were shown to include numerous cryptic species, with possibly four occurring in Angola, and the *Hyperolius cinnamomeoventris* complex was split into different sister clades (Schick et al. 2010).

Another Angolan international expedition, again organised by Brian Huntley in 2011, visited the unexplored Lagoa Carumbo, Angola's second largest freshwater lake, in Lunda Norte province. Preliminary findings revealed a complex herpetofauna (Branch and Conradie 2015), with the description of the new *Hyperolius raymondi* (Conradie et al. 2013), and the addition of two new country records: *Amnirana* cf. *lepus* and *Hyperolius pardalis.*

Two books, *Treefrogs of Africa* (Schiøtz 1999) and *Amphibians of Central and Southern Africa* (Channing 2001) address the Angolan territory, providing species identification keys, colour photographs, and distribution maps. In 2011, a book on the central African and Angolan amphibians was released (Frétey et al. 2011). It addressed Angolan fauna only briefly, providing a species list (without discussion), and synthesis of species and habitat/biogeographical associations. In *Tadpoles of Africa* (Channing et al. 2012), the larvae of several species occurring in the country are described, and the description of *Leptopelis anchietae* and *Ptychadena porosissima* tadpoles are based on Angolan specimens. The popular book *Frogs of Southern Africa – A Complete Guide* (Du Preez and Carruthers 2009, 2017) provides descriptions of species, morphology, distribution, behaviour, and has advertisement calls available for many species. It has been recently updated to a cell phone app. "Frogs of Southern Africa" and has relevant information about species that also occur in Angola.

In 2012 and 2013, studies of the lower catchments of the Cubango, Cuito and Cuando rivers in southeastern Angola were organised by the Southern Africa Regional Environmental Program (SAREP), funded by the USAID, and included herpetological surveys. Preliminary results have been published (Brooks 2012, 2013), as well as an annotated checklist of the herpetofauna of the region (Conradie et al. 2016).

In 2013, a partnership between the Kimpa Vita University in Uíge, the Technical University Dresden and Senckenberg Natural History Collections, Dresden, promoted herpetological surveys in the extremely poorly known Serra do Pingano ecosystem and surrounding forest fragments in Uíge Province. Two forest species, *Trichobatrachus robustus* and *Xenopus andrei*, typical of the Congo Basin, were added to the country's list (Ernst et al. 2014, 2015). Both these observations represented southern range extensions of hundreds of kilometers. Additional important discoveries from this survey await formal publication, and will certainly increase current knowledge of the taxonomy and biogeography of Angolan amphibians, as well as highlight the exceptional biodiversity of northern Angola (Ernst pers. comm.).

Since 2013, a project of the California Academy of Sciences in collaboration with the National Institute of Biodiversity and Conservation Areas (INBAC), Angola, initiated a study of the Angolan herpetofauna, including the development of an atlas of the Angolan amphibians and reptiles, based on literature, analysis of museum collections from several countries, and new findings (Marques et al. 2014, 2018). The Angolan type material deposited in the Porto Museum was studied, and the nomenclature and taxonomy of hyperoliids, *Leptopelis* and *Arthroleptis* described by Ferreira were discussed (Ceríaco et al. 2014a). Analysis of amphibians collected in the Capanda Dam surroundings in Malanje (Ceríaco et al. 2014a) included a possible record of *Kassina maculosa,* which if confirmed would be the first for the country. In a study of the Namibe Province herpetofauna, *Tomopterna damarensis* was recorded for the first time for Angola (Ceríaco et al. 2016a; Heinicke et al. 2017), and a new species of pygmy toad has been described from Serra da Neve (Ceríaco et al. 2018a). A booklet on the herpetofauna of the Cangandala National Park in Malanje (Ceríaco et al. 2016c) was also released, followed by a scientific publication on the same subject (Ceríaco et al. 2018b). Research on the project's findings and surveys to additional regions in Angola are ongoing.

In 2015 the Wild Bird Trust, supported by the National Geographic Society, organised Angolan expeditions associated with the Okavango Wilderness Project. Herpetological surveys took place in the headwaters of the Cuito, Cuanavale, Cubango and Cuando rivers and other river sources in the region in both wet and dry seasons. Whilst some of these results have been published (Conradie et al. 2016), the project is ongoing but already two new country records (*Kassinula wittei* and *Leptopelis* cf. *parvus*), numerous range extensions for Angolan herpetofauna, and a number of candidate new species of amphibians have been identified.

Within the Southern African Science Service Centre for Climate Change and Adaptive Land Management (SASSCAL) project, research on herpetology is being undertaken by the *Instituto Superior de Ciências da Educação* (ISCED)-Huíla. Observatories have been implemented in Tundavala, Bicuar National Park, Cameia National Park, Iona National Park, Candelela and Cusseque (Jürgens et al. 2018). Opportunistic surveys of herpetofauna are made at all observatories (SASSCAL ObservationNet 2017), herpetofauna monitoring has been carried out at the Tundavala observatory since 2016 (Baptista et al. 2018), and a checklist of Bicuar National Park herpetofauna compiled (Baptista et al. in press). Additionally, in collaboration with Fundação Kissama, herpetological surveys have been made at several sites in Huíla Province, and throughout Angola, with emphasis along the Angolan escarpment: Cuanza-Norte, Cuanza-Sul (Cumbira) and Huíla Provinces. An Angolan herpetofauna archive is being developed at ISCED Huíla, and research undertaken in conjunction with these projects.

International and National Resources

Given the scarcity and the difficulties in obtaining information about Angolan amphibians, the compilation and listing of existing information sources is relevant. Table 12.1 lists generalist on-line platforms with relevant information about amphibians that include Angolan species, as well as a list of institutions known to have significant Angolan material in their assets.

The Current State of Knowledge on Angolan Amphibians

Despite some progress made during the last decade, the Angolan herpetofauna remains one of the most poorly known in Africa (Conradie et al. 2016). This lack of information becomes more evident when contrasted with the comprehensive information compilations regarding adjacent Namibia, which include updated lists of species (Herrmann and Branch 2013) and analysis of habitat availability, species richness and conservation (Curtis et al. 1998). For Angola, even basic information, such as accurate species checklists for the country, is absent. Existing information is scattered in recent and historical publications, many of which are not easily accessible. The recent Atlas of Angolan herpetofauna (Marques et al. 2018) contributes to filling this gap. Figure 12.1 shows the localities where amphibians have been collected before and after independence. Although recent surveys have filled some gaps, many areas remain unsurveyed. Figure 12.2 depicts some of the amphibian diversity present in Angola.

Checklist of Angolan Amphibians

Currently only 111 species are recorded from Angola (Appendix). Marques et al. (2018) list 117 species for the country. This discrepancy results from the use of different criteria for synonymies, and of a conservative approach of the present authors not incuding unconfirmed records, which are discussed elsewhere in this chapter. Both these totals are considered to be underestimates, given the country's size and habitat richness, including the southern desert, the tropical northern forests, the unique escarpment and the extensive plateau, many areas of which remain unsurveyed. This becomes more evident when compared with a country of similar size such as South Africa, whose herpetofauna is the best studied in Africa and which is considerably drier and cooler (and therefore less suitable for amphibians) than Angola, and yet it has 128 species (Frost 2018), and new species continue to be discovered (Turner and Channing 2017; Minter et al. 2017).

Table 12.1 List of relevant websites with information regarding Angolan amphibians, and collections where Angolan amphibian specimens are deposited, according to available literature

On-line platforms and mobile phone apps

Amphibian Species of the World: http://research.amnh.org/vz/herpetology/amphibia/

AmphibiaWeb http://amphibiaweb.org/

IUCN Red List http://www.iucnredlist.org/initiatives/amphibians

Frogs of Southern Africa https://play.google.com/store/apps/details?id=com.coolideas.eproducts.safrogs

Collections where amphibians from Angola are deposited

Angola	Instituto Nacional para a Biodiversidade e áreas de Conservação, Ministério do Ambiente (INBAC/MINAMB)[a]
	Museu do Dundo (MD)
	Museu Nacional de História Natural (Luanda)[a]
	Southern African Science Service Centre for Climate Change and Adaptive Land Management (SASSCAL) / Instituto Superior de Ciências da Educação da Huíla (ISCED-Huíla)[a]
Austria	Imperial Natural History Museum (K.K. Museum) / Natural History Museum of Vienna (NHMW)
France	National Museum of Natural History (Paris) (MNHNP)
Germany	Berlin Zoological Museum (ZMB – Zoologisches Museum)[a]
	Forschungsinstitut und Naturmuseum Senckenberg (SMF)
	Hamburg Museum (ZMH – Zoologisches Museum für Hamburg)
	Senckenberg Natural History Collections Dresden (MTD – Museum für Tierkunde Dresden)[a]
Portugal	Centro de Zoologia do Instituto de Investigação Científica Tropical, Lisbon (IICT)
	Museu de História Natural na Universidade do Porto (MUP)
	Museu Nacional de História Natural e da Ciência, formerly Museu Bocage, Lisbon (MBL) – collections destroyed on the 1978 fire
South Africa	Ditsong National Museum of Natural History (formerly Transvaal Museum) (TMP), Pretoria
	Port Elizabeth Museum at Bayworld (PEM)[a]
	South African Institute for Aquatic Biodiversity (SAIAB)[a], Grahamstown
Spain	Estación Biológica de Doñana (EBD-CSIC), Sevilla
Switzerland	Musée de la Chaux-de-Fonds (LCFM)
	Museum d'histoire naturelle de la Ville de Genève (MHNG – Geneva Natural History Museum)
United Kingdom	Natural History Museum, London (NHMUK, formerly British Museum)
	Natural History Museum at Tring
United States of America	Carnegie Museum of Natural History (CM), Pittsburgh
	California Academy of Sciences (CAS), San Francisco[a]
	American Museum of Natural History (AMNH), New York[a]
	Academy of Natural Sciences of Philadelphia (ANSP), Philadelphia
	Field Museum of Natural History (FMNH), Chicago
	Museum of Comparative Zoology (MCZ), Harvard University, Cambridge, Massachusets
	National Museum of Natural History, Smithsonian Institution (NMNH), Washington, D.C.

[a]indicates the institutions containing specimens from recent (post-1975) surveys

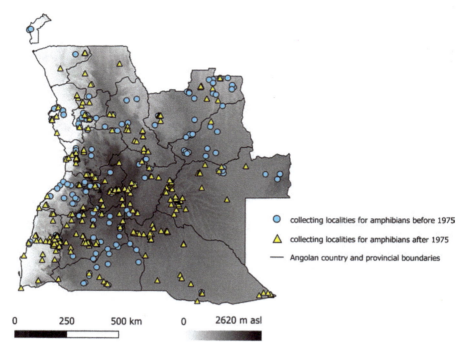

Fig. 12.1 Map with collecting localities for amphibians. Blue circles represent surveys before 1975 (based on literature records), and yellow triangles represent surveys after 1975 (literature records, localities from the 2009 and 2011 expeditions, SAREP and NGOWP trips to southeastern Angola, surveys in the scope of the SASSCAL Project and Fundação Kissama work, and Senckenberg Technical University, Dresden)

Records That Require Confirmation

A number of unconfirmed records for Angola require further investigation. These include *Leptopelis notatus* (Laurent 1964), *Ptychadena schillukorum* (Channing 2001), and *Kassina maculosa* (Ceríaco et al. 2014a). Monard (1937) noted one specimen of *Aubria subsigillata* from Caquindo that Perret (1996) could not confidently associate with any known genus, but that Channing (2001) considered to be *A. masako*. However, the latter is a closed-canopy forest species that is not expected to occur in southern Angola. The stated locality is either in error or the specimen deserves further investigation. *Phrynobatrachus dispar* was recorded from Cabinda by Peters (1877, as *Arthroleptis dispar*), but this species originates from São Tomé and Príncipe islands (Uyeda et al. 2007; Frost et al. 2018), and it is therefore likely that the Angolan record refers to another species. *Hyperolius nitidulus* was also recorded from Angola (Peters 1877), but was described from Nigeria and is currently considered to extend south only to Cameroon (Amiet 2012). *Hyperolius ocellatus* has been described both from Angola and Fernando Pó, but the type locality was later restricted to Fernando Pó (Perret 1975) which leaves Angolan specimens

Fig. 12.2 Representative of some of the families of frogs present in Angola. **1** Congulo Forest Tree Frog (*Leptopelis jordani*) from Congulo. **2** Dombe Pigmy Toad (*Poyntonophrynus dombensis*) from Meva. **3** Kuvango Kassina (*Kassina kuvangensis*) from Cuanavale River Source. **4** Spot-bellied Grass Frog (*Ptychadena subpunctata*) from Cameia National Park. **5** Marbled Rubber Frog (*Phrynomantis annectens*) from Meva. **6** Marbled Snout-Burrower (*Hemisus marmoratus*) from Bicuar National Park. **7** Angolan Reed Frog (*Hyperolius* cf. *parallelus*) from Quilengues. **8** Rain frog (*Breviceps* sp. nov.) from Cuando River Source. (Photo credits – N Baptista: **4,6,7**; P Vaz Pinto: **1,2,5**; W Conradie: **3,8**)

with no appliable name. *Phrynobatrachus auritus* was recorded from Cabinda by Peters (1877) as *Arthroleptis plicatus*, but the validity of this synonymy for Cabinda requires further study. A number of species recorded from Angola were presumably misidentified as the currently known species' range does not include Angola, including: *Phrynobatrachus minutus* recorded by Ruas (1996), but which is now restricted to Ethiopia; *Hyperolius microps* recorded by Bocage (1895) and Monard (1937), now restricted to Eastern Africa; *Hyperolius multifasciatus* Ahl 1931 which was included provisionally by Monard (1937), but placed in the synonymy of *H. kivuensis* Ahl 1931, by Pickersgill (2007); and *Xenopus calcaratus* recorded by Peters (1877), but now restricted to western Africa. Specimens of *Ptychadena* cf. *aequiplicata*, which occurs approximately 50 km from the Cabinda Enclave (Nagy et al. 2013), exist in the AMNH collection, but their identity requires confirmation (Ernst pers. comm.).

Species Likely to Occur in Angola But Not Yet Recorded

The ranges of many species occurring in adjacent countries (Namibia, Zambia and the Democratic Republic of the Congo, DRC) are likely to extend into Angola and are discussed below. A significant example is that of caecilians (Order Gymnophiona), which are known from the Congo Basin but have not been recorded in Angola, including Cabinda. Species that have been recorded close to the Angolan border and that are likely to occur in the country are listed below.

Caecilians (Gymnophiona)

The Gaboon Caecilian (*Geotrypetes seraphini* (Duméril, 1859)) and the Congo Caecilian (*Herpele squalostoma* (Stutchbury, 1836)) have both been recorded from the extreme western DRC, in Mayombe, River Minkala, Vemba-Minionzi, around Kidima, around 40 km from the Angolan border (Scheinberg and Fong 2017), and are likely to occur in this poorly known region.

Frogs and Toads (Anura)

Arthroleptidae

Cryptic Tree Frog (*Leptopelis parbocagii* Poynton and Broadley, 1987). This tree frog occurs in northern Mwinilunga district, northwest Zambia, less than 50 km from Cazombo, eastern Angola (Schiøtz and Van Daele 2003), and may occur on the Angolan side of the border.

Breviciptidae

Power's Rain Frog (*Breviceps poweri* Parker, 1934). This rainfrog was found in southwestern Zambia, less than 100 km from the Angolan border (Pietersen et al. 2017), and can be expected in Angola.

Bufonidae

Beira Pygmy Toad (*Poyntonophrynus beiranus* (Loveridge, 1932)). Recorded from southwestern Zambia near the Angolan border (Poynton and Broadley 1991) and may occur in Angola.
Northern Pygmy Toad (*Poyntonophrynus fenoulheti* (Hewitt and Methuen, 1913)). This pigmy toad is recorded from Caprivi Strip in northeastern Namibia (Channing and Griffin 1993) and southwestern Zambia (Pietersen et al. 2017), less than 100 km from the Angolan border, and its presence is expected in southeastern Angola.

Hemisotidae

Perret's Snout-burrower (*Hemisus perreti* Laurent, 1972). Recorded in Singa Mbamba, Mayumbe (Royal Museum for Central Africa 2017) and in the region of Kipanzu, Tshela (MHNG 2017) both in the Bas-Congo Province, DRC, in close proximity to the Cabinda enclave, and it is therefore expected to occur in Cabinda.
Barotse Snout-burrower (*Hemisus barotseensis* Channing and Broadley 2002). Described from the Barotse floodplain near Mongu, in southwestern Zambia, 120 km east of the Angolan border, but may occur in suitable floodplain habitat along the eastern Zambezi drainage.

Hyperoliidae

Foulassi Spiny Reed Frog (*Afrixalus paradorsalis* (Perret, 1960)). This hyperoliid was found in Luango-Nzambi, DRC, around 50 km from the Cabinda Enclave (Nagy et al. 2013) and is likely to occur in Angola.
Rainforest Reed Frog (*Hyperolius tuberculatus* (Mocquard, 1897)). Also found in Luango-Nzambi, DRC (Nagy et al. 2013) and likely to occur at least in Cabinda.
Kachalola Reed Frog (*Hyperolius kachalolae* Schiøtz, 1975). Known from Mwinilunga district, in northwestern Zambia (Schiøtz and Van Daele 2003), less than 50 km from the eastern Angolan border.
Hyperolius major Laurent, 1957. This reed frog occurs in Mwinilunga district, in northwestern Zambia, less than 50 km of Cazombo (Poynton and Broadley 1991; Schiøtz and Van Daele 2003), eastern Angola.

Phrynobatrachidae

Golden Puddle Frog (*Phrynobatrachus auritus* Boulenger, 1900). This species of puddle frog was found in Luki, DRC, only 20 km north of Angola (Nagy et al. 2013) and may occur in the country.

Horned Puddle Frog (*Phrynobatrachus* sp. aff. *cornutus* (Boulenger, 1906)), also found in Luki, DRC (Nagy et al. 2013) and likely to occur in Angola.

Pipidae

Gaboon Dwarf Clawed Frog (*Hymenochirus* sp. aff. *feae* Boulenger 1906), and *Xenopus (Silurana)* sp. This Dwarf Clawed Frog and an unidentified species of clawed frog were found in Luki, DRC, 20 km north of the Angolan border (Nagy et al. 2013) and are expected in Angolan territory.

Fraser's Clawed Frog (*Xenopus* cf. *fraseri* Boulenger, 1905). This clawed frog has been found in Luki, DRC, 20 km north of the Angolan border (Nagy et al. 2013) and is expected in Angola, although the records of these species are considered to need a critical revision (Ernst et al. 2015).

Common Platanna (*Xenopus laevis* (Daudin, 1802)). Recorded in Luki, DRC, 20 km north of the Angolan border, and in Tsumba-Kituti (Nagy et al. 2013) and might occur in Angola.

Ptychadenidae

Dark Grass Frog (*Ptychadena obscura* (Schmidt and Inger, 1959)). This species has been recorded in the Ikelenge pedicle, northern Mwinilunga district, northwestern Zambia, in close proximity to the Angolan eastern border (Poynton and Broadley 1991).

Mapacha Grass Frog (*Ptychadena* cf. *mapacha* Channing, 1993). This Grass Frog is described from the Caprivi Strip in Namibia, near southeastern Angola (Channing 1993). It has also been recorded about 80 km east of Rundu (Haacke 1999), near Vicota, around 30 km south of the Angolan border (Ceríaco et al. 2016a), and in southwestern Zambia (Pietersen et al. 2017). Conradie et al. (2016) collected a series of *Ptychadena* at Jamba provisionally assigned to *P.* cf. *mossambica*, but mentioned that the specimens might be referable to *P. mapacha*. All these records suggest that this species may occur in southeastern Angola.

Perret's Grass Frog (*Ptychadena* cf. *perreti* Guibé and Lamotte, 1958). This grass frog was found in Nkamuna, in the Bas-Congo province of DRC, near Angola (Nagy et al. 2013).

Pyxicephalidae

Boettger's Dainty Frog (*Cacosternum boettgeri* (Boulenger, 1882)). This species has been recorded near the Angolan border in northern Namibia in Caprivi Strip and in Omusati province (Channing and Griffin 1993), and Southern Province in Zambia (Broadley 1971) and may occur in Angolan territory.

Knocking Sand Frog (*Tomopterna krugerensis* Passmore and Carruthers, 1975). This frog has been recorded in northern Namibia close to the Angolan border (Channing and Griffin 1993).

Tandy's Sand Frog (*Tomopterna tandyi* Channing and Bogart, 1996). Recorded from northern Namibia near the Angolan border (Coetzer 2017), and may occur in southwestern Angola. A recent fing of *Tomopterna* has been made in Bicuar National Park and its identification as *T. tandyi* is under discussion (Baptista et al. in press).

Rhacophoridae

Southern Foam Nest Frog (*Chiromantis xerampelina* Peters, 1854). Recorded from Caprivi Strip in northern Namibia (Channing and Griffin 1993) and from southeastern Zambia (Broadley 1971; Pietersen et al. 2017), and therefore expected in southeastern Angola. It is recorded from southwestern Angola (Schiøtz 1999), but the original source of this record is unknown. This odd distribution record requires confirmation.

Western Foam-nest Tree Frog (*Chiromantis rufescens* (Günther, 1869)). This species is known from near Boma, close to the northern bank of the Congo River (Royal Belgian Institute of Natural Sciences 2017) and may occur in Angola.

According to Frost (2018), based on distribution and species' habitat affinities, around 20 additional species are expected in the country, mostly from the northern forests and expected in the Cabinda enclave in northern Angola. These are generalist assumptions that do not necessarily take into account actual proximity to the Angolan border. These include arthroleptids Silver Long-fingered Frog (*Cardioglossa leucomystax* (Boulenger, 1903)), Kala Forest Treefrog (*Leptopelis aubryioides* (Andersson, 1907)), Victoria Forest Treefrog (*Leptopelis boulengeri* (Werner, 1898)), Red Treefrog (*Leptopelis rufus* Reichenow, 1874)); bufonids [High Tropical Forest Toad (*Sclerophrys latifrons* (Boulenger, 1900))], hyperoliids [African Wart Frog (*Acanthixalus spinosus* (Buchholz and Peters, 1875)); Greshoff's Wax Frog (*Cryptothylax greshoffii* (Schilthuis 1889) with an unconfirmed record from northwestern Angola (Schiøtz 1999)), Olive Striped Frog (*Phlyctimantis leonardi* (Boulenger, 1906), ptychadenids [Savanna Grass Frog (*Ptychadena superciliaris* (Günther, 1858)], and pipids [Western Dwarf Clawed Frog (*Hymenochirus curtipes* Noble, 1924), False Fraser's Clawed Frog (*Xenopus allofraseri* Evans, Carter, Greenbaum, et al., 2015)].

Hidden Among the Unknown – Angolan Tadpoles

An important and often neglected component of studying amphibians is knowledge of their larvae. Unlike adult frogs, whose activity is quite dependent on appropriate weather conditions, breeding season, and nocturnal activity for most species, tadpoles can be easily found in water bodies, during the day, and throughout the year in some cases. The study of tadpoles includes not only morphology, but also microhabitat requirements, ecology, behaviour, feeding habits, predator-prey interactions, etc. Although they look similar at first glance, tadpole morphology usually allows the identification to genus, and a more precise analysis can often lead to species identification.

Early literature regarding southern African tadpoles often covers species occurring in Angola (Van Dijk 1966, 1971). Channing et al. (2012) provide a comprehensive review of the knowledge about African tadpoles with keys to the identification of genera and detailed description of species. Given the poorly known status of Angolan amphibians, it is not surprising that very little is known about Angolan tadpoles. Out of the 99 Angolan species that have tadpoles (i. e., *Breviceps* and *Arthroleptis* genera not included), the tadpoles of only 44 species have been described, and only those of *Ptychadena porosissima* (Channing et al. 2012), *Amietia angolensis* (Channing et al. 2016) and of the recent discoveries of the endemic *Hyperolius chelaensis* (Conradie et al. 2012), *H. cinereus* and *H. raymondi* (Conradie et al. 2013) are based on Angolan material. A recent description of *Leptopelis anchietae* tadpoles is also based on Angolan material (Channing et al. 2012), but it was not encountered with adult specimens, and was based on the association with the first description of that tadpole (Lamotte and Perret 1961), which was based on a specimen from Cameroon that may involve another species. A list of the Angolan frogs with undescribed tadpoles (Table 12.2) includes some of the more common local species.

Comments on Selected Groups

As a consequence of the current poor knowledge of Angola's amphibians, the taxonomic status of many species in the checklist remains unresolved. Some of these are discussed in this section, as well as recent discoveries from ongoing studies.

Species Complexes and Species with Unclear Boundaries

Some morphologically similar species display variation in calls or habitat and are considered to form a complex of closely-related species, and the resolution of their taxonomic status and distribution requires comprehensive investigation. This is

Research on Angolan Amphibians: Past and Present   251

exemplified by the *Hyperolius marmoratus/viridiflavus* complex in Africa, in which 15 names from Angola have been synonymised *(Hyperolius cinctiventris, H. decoratus, H. huillensis, H. insignis, H. marungaensis, H. microstictus, H. pliciferus, H. vermiculatus, Rappia cinctiventris, R. marmorata marginata, R. m. paralella, R. m. variegata, R. plicifera, R. toulsonii, H. m. alborufus). Hyperolius parallelus* is closely related to this complex, and has several Angolan taxa in its synonymy *(H. angolensis, H. marmoratus* var. *angolensis, H. erythromelanus, H. toulsonii, Rappia marmorata huillensis, R. m. insignis, R. m. taeniolata).* Other difficult groups are the *Hyperolius platyceps* complex, with four names currently subsumed within it *(Hyperolius angolanus, Rappia platyceps* var. *angolensis,*

Table 12.2 Angolan frog species with undescribed tadpoles

Leptopelis bocagii (Günther, 1865)	*Hyperolius platyceps* (Boulenger, 1900)
Leptopelis cynnamomeus (Bocage, 1893)	*Hyperolius polli* (Laurent, 1943)
Leptopelis jordani (Parker, 1936)	*Hyperolius protchei* (Rochebrune, 1885)
Leptopelis marginatus (Bocage, 1895)	*Hyperolius quinquevittatus* (Bocage, 1866)
Leptopelis parvus (Schmidt and Inger, 1959)	*Hyperolius rhizophilus* (Rochebrune, 1885)
Mertensophryne melanopleura (Schmidt and Inger, 1959)	*Hyperolius steindachneri* (Bocage, 1866)
	Hyperolius vilhenai (Laurent, 1964)
Mertensophryne mocquardi (Angel, 1924)	*Kassinula wittei* (Laurent, 1940)
Poyntonophrynus grandisonae (Poynton and Haacke, 1993)	*Phrynomantis affinis* (Boulenger, 1901)
	Phrynobatrachus brevipalmatus (Ahl, 1925)
Poyntonophrynus kavangensis (Poynton and Broadley, 1988)	*Phrynobatrachus cryptotis* (Schmidt and Inger, 1959)
Poyntonophrynus pachnodes. (Ceríaco, Marques, Bandeira et al. 2018a)	*Phrynobatrachus parvulus* (Boulenger, 1905)
	Xenopus andrei (Loumont, 1983)
Sclerophrys buchneri (Peters, 1882)	*Xenopus petersii* (Bocage, 1895)
Afrixalus osorioi (Ferreira, 1906)	*Xenopus epitropicalis* (Fischberg, Colombelli, and Picard, 1982)
Afrixalus fulvovittatus (Cope, 1861)	
Afrixalus wittei (Laurent, 1941).	*Hildebrandtia ornatissima* (Bocage, 1879)
Hyperolius adspersus (Peters, 1877)	*Ptychadena ansorgii* (Boulenger, 1905)
Hyperolius benguellensis (Bocage, 1893)	*Ptychadena bunoderma* (Boulenger, 1907)
Hyperolius bicolor (Ahl, 1931)	*Ptychadena grandisonae* (Laurent, 1954)
Hyperolius bocagei (Steindachner, 1867)	*Ptychadena guibei* (Laurent, 1954)
Hyperolius cinnamomeoventris (Bocage, 1866)	*Ptychadena keilingi* (Monard, 1937)
Hyperolius fasciatus (Ferreira, 1906)	*Ptychadena perplicata* (Laurent, 1964)
Hyperolius ferreirai (Noble, 1924)	*Ptychadena taenioscelis* (Laurent, 1954)
Hyperolius fuscigula (Bocage, 1866)	*Ptychadena upembae* (Schmidt and Inger, 1959)
Hyperolius gularis Ahl, 1931	*Ptychadena uzungwensis* (Loveridge, 1932)
Hyperolius langi (Noble, 1924)	*Tomopterna damarensis* (Dawood and Channing, 2002)
Hyperolius lucani (Rochebrune, 1885)	
Hyperolius maestus (Rochebrune, 1885)	*Tomopterna tuberculosa* (Boulenger, 1882)
Hyperolius nobrei (Ferreira, 1906)	*Amnirana parkeriana* (Mertens, 1938)
Hyperolius parallelus (Günther, 1858)	

Hyperolius fasciatus, Hyperolius ferreirai (originally *Rappia bivittata*)), and the super-cryptic *Hyperolius nasutus* complex. Currently this is represented in Angola by at least four species *(H. adspersus, H. benguellensis, H. dartevellei, H. nasutus)* (Channing et al. 2013) and additional names that have been synonymised *(H. punctulatus, Rappia punctulata)* (Channing et al. 2013) or not assigned to any known species occurring in Angola *(H. microps)*.

Typical toads are another problematic group. Formerly known as *Bufo,* which was cosmopolitan in distribution and included the majority of bufonids, the genus was partitioned with African typical toads transferred to *Amietophrynus* (Frost et al. 2006), and more recently renamed in the reinstated genus *Sclerophrys* (Poynton et al. 2016). Seven species of typical toad occur in Angola (see Table 12.2). The mysterious *S. buchneri,* known only from the holotype from northeastern Angola, is considered as a valid species (Frost 2018), but synonymy with *S. funerea* has been suggested and requires further studies (Tandy and Keith 1972). Apart from *S. lemairii* which is easily distinguishable morphologically from the remaining species, distinction between the other *Sclerophrys* is difficult, even between the most common species. Hybridisation between *Sclerophrys* species has been documented and discussed (Guttman 1967; Passmore 1972; Cunningham and Cherry 2004) and may further complicate identification. The red coloration of the interior thigh and parotid gland development are features commonly used to distinguish the often sympatric *S. pusilla, S. gutturalis* and *S. regularis* (Du Preez and Carruthers 2017), but do not distinguish these species in Angola. It is likely that cryptic diversity exists, and understanding of the genus and delimitation of species boundaries requires an integrative approach with comprehensive surveys, analysis of advertisement calls and genetic studies.

Grass frogs, *Ptychadena* spp., are a challenging genus. At least 15 species of this specious genus are represented in Angola (Appendix). *P. mascareniensis,* a large species complex widespread in Africa and Madagascar, has been recently partitioned (Dehling and Sinsch 2013b) with *Ptychadena nilotica* in much of continental Africa, including Angola (Zimkus et al. 2017). Difficulties in distinguishing *Ptychadena* species have been discussed (Poynton and Broadley 1985b; Dehling and Sinsch 2013a, b), although coloration features such as triangular patch on the head, pattern of the interior thigh (Poynton 1970) and several morphometric and morphological features enable species identification (Dehling and Sinsch 2013a, b). Species distinction in Angola is not clear, and in a recent study as many as six different species of *Ptychadena* were found in the same region (Conradie et al. 2016).

Rainfrogs in Angola are known from a single species, *Breviceps adspersus.* However, analysis of material from Angola and adjacent regions has revealed that the Angolan form has features of *B. mossambicus* and may indicate an undescribed Angolan species (Poynton and Broadley 1985a, 1991).

Groups that remain not fully understood such as *Phrynobatrachus* (Zimkus et al. 2010), *Xenopus* (Furman et al. 2015) and *Amnirana* (Jongsma et al. 2018), all have species widespread in Africa with type localities from Angola, and the resolution of their taxonomy depends on detailed studies in Angola.

Species Synonymised with No Clear Justification

A number of putative Angolan species currently placed in synonymy require reassessment as they may represent hidden diversity currently placed under a different name. Cases are mentioned in the previous section, especially in the *Hyperolius* genus. Other examples of this include the placement of *Hylambates (=Leptopelis) angolensis* in the synonymy of *Leptopelis bocagii*. This resulted from comparison between adult and juvenile specimens (Perret 1976) that may not be comparable. *Hylambates bocagei* var. *leucopunctata* Ferreira 1904, has also been placed in the synonymy of *Leptopelis bocagii* (Ceríaco et al. 2014b) and this also requires further investigation as the well developed finger pads in the type specimen of *H. b. leucopunctata* suggests an arboreal habit, very different from the ground-dwelling habits of *L. bocagii,* which lacks pads on fingers or toes.

Species with Questionable Distributions

Some species described from Angola have widespread distributions throughout Africa and inhabit diverse habitats, suggesting that cryptic diversity may be involved (see examples in Endemism section, below). The classic example of this is the Common River Frog, *Amietia angolensis,* which was considered widespread in the continent, but which was discovered to be in fact a complex of cryptic species, with true *A. angolensis* being restricted to Angola (Channing and Baptista 2013). Another potential example is *Afrixalus osorioi,* which was described from western Angola and remains known in the country only from the type locality, whereas the closest other records are in DRC, nearly 1000 km away from the type locality. Other examples include *Ptychadena porosissima, Leptopelis cynnamomeus, L. bocagii, Hyperolius bocagei,* and highlight the earlier comments that study of Angolan amphibians is crucial for solving many problems in African amphibian taxonomy.

Recent Discoveries and Ongoing Studies

The endemic Anchieta's Treefrog, *Leptopelis anchietae* and Congulo Forest Treefrog, *Leptopelis jordani* have been rediscovered in the Angolan escarpment (Baptista et al. 2017), and together with other frogs belonging to the genus *Kassina, Arthroleptis* and *Amnirana* found in the region, their conservation and taxonomic status are being investigated (Baptista et al. in prep.). Further to this, ongoing studies (Baptista et al. in prep.) are assessing: a candidate new species of *Schismaderma*; the taxonomic status of *Hildebrandtia ornatissima* from the Angolan central plateau, previously discussed by Boulenger (1919); the status of *Hyperolius punctulatus* from the Cuanza River (currently in the synonymy of *Hyperolius nasutus*); and the status of various populations of morphologically distinct pygmy toads that cannot be assigned to known *Poyntonophrynus* species. During the 2011 expedition to Lagoa Carumbo, a large white-lipped frog was morphologically assigned to the

Amnirana lepus group (Branch and Conradie 2015). This assigment has been confirmed in a phylogeny of the genus (Jongsma et al. 2018), and further studies are underway to address the taxonomical status of the Angolan population (Conradie pers. comm.). On the SAREP (2012/3) and the NGOWP (2016/7) expeditions to southeastern Angola, numerous candidate new species were discovered, in the genera *Phrynobatrachus, Breviceps* and *Amnirana,* and are currently under investigation (Conradie pers. comm.). The new country records of *Kassinula wittei* and *Leptopelis* cf. *parvus* are being studied to determine if they conform to the nominal forms from northern Zambia and southern DRC, respectively (Conradie pers. comm.). During recent independent surveys conducted in the northern Angolan provinces of Uíge (Ernst et al. 2014, 2015) and Zaire (Vaz Pinto and Baptista Unpublished Data), two different *Alexteroon* spp. were discovered. The taxonomic status of these, the first Angolan records for this poorly-known hyperoliid genus, are under investigation with the Uíge species tentatively assigned to the nominal species *A. hypsiphonus,* whilst the Zaire discovery has affinities to *A. obstetricans.* The material awaits formal taxonomic assignment pending analysis of type material.

Biogeography

Angola is one of the most biogeographically rich countries in Africa (Huntley 1974, 2019). Geomorphologically, the country can be divided into various regions, including the western lowlands of the Coast Belt, the Transition Zone which includes the escarpment, the Marginal Mountain Chain, the Old (Highland) Plateau, also known as central plateau, which progressively decreases in altitude to the east, where the Congo Basin in the north and the Zambezi-Cubango Basin in the south are located (Huntley 1974). Each of these regions have several biome associations, with habitats ranging from the tropical rainforests on the Maiombe region in the north, to the Namib Desert in the south, one of the oldest deserts in the world (Huntley 1974, 2019). This complexity is reflected in the country's diverse fauna.

The difficulties in establishing clear biogeographic regions for amphibians is demonstrated by Poynton and Broadley (1991), in their thorough analysis of the biogeography of the Zambezian amphibians. For Angolan amphibians, which are much more poorly known, this difficulty is immensely increased. The biogeography of Angolan amphibians can only be assessed after major taxonomic issues are resolved, which in some cases requires the revision of entire genera (Cei 1977; Blanc and Frétey 2000). In early studies of the Angolan herpetofauna, several attempts were made to group species according to the distributions known at the time, and these will be summarised below.

Bocage (1895a) made the first grouping, distinguishing a northern and southern region, each divided into coastal, intermediate and high-altitude zones, and listing species occurring in each block. Monard (1937) used humidity to explain the higher diversity of amphibian species in central Angola (a high-rainfall region), compared to the south. He divided Angolan amphibians into four groups: (i) pan-African species (4% of the country's species; such as *Rana mascareniensis (=Ptychadena nilotica)*, and *Bufo (=Sclerophrys) regularis;* (ii) southern species (10%) which reached their northern limit in Angola, such as *Pyxicephalus adspersus*; (iii) tropical species (40%), from western, central and eastern Africa, highlighting the central African tropics as the most significant influence, and including *Rana (=Amnirana) albolabris* and *Rana (=Hoplobatrachus) occipitalis*; and (iv) endemic species (46%), most of which are no longer considered endemic (see Endemism section).

Based on the species known from Angola at the time, Cei (1977) organised Angolan amphibians in three questionable groups, each with affinities to different habitats and regions: (i) the northern and northeastern forests and savannas, (ii) the plateau, and (iii) the arid and semi-arid regions of the coast and of the south, providing a map to delineate those areas. The first area is wide, with northern and northeastern limits in the Congo, Cuanza and Cassai rivers (in Zaire, Uíge, Malanje and Lunda-Norte), and extending to the southeast through Moxico and Cuando Cubango. Examples of species within this group are *Arthroleptis carquejai* and *Hyperolius steindachneri*. The second region corresponds to the south of Congo and Cuango rivers and comprises the southern tropics: Cuanza-Norte, Cuanza-Sul, Huambo, Bié, Malanje and Huíla provinces. Characteristic species in this group include *Hildebrandtia ornatissima, Hyperolius cinereus, Hyperolius quinquevittatus, Leptopelis anchietae*. The third and southernmost region comprises the arid sections of Benguela, Namibe, and Cunene provinces. The fauna on this group is related to that of the Namib, Kalahari, and Namaqualand regions, and can be exemplified by *Pyxicephalus adspersus* and *Poyntonophrynus dombensis*.

Surprisingly, the Great Escarpment of Angola has not been considered in any of these studies. This escarpment is part of a much larger geomorphological unit that dominates the African subcontinent and extends into western Angola, where it acts as a barrier between the dry coastline and the inland plateau. Due to its climatic and topographic peculiarities, it promotes isolation and thus speciation (Huntley 1974). It is a well-documented center of endemism for birds (Hall 1960), and although the escarpment herpetofauna is poorly understood, its endemism potential for herpetofauna has been highlighted (Laurent 1964, Clark et al. 2011, Baptista et al. 2018, Branch et al. 2019), and endemic amphibian species are known from the region *(Leptopelis jordani, L. marginatus, Amnirana parkeriana* and *Hyperolius chelaensis)*. Bordering the Angolan escarpment to the east, the highlands of the ancient massif include patches of Afromontane forest. These consist of islands of relic cool moist Afromontane forest with great biogeographic interest (Huntley 1974), and also potential for endemism.

Inland to the escarpment zone, the plateau is broadly dominated by miombo woodlands, and its fauna often has influences from adjacent regions. Boundaries between regions are not always clear or well understood. Some of these uncertainties have been mentioned in early studies and still require explanation. Hellmich (1957b) referred to the difficulty in establishing geographical limits between the moist forests of the north and the central plateau. An example of this is the penetration of forest species in association with riverine habitats along the northern rivers of Angola. He also noted that faunal boundaries between the slopes of the eastern plateau and the flatlands between Cassai and Cuando were not clear, with the presence of 'pockets' of herpetological elements typical of the south on the central plateau. Laurent (1964) referred to the known affinities between the species of Katanga, in southeastern DRC, with the species of the Lundas and Moxico in Angola.

All of these early biogeographic regions and the species assigned to them need to be re-evaluated with updated taxonomy, accurate species distributions, and in association with the study of phylogenetic relationships among the various amphibian families and genera occurring in Angola. The confirmation of ancestral relationships within these groups is a prerequisite for testing hypotheses about the timing and environmental correlates of amphibian movement and speciation across the Angolan landscape.

Endemism

The originality of the Angolan amphibians due to the richness of endemic species has been highlighted (Blanc and Frétey 2000). Angola's more unique amphibians are also the most poorly known. There are 21 species of amphibians endemic to Angola, of which about 75% are only known from the type locality or type specimens (Table 12.3). Many have not been found for decades, and in some cases for over 100 years. Most of these species are classified as Data Deficient in the IUCN Red List (IUCN 2017).

A number of endemic taxa have been mentioned in the literature but still await formal description: *Hyperolius* sp. I, *Hyperolius* sp. II, *Hyperolius* sp. III (Monard 1937), possibly unknown genus (Perret 1996), and as the taxonomic studies on Angolan amphibians progress, more endemic species will very likely be discovered. In contrast, many early species that were considered endemic have now been relegated to the synonymy of wide-ranging species. Monard (1937), for instance, considered nearly half (46%) of the 80 Angolan species he considered to occur in the country as endemic. However, of the 37 endemic species he identified, only eight are still recognised. Sixteen of these former 'endemics' have been synonymised with other species; e.g. *Leptopelis angolensis* (= *L. bocagii*), *Rana buneli* (= *Ptychadena bunoderma*), *Hyperolius seabrai* (= *H. bocagei*), *Hyperolius angolanus, H. ferreirai, H. fasciatus* (all =*Hyperolius platyceps*), *H. pliciferus, H. vermiculatus, H. marungaensis* (all =*Hyperolus marmoratus*), *H. angolensis, H. erythromelanus, H. toulsonii* (all =*Hyperolius parallelus*), *H. punctulatus* (=*Hyperolius nasutus*), *Rana myotympanum* (=*Hildebrandtia ornatissima*), *Rana cacondana* and *R. signata* (=*Tomopterna tuberculosa*). Many of these synonymies have poor justification, and whilst some names may reflect regional variation, others

Table 12.3 List of amphibian species endemic to Angola, with IUCN Red List Category (*LC* least concern, *DD* data deficient, *N/A* not assessed), and marked (X) when known only from the type locality

Common name	Scientific name	IUCN	TYPE
Angola River frog	*Amietia angolensis* (Bocage, 1866)	LC	
Parker's white-lipped frog	*Amnirana parkeriana* (Mertens, 1938)	DD	X
Cambondo squeaker	*Arthroleptis carquejai* (Ferreira, 1906)	DD	X
Angola ornate frog	*Hildebrandtia ornatissima* (Bocage, 1879)	DD	
Two-colored reed frog	*Hyperolius bicolor* (Ahl, 1931)	DD	X
Chela Mountain Reed Frog	*Hyperolius chelaensis* (Conradie et al., 2012)	N/A	X
Monard's Reed Frog	*Hyperolius cinereus* (Monard, 1937)	LC	
Brown-throated Reed Frog	*Hyperolius fuscigula* (Bocage, 1866)	DD	X
Loanda Reed Frog	*Hyperolius gularis* (Ahl, 1931)	DD	X
Landana Reed Frog	*Hyperolius lucani* (Rochebrune, 1885)	DD	X
Cabinda Reed Frog	*Hyperolius maestus* (Rochebrune, 1885)	DD	X
Nobre's Reed Frog	*Hyperolius nobrei* (Ferreira, 1906)	N/A	X
Rochebrune's Reed Frog	*Hyperolius protchei* (Rochebrune, 1885)	DD	X
Raymond's Reed Frog	*Hyperolius raymondi* (Conradie et al., 2013)	N/A	
African Reed Frog	*Hyperolius rhizophilus* (Rochebrune, 1885)	DD	X
Luita River Reed Frog	*Hyperolius vilhenai* (Laurent, 1964)	DD	X
Congulo Forest Treefrog	*Leptopelis jordani* (Parker, 1936)	DD	X
Quissange Forest Treefrog	*Leptopelis marginatus* (Bocage, 1895)	DD	X
Ahl's Puddle Frog	*Phrynobatrachus brevipalmatus* (Ahl, 1925)	DD	X
Grandison's Pygmy Toad	*Poyntonophrynus grandisonae* (Poynton and Haacke, 1993)	DD	X
Serra da Neve Pygmy Toad	*Poyntonophrynus pachnodes* (Ceríaco, Marques, Bandeira et al., 2018a)	N/A	X

Taxonomy follows Frost (2018)

referred to species found in other countries may not be conspecific (see Comments on selected groups). All deserve careful re-examination.

At least four species (*Leptopelis marginatus, L. jordani, Amnirana parkeriana,* and *Hyperolius chelaensis*) are escarpment-endemics, and others are plateau-endemics (*Hildebrandtia ornatissima, H. cinereus*). However, in order to effectively protect Angolan endemic amphibians and their habitats, further studies are needed to reveal the relations between endemic amphibians and particular habitat, and also the importance of other potential areas of endemism (e.g. relic Afromontane forest patches, isolated mountains such as Serra da Neve, the Angolan escarpment).

Directions for Future Research in Angola

Detailed species lists for a country are an essential baseline tool for understanding biodiversity, its distribution and conservation status. The confusing status of Angolan amphibian taxonomy has been discussed in previous sections and demonstrates how studying taxonomy forms the bedrock for resolving the many pressing questions regarding Angolan amphibian conservation and biology.

A critical first taxonomic step is to revisit the type localities of all the species described from the country to obtain new topotypical material. This is particularly important for the 15 species described by Bocage *(Amietia angolensis, Hyperolius benguellensis, H. cinnamomeoventris, H. fuscigula, H. quinquevittatus, H. steindachneri, Ptychadena anchietae, P. subpunctata, Sclerophrys funerea, Leptopelis anchietae, L. cynnamomeus, L. marginatus, Hildebrandtia ornatissima, Poyntonophrynus dombensis, Xenopus petersii)*, for which many of the type specimens were lost in the fire that destroyed the collections of the Natural History Museum of Lisbon, and for which the original descriptions are the only available source of information. Possibly also lost are the type specimens of several Angolan endemics described by Rochebrune *(Hyperolius lucani, H. maestus, H. protchei, H. rhizophilus)* (Frost 2018), which have very vague descriptions. For many species, neotypes may need to be designated in order to stabilise their taxonomy. Integrative taxonomic studies, including analysis of genetic material, advertisement calls, adult and larval morphology, habitat associations and natural history are crucial to bring Angolan studies into the new millennium.

Many regions of Angola have never been surveyed for amphibians (see Fig. 12.1). Surveying these areas would greatly improve understanding of amphibian distributions, habitat associations and relative abundance, but are also critical for making assessments on their conservation status in terms of IUCN criteria. Priority areas include the northwestern provinces (Uíge and Zaire), the extensive wetlands of Moxico, the escarpment and the adjacent Afromontane forest patches that are rich in endemic birds (Hall 1960), other vertebrates (Crawford-Cabral 1966; Clark et al. 2011) and also probably amphibians, and for which the urgent need of studies has been highlighted (Laurent 1964; Clark et al. 2011).

The controversial frog from Caquindo (Perret 1996) for which genus assignment lacks consensus (see *Records that require confirmation*), still has to be recollected and its true affinities resolved. This could enrich Angolan herpetology possibly with a new endemic genus. This begs the question – how much remains to be discovered about Angolan amphibians? It also shows how the analysis of extant collections can contribute significantly to the knowledge of the country's fauna. Collections that remain to be studied include those from the Rohan-Chabot Mission, the Vernay Angola Expedition, and the Leptopelinae and Hyperoliidae from the Portuguese Mission of Apiarian Studies of the Overseas.

Another important step to furthering amphibian knowledge is studying the biology of individual species. Some studies are available for iconic species such as the Dombe Pigmy toad *Poyntonophrynus dombensis* (Channing and Vences 1999), based on individuals from Namibia, and Lemaire's toad *Sclerophrys lemairii,* the first study of this kind made in Angolan territory (Conradie and Bills 2017). However, this is still missing for many species, and understanding their natural history, reproduction strategies, breeding sites, breeding seasons, behaviour, habitat and microhabitat requirements, both for adults and tadpole stages, are key for an effective planning of species conservation. All of this is even more relevant for the extremely poorly known Angolan endemics.

Conservation-driven studies about Angolan amphibians require awareness of potential threats to biodiversity, particularly those resulting from habitat loss and climate change. Habitat degradation as a result of exploitation of natural resources and associated with industrialisation have increased dramatically in Angola in recent decades and will affect amphibians. The implementation of monitoring programmes are crucial for documenting and understanding this relation. Research about the appearance and effect of global amphibian diseases such as the chytrid fungus *(Batrachochytrium dendrobatidis),* viruses *(Ranavirus* spp.*),* and other pathogens, are lacking in Angola, even though they are threatening amphibians around the world and are reported from neighbouring countries (Greenbaum et al. 2014).

The study of Angolan amphibians is a broad and important component of biodiversity studies, for which many baseline questions remain unanswered, and exciting discoveries are still to be made. This becomes more evident when confronted with the fact that Angolan fauna is among the least studied in Africa. Increasing public awareness about amphibians and their importance is necessary for their conservation, and requires developing local knowledge and expertise, as well as constructing functional amphibian collections in national archives. These are essential steps for understanding and protecting this rich, diverse and ecologically important group. This is even more urgent in an era where an "amphibian decline crisis" is happening around the world (Beebee and Griffiths 2005), and where this decline is known to have major consequences in ecosystem function (Whiles et al. 2006).

Acknowledgements The writing of this chapter was made possible through a convergence of efforts and projects. SASSCAL Project (sponsored by the German Federal Ministry of Education and Research (BMBF) under promotion number 01LG1201M); Conservation Leadership Programme (Project CLP ID: F01245015: Conserving Angolan Scarp Forests: a Holistic Approach for Kumbira Forest); National Geographic/Okavango Wilderness Project (NGOWP); South Africa's National Research Foundation (2009–2017, WRB), National Geographic Society (Explorer Grant 2011, WRB); Fundação para a Ciência e Tecnologia (contract SFRH/PD/BD/140810/2018, NB); and Wild Bird Trust 2015–2018. Particular thanks go to Fernanda Lages (ISCED Huíla), Brian Huntley (South Africa), John Hilton and Rainer Von Brandis (Wild Bird Trust) for logistical and administrative support.

Appendix 1

Checklist of the amphibians recorded in Angola, based on historical records and on confirmed records from recent surveys. Taxonomy follows Frost (2018). Unconfirmed records are not included. To avoid redundancy, records included in existing compilations (e.g. Monard 1937; Ruas 1996) are mentioned under the compilation reference, and the original reference(s) is not included in the list

Common name	Species	References
Family Arthroleptidae		
Carqueja's Squeaker	*Arthroleptis carquejai* (Ferreira, 1906)	Ferreira (1906)
Lameer's Squeaker	*Arthroleptis lameerei* (De Witte, 1921)	Laurent (1964) and Ruas (1996)
Tanganyika Screeching Frog	*Arthroleptis spinalis* (Boulenger, 1919)	Laurent (1950)
Common Squeaker	*Arthroleptis stenodactylus* (Pfeffer, 1893)	Laurent (1964), Ruas (1996) and Conradie et al. (unpub. data)
Variable Squeaker	*Arthroleptis variabilis* (Matschie, 1893)	Baptista and Vaz Pinto (unpub. data)
Plain Squeaker	*Arthroleptis xenochirus* (Boulenger, 1905)	Monard (1937), Laurent (1964), Ruas (1996), Ceríaco et al. (2018b), Conradie et al. (unpub. data), Baptista and Vaz Pinto (unpub. data), and Ernst (unpub. data)
Anchieta's Treefrog	*Leptopelis anchietae* (Bocage, 1873)	Bocage (1895), Boulenger (1905), Schmidt (1936), Monard (1937), Laurent (1964), Conradie et al. (2016), Baptista et al. (2018), (in prep.) and Ernst (unpub. data)
Gaboon Forest Treefrog	*Leptopelis aubryi* (Duméril, 1856)	Peters (1887) and Laurent (1954)
Bocage's Burrowing Treefrog	*Leptopelis bocagii* (Günther, 1865)	Bocage (1895), Monard (1937), Hellmich (1957b), Laurent (1954, 1964), Ceríaco et al. (2018b), Baptista et al (2018, in prep), Baptista and Vaz Pinto (unpub. data) and Conradie et al. (unpub. data)
Efulen Forest Treefrog	*Leptopelis calcaratus* (Boulenger, 1906)	Baptista and Vaz Pinto (unpub. data)
Cinnamon Treefrog	*Leptopelis cynnamomeus* (Bocage, 1893)	Bocage (1895), Monard (1937) and Laurent (1964)
Congulo Forest Treefrog	*Leptopelis jordani* (Parker, 1936)	Parker (1936) and Baptista et al. (2017)
Quissange Forest Treefrog	*Leptopelis marginatus* (Bocage, 1895)	Bocage (1895)
Kanole Forest Treefrog	*Leptopelis* cf. *parvus* (Schmidt and Inger, 1959)	Conradie et al. (unpub. data)
Rusty Forest Treefrog	*Leptopelis viridis* (Günther, 1869)	Boulenger (1882) and Bocage (1895)
Hairy Frog	*Trichobatrachus robustus* (Boulenger, 1900)	Ernst et al. (2014)

(continued)

Research on Angolan Amphibians: Past and Present

Common name	Species	References
Lang's Reed Frog	*Hyperolius langi* (Noble, 1924)	Monard (1937)
Landana Reed Frog	*Hyperolius lucani* (Rochebrune, 1885)	Rochebrune (1885)
Cabinda Reed Frog	*Hyperolius maestus* (Rochebrune, 1885)	Rochebrune (1885)
Marbled Reed Frog	*Hyperolius marmoratus* (Rapp, 1842)	Boulenger (1882), Bocage (1895) and Monard (1937)
Large-nosed Long Reed Frog	*Hyperolius nasutus* (Günther, 1865)	Bocage (1895), Monard (1937), Laurent (1950, 1954, 1964), Hellmich (1957b), Baptista and Vaz Pinto (unpub. data) and Ceríaco et al. (2018b)
Nobre's Reed Frog	*Hyperolius nobrei* (Ferreira, 1906)	Ferreira (1906)
Angolan Reed Frog	*Hyperolius parallelus* (Günther, 1858)	Monard (1937), Laurent (1950, 1954, 1964), Ceríaco et al. (2018b), Conradie et al. (unpub. data), Baptista et al. (2018) and Baptista and Vaz Pinto (unpub. data)
Leopard Reed Frog	*Hyperolius pardalis* (Laurent, 1948)	Conradie (unpub. data)
Rio Luinha Reed Frog	*Hyperolius platyceps* (Boulenger, 1900)	Monard (1937), Laurent (1950, 1954) and Baptista and Vaz Pinto (unpub. data)
Tshimbulu Reed Frog	*Hyperolius polli* (Laurent, 1943)	Laurent (1954)
Rochebrune's Reed Frog	*Hyperolius protchei* (Rochebrune, 1885)	Rochebrune (1885)
Five-striped Reed Frog	*Hyperolius quinquevittatus* (Bocage, 1866)	Bocage (1895), Laurent (1950, 1954) and Baptista and Vaz Pinto (unpub. data)
Raymond's Reed Frog	*Hyperolius raymondi* (Conradie et al. 2013)	Conradie et al. (2013)
African Reed Frog	*Hyperolius rhizophilus* (Rochebrune, 1885)	Rochebrune (1885)
Steindachner's Reed Frog	*Hyperolius steindachneri* (Bocage, 1866)	Bocage (1895), Monard (1937), Laurent (1950, 1954, 1964), Poynton and Haacke (1993) and Channing and Vaz Pinto (unpub. data)
Luita River Reed Frog	*Hyperolius vilhenai* (Laurent, 1964)	Laurent (1964)
Kuvangu Kassina	*Kassina kuvangensis* (Monard, 1937)	Monard (1937) and Conradie et al. (2016, unpub. data)
Family Hemisotidae (cont.)		
Bubbling Kassina	*Kassina senegalensis* (Duméril and Bibron, 1841)	Monard (1937), Laurent (1954, 1964), Poynton and Haacke (1993), Conradie et al. (2016), Baptista et al. (2018), Baptista and Vaz Pinto (unpub. data), Conradie et al. (unpub. data) and Ernst (unpub. data)
De Witte's Clicking Frog	*Kassinula wittei* (Laurent, 1940)	Conradie et. al (unpub. data)

(continued)

Common name	Species	References
Family Microhylidae		
Spotted Rubber Frog	*Phrynomantis affinis* (Boulenger, 1901)	Laurent (1964)
Marbled Rubber Frog	*Phrynomantis annectens* (Werner, 1910)	Ruas (1996) and Vaz Pinto and Branch (unpub. data)
Banded Rubber Frog	*Phrynomantis bifasciatus* (Smith, 1847)	Boulenger (1882), Monard (1937), Ruas (1996), Channing (unpub. data) and Baptista et al. (unpub. data)
Family Phrynobatrachidae		
Ahl's Puddle Frog	*Phrynobatrachus brevipalmatus* (Ahl, 1925)	Ahl (1925)
Cryptic Puddle Frog	*Phrynobatrachus cryptotis* (Schmidt and Inger, 1959)	Laurent (1964)
Mababe Puddle Frog	*Phrynobatrachus mababiensis* (FitzSimons, 1932)	Poynton and Haacke (1993), Conradie et al. (2016, unpub. data) and Baptista and Vaz Pinto (unpub. data)
Snoring Puddle Frog	*Phrynobatrachus natalensis* (Smith, 1849)	Bocage (1895), Monard (1937), Hellmich (1957b), Ruas (1996), Conradie et al. (2016, unpub. data), Ceríaco et al. (2018b), Baptista et al. (2018, in press) and Baptista and Vaz Pinto (unpub. data)
Loanda River Frog	*Phrynobatrachus parvulus* (Boulenger, 1905)	Ruas (1996), Baptista and Vaz Pinto (unpub. data) and Conradie et al. (unpub. data)
Family Pipidae		
Andre's Clawed Frog	*Xenopus andrei* (Loumont, 1983)	Ernst et al. (2015)
Southern Tropical Clawed Frog	*Xenopus epitropicalis* (Fischberg et al., 1982)	Laurent (1950, 1954) and Klein (unpub. data)
Müller's Clawed Frog	*Xenopus muelleri* (Peters, 1844)	Conradie et al. (2016)
Peters' Clawed Frog	*Xenopus petersii* (Bocage, 1895)	Bocage (1895), Monard (1937), Hellmich (1957b), Ruas (1996), Baptista et al. (2018), Baptista and Vaz Pinto (unpub. data), Ceríaco et al. (2018b) and Ernst (unpub. data)
Power's Clawed Frog	*Xenopus poweri* (Hewitt, 1927)	Conradie et al. (2016)
Family Pipidae (cont.)		
Clawed Frog	*Xenopus* sp.	Laurent (1950)
Family Ptychadenidae		
Common Ornate Frog	*Hildebrandtia ornata* (Peters, 1878)	Poynton and Haacke (1993)
Angola Ornate Frog	*Hildebrandtia ornatissima* (Bocage, 1879)	Bocage (1895), Monard (1937) and Baptista and Vaz Pinto (unpub. data)
Anchieta's Grass Frog	*Ptychadena anchietae* (Bocage, 1868)	Ruas (1996), Ceríaco et al. (In press.), Baptista et al (2018) and Baptista and Vaz Pinto (unpub. data)
Ansorge's Grass Frog	*Ptychadena ansorgii* (Boulenger, 1905)	Monard (1937) and Ruas (1996)

(continued)

Common name	Species	References
Rough Grass Frog	*Ptychadena bunoderma* (Boulenger, 1907)	Monard (1937), Ruas (1996) and Conradie et al. (unpub. data)
Grandison's Grass Frog	*Ptychadena grandisonae* (Laurent, 1954)	Ruas (1996)
Guibe's Grass Frog	*Ptychadena guibei* (Laurent, 1954)	Ruas (1996), Ceríaco et al. (in press), Conradie et al. (2016) and Baptista and Vaz Pinto (unpub. data)
Keiling's Grass Frog	*Ptychadena keilingi* (Monard, 1937)	Ruas (1996) and Conradie et al. (unpub. data)
Mozambique Grass Frog	*Ptychadena* cf. *mossambica* (Peters, 1854)	Conradie et al. (2016) and Conradie (unpub. data)
Nile Grass Frog	*Ptychadena nilotica* (Seetzen, 1855)	Monard (1937), Schmidt and Inger (1959), Ruas (1996), Conradie et al. (2016), Dehling and Sinsch (2013b) and Zimkus et al. (2017)
Sharp-nosed Grass Frog	*Ptychadena oxyrhynchus* (Smith, 1849)	Monard (1937), Hellmich (1957b), Ruas (1996), Ceríaco et al. (2018b), Conradie et al. (2016) and Baptista (unpub. data)
Many-Grass Frog	*Ptychadena perplicata* (Laurent, 1964)	Laurent (1964)
Striped Grass Frog	*Ptychadena porosissima* (Steindachner, 1867)	Ruas (1996), Conradie et al. (unpub. data) and Channing et al. (2012)
Spot-bellied Grass Frog	*Ptychadena subpunctata* (Bocage, 1866)	Ruas (1996), Conradie et al. (2016) and Baptista (unpub. data)
Small Grass Frog	*Ptychadena taenioscelis* (Laurent, 1954)	Ruas (1996) and Conradie et al. (2016)
Upemba Grass Frog	*Ptychadena upembae* (Schmidt and Inger, 1959)	Ruas (1996)
Udzungwa Grass Frog	*Ptychadena uzungwensis* (Loveridge, 1932)	Ruas (1996) and Conradie et al. (2016, unpub. data)
Family Pyxicephalidae		
Angola River Frog	*Amietia angolensis* (Bocage, 1866)	Bocage (1895), Monard (1937), Ruas (1996), Channing and Baptista (2013), Ceríaco et al. (2016b), Channing et al. (2016), Baptista et al. (2018) and Baptista and Vaz Pinto (unpub. data)
African Bullfrog	*Pyxicephalus adspersus* (Tschudi, 1838)	Monard (1937) and Ruas (1996)
Cryptic Sand Frog	*Tomopterna cryptotis* (Boulenger, 1907)	Monard (1937), Ruas (1996), Conradie et al. (2016) and Baptista et al. (in press)
Damaraland Sand Frog	*Tomopterna damarensis* (Dawood and Channing, 2002)	Ceríaco et al. (2016a) and Heinicke et al. (2017)
Rough Sand Frog	*Tomopterna tuberculosa* (Boulenger, 1882)	Bocage (1895), Monard (1937), Ruas (1996), Baptista et al (2018, unpub. data) and Conradie et al. (unpub. data)
Family Ranidae		
Forest White-lipped Frog	*Amnirana albolabris* (Hallowell, 1856)	Bocage (1895), Monard (1937), Ruas (1996) and Jongsma et al. (2018)

(continued)

Common name	Species	References
Darling's White-lipped Frog	*Amnirana darlingi* (Boulenger, 1902)	Monard (1937), Laurent (1964), Ruas (1996), Ceríaco et al. (2018b) Branch and Conradie (2015) and Conradie et al. (unpub. data)
Lemaire's White-lipped Frog	*Amnirana lemairei* (De Witte, 1921)	Laurent (1964), Ruas (1996) and Baptista and Vaz Pinto (unpub. data)
Andersson's White-lipped Frog	*Amnirana* cf. *lepus* (Andersson, 1903)	Branch and Conradie (2015)
Parker's White-lipped Frog	*Amnirana parkeriana* (Mertens, 1938)	Mertens (1938)

References

Ahl E (1925 "1923") Ueber neue afrikanische Frösche der Familie Ranidae. Sitzungsberichte der Gesellschaft Naturforschender Freunde zu Berlin 1923:96–106

Ahl E (1931) Amphibia, Anura III, Polypedatidae. Das Tierreich 55: xvi + 477

Amiet JL (2012) Les Rainettes du Cameroun (Amphibiens Anoures). La Nef des Livres, Saint-Nazaire, 591 pp

Baptista N, António T, Branch WR The herpetofauna of Bicuar National Park and surrounds, southwestern Angola: a first description and preliminary checklist. Amphibian & Reptile Conservation. (in press).

Baptista N, Vaz Pinto P, Ernst R et al (2017) Cryptic diversity in treefrogs (*Leptopelis*) of the Angolan escarpment – fitting the pieces together. 13th conference of the Herpetological Association of Africa, Bonamanzi, South Africa

Baptista N, António T, Branch WR (2018) Amphibians and reptiles of the Tundavala region of the Angolan Escarpment. In: Revermann R, Krewenka KM, Schmiedel U et al (eds) Climate change and adaptive land management in Southern Africa – assessments, changes, challenges, and solutions, Biodiversity & ecology, vol 6, pp 397–403

Beard KH, Vogt KA, Kulmatiski A (2002) Top-down effects of a terrestrial frog on forest nutrient dynamics. Oecologia 133(4):583–593

Beebee TJ, Griffiths RA (2005) The amphibian decline crisis: a watershed for conservation biology? Biol Conserv 125(3):271–285

Blanc CP, Frétey T (2000) Biogeographie des Amphibiens d'Afrique Centrale et d'Angola. Biogeographica 76(3):107–118

Bocage JVB (1866a) Lista dos reptis das possessões portuguesas d' Africa occidental que existem no Museu de Lisboa. Jornal de Sciências Mathemáticas, Physicas e Naturaes. Lisboa 1:37–56

Bocage JVB (1866b) Reptiles nouveaux ou peu connus recueillis dans les possessions portugaises de l'Afrique occidentale, qui se trouvent au Muséum de Lisbonne. Jornal de Sciencias Mathematicas, Physicas e Naturaes. Lisboa I(1):57–78

Bocage JVB (1867) Batraciens nouveaux de l'Afrique occidentale (Loanda et Benguella). Proc Zool Soc London 35:843–846

Bocage JVB (1873) Mélanges erpétologiques. Sur quelques Reptiles et Batraciens nouveux, rares ou peu connues de l'Afrique occidentale. Jornal de Sciências Mathemáticas, Physicas e Naturaes. Lisboa 4(1):209–227

Bocage JVB (1879a) Reptiles et batraciens nouveaux d' Angola. Jornal de Sciências Mathemáticas, Physicas e Naturaes. Lisboa 7(26):97–99

Bocage JVB (1879b) Subsidio para a fauna das possessões portuguezas d'África occidental. Jornal de Sciências Mathemáticas, Physicas e Naturaes. Lisboa (1) 7:85–95

Bocage JVB (1882) Reptiles rares ou nouveaux d'Angola. Jornal de Sciencias, Mathemáticas,

Physicas e Naturaes. Lisboa (1) 8:299–304

Bocage JVB (1893) Diagnose de quelques nouvelles espéces de reptiles et batraciens d' Angola. Jornal de Sciências Mathemáticas, Physicas e Naturaes. Lisboa (2) 10:115–121

Bocage JVB (1895a) Herpétologie d'Angola et du Congo. Lisbonne, Imprimerie Nationale, 203 pp, 19 pls

Bocage JVB (1895b) Sur une espêce de Crapaud à ajouter à la faune herpétologique d'Angola. Jornal de Sciências Mathemáticas, Physicas e Naturaes. Lisboa (2) 4:51–53

Bocage JVB (1896a) Mamíferos, aves e réptis da Hanha, no sertão de Benguella. Jornal de Sciências Mathemáticas, Physicas e Naturaes. Lisboa (2) 14:105–114

Bocage JVB (1896b) Répteis de algumas possessões portuguesas de Africa que existem no Museu de Lisboa. Jornal de Sciências Mathemáticas, Physicas e Naturaes 2:65–104

Bocage JVB (1897a) Mamíferos, aves e reptis da Hanha, no sertão de Benguella. Segunda lista. Jornal de Sciências Mathemáticas, Physicas e Naturaes. Lisboa (2):207–211

Bocage JVB (1897b) Mamíferos, réptis e batrachios d'África de que existem exemplares típicos no Museu de Lisboa. Jornal de Sciências Mathemáticas, Physicas e Naturaes. Lisboa (2) 4:187–206

Boulenger GA (1882) Catalogue of the Batrachia Salientia s. Ecaudata in the collection of the British museum, 2nd edn. Taylor and Francis, London

Boulenger GA (1905) A list of the batrachians and reptiles collected by Dr. W. J. Ansorge in Angola, with descriptions of new species. Ann Mag Nat Hist Ser 7 16(92):8–115

Boulenger GA (1907a) Descriptions of three new lizards and a new frog, discovered by Dr. W J. Ansorge in Angola. Ann Mag Nat Hist Ser 7 19:212–214

Boulenger GA (1907b) Description of a new frog discovered by Dr. W. J. Ansorge in Mossamedes, Angola. Ann Mag Nat Hist Ser 7 20:109

Boulenger GA (1919) On *Rana ornatissima*, Bocage, and *R. ruddi*, Blgr. Trans Royal Soc S Afr 8:33–37

Branch WR, Bauer AM (2005) The life and herpetological contributions of Andrew Smith. pp. 1-19 in *Smith, A. The Herpetological Contributions of Sir Andrew Smith*. Society for the Study of Amphibians and Reptiles, Villanova, PA. iv + 84 pp

Branch WR, Conradie WC (2015) Herpetofauna da região da Lagoa Carumbo (Herpetofauna of the Carumba Lagoon Area). In: Huntley BJ (ed), Relatório sobre a Expedição Avaliação rápida da Biodiversidade de região da Lagoa Carumbo, Lunda-Norte – Angola, República de Angola. Ministério do Ambiente, pp 194–209, 219p

Branch WR, Baptista N, Keates CW, et al (2019) Rediscovery, taxonomic status, and phylogenetic relationships of two rare and endemic snakes (Serpentes: Psammophinae) from the Angolan Escarpment. Zootaxa, in press

Brito D (2010) Overcoming the Linnean shortfall: data deficiency and biological survey priorities. Basic Appl Ecol 11(8):709–713

Broadley DG (1971) The reptiles and amphibians of Zambia. The Puku, Occ Pap Dept Game Fish Zambia 7:1–143

Brooks C (2012) Biodiversity survey of the upper Angolan Catchment of the Cubango-Okavango River Basin. USAid-Southern Africa. 151 pp

Brooks C (2013) Trip report: aquatic biodiversity survey of the lower Cuito and Cuando river systems in Angola. USAid-Southern Africa. 43 pp

Cei JM (1977) Chaves para uma identificação preliminar dos batráquios anuros da R. P. de Angola. Boletim da Sociedade Portuguesa de Ciências Naturais 17:5–26

Ceríaco LMP, Bauer AM, Blackburn et al (2014a) The herpetofauna of the Capanda Dam region, Malanje, Angola. Herpetol Rev 45(4):667–674

Ceríaco LMP, Blackburn DC, Marques MP et al (2014b) Catalogue of the amphibian and reptile type specimens of the Museu de História Natural da Universidade do Porto in Portugal, with some comments on problematic taxa. Alytes 31(1):13–36

Ceríaco LMP, Bauer AM, Heinicke MP et al (2016a) Geographical distributions: Ptychadenidae, *Ptychadena mapacha* Channing, 1993 – Mapacha ridged frog in Namibia. Afr Herp News 63:19–20

Ceríaco LMP, de Sá SAC, Bandeira SA et al (2016b) Herpetological survey of Iona National Park and Namibe regional natural park, with a synoptic list of the amphibians and reptiles of Namibe

Province, Southwestern Angola. Proc Calif Acad Sci 63(2):15–61

Ceríaco LMP, Marques MP, Bandeira SA et al (2016c) Anfíbios e répteis do Parque Nacional da Cangandala. Instituto Nacional da Biodiversidade e Áreas de Conservação & Museu Nacional de História Natural e da Ciência, 96 pp

Ceríaco LMP, Marques MP, Bandeira S et al (2018a) A new earless species of Poyntonophrynus (Anura: Bufonidae) from the Serra da Neve Inselberg, Namibe Province, Angola. Zookeys 780:109–136

Ceríaco LMP, Marques MP, Bandeira S et al (2018b) Herpetological survey of Cangandala National Park, with a synoptic list of the amphibians and reptiles of Malanje Province, Central Angola. Herpetological Review 49(3):408–431

Channing A (1993) A new grass frog from Namibia. S Afr J Zool 28:142–145

Channing A (1999) Historical overview of amphibian systematics in Southern Africa. Trans Royal Soc S Afr 54(1):121–135

Channing A (2001) Amphibians of central and southern Africa. Cornell University Press, New York, 470 pp

Channing A, Baptista N (2013) Amietia angolensis and A. fuscigula (Anura: Pyxicephalidae) in southern Africa: a cold case reheated. Zootaxa 3640(4):501–520

Channing A, Griffin M (1993) An annotated checklist of the frogs of Namibia. Modoqua 18:101–116

Channing A, Vences M (1999) The advertisement call, breeding biology, description of the tadpole and taxonomic status of Bufo dombensis, a little-known dwarf toad from southern Africa. S Afr J Zool 34:74–79

Channing, A., and D. G. Broadley. 2002. A new snout-burrower from the Barotse Floodplain (Anura: Hemisotidae: Hemisus). Journal of Herpetology 36: 367–372.

Channing A, Rödel MO, Channing J (2012) Tadpoles of Africa: the biology and identification of all known tadpoles in sub-Saharan Africa. Edition Chimaira, Frankfurt, 401pp

Channing A, Hillers A, Lötters S et al (2013) Taxonomy of the super-cryptic Hyperolius nasutus group of long reed frogs of Africa (Anura: Hyperoliidae), with descriptions of six new species. Zootaxa 3620(3):301–350

Channing A, Dehling JM, Lötters S et al (2016) Species boundaries and taxonomy of the African river frogs (Amphibia: Pyxicephalidae: Amietia). Zootaxa 4155(1):1–76

Clark VR, Barker NP, Mucina L (2011) The great escarpment of southern Africa: a new frontier for biodiversity exploration. Biodivers Conserv 20(12):2543–2561

Coetzer W (2017) Occurrence records of southern African aquatic biodiversity. Version 1.10. The south African Institute for Aquatic Biodiversity. Occurrence dataset https://doi.org/10.15468/pv7vds. Accessed via GBIF.org

Conradie W, Bills R (2017) Wannabe Ranid: notes on the morphology and natural history of the Lemaire's toad (Bufonidae: Sclerophrys lemairii). Salamandra 53(3):439–444

Conradie W, Branch WR, Measey GJ et al (2012) A new species of Hyperolius Rapp, 1842 (Anura: Hyperoliidae) from the Serra da Chela mountains, South-Western Angola. Zootaxa 3269(1):1–17

Conradie W, Branch WR, Tolley KA (2013) Fifty Shades of Grey: giving colour to the poorly known Angolan Ashy reed frog (Hyperoliidae: Hyperolius cinereus), with the description of a new species. Zootaxa 2636(3):201–223

Conradie W, Bills R, Branch WR (2016) The herpetofauna of the Cubango, Cuito, and lower Cuando river catchments of South-Eastern Angola. Amphibian Reptile Conserv 10(2):6–36

Crawford-Cabral JC (1966) Some new data on Angolan Muridae. Zool Afr 2:193–203

Crawford-Cabral J, Mesquitela LM (1989) Índice toponímico de colheitas zoológicas em Angola. Instituto de Investigação Cientifica Tropical, Centro de Zoologia, Lisbon, 206

Cunningham M, Cherry MI (2004) Molecular systematics of African 20-chromosome toads (Anura: Bufonidae). Mol Phylogenet Evol 32(3):671–685

Curtis B, Roberts KS, Griffin M et al (1998) Species richness and conservation of Namibian freshwater macro-invertebrates, fish and amphibians. Biodivers Conserv 7(4):447–466

Davic RD, Welsh JHH (2004) On the ecological roles of salamanders. Annu Rev Ecol Evol Syst 35:405–434

Dehling JM, Sinsch U (2013a) Diversity of *Ptychadena* in Rwanda and taxonomic status of *P. chrysogaster* Laurent, 1954 (Amphibia, Anura, Ptychadenidae). Zoo Keys 356:69–102

Dehling JM, Sinsch U (2013b) Diversity of Ridged Frogs (Anura: Ptychadenidae: *Ptychadena* spp.) in wetlands of the upper Nile in Rwanda: morphological, bioacoustic, and molecular evidence. Zoologischer Anzeiger 253(2):143–157

Du Preez L, Carruthers V (2009) A complete guide to the frogs of Southern Africa. Struik Publishers, Cape Town, 488 pp

Du Preez L, Carruthers V (2017) Frogs of Southern Africa: a complete guide. Struik Publishers, Cape Town, 520 pp

Ernst R, Nienguesso ABT, Lautenschlaeger T et al (2014) Relicts of a forested past: southernmost distribution of the hairy frog genus *Trichobatrachus* Boulenger, 1900 (Anura: Arthroleptidae) in the Serra do Pingano region of Angola with comments on its taxonomic status. Zootaxa 3779(2):297–300

Ernst R, Schmitz A, Wagner P, Branquima MF et al (2015) A window to Central African forest history: distribution of the *Xenopus fraseri* subgroup south of the Congo Basin, including a first country record of *Xenopus andrei* from Angola. Salamandra 52(1):147–155

Ferreira JB (1897) Lista dos reptis e amphibios que fazem parte da última remessa de J. d'Anchieta. Jornal de Sciências Mathemáticas, Physicas e Naturaes 5(2):240–246

Ferreira JB (1900) Sobre alguns exemplares pertencentes à fauna do norte de Angola (Reptis, Batrachios, Aves e Mammiferos). Jornal de Sciências Mathemáticas, Physicas e Naturaes, Lisboa 2(6):48–54

Ferreira JB (1904) Reptis e amphibios de Angola da região ao norte do Quanza (Collecção Newton – 1903). Jornal de Sciências Mathemáticas, Physicas e Naturaes, Segunda Série 7(26):111–117

Ferreira JB (1906) Algumas espécies novas ou pouco conhecidas de amphibios e reptis de Angola (Collecção Newton – 1903). Jornal de Sciências Mathemáticas, Physicas e Naturaes, Segunda Série 7(26):159–171

Frétey T, Dewynter M, Blanc CP (2011) Amphibiens d'Afrique central et d'Angola. Clé de détermination ilustrée desamphibiens du Gabo et du Mbini/Illustrated identification key of the amphibians from Gabon and Mbini. Biotope, Mèze/Muséum national d'Histoire naturelle, Paris, 232 pp

Frost DR (2018) Amphibian species of the world: an online reference. Version 6.0. Electronic database accessible at http://research.amnh.org/herpetology/amphibia/index.html. American Museum of Natural History, New York

Frost DR, Grant T, Faivovich J et al (2006) The amphibian tree of life. Bull Am Mus Nat Hist 297:1–370

Furman BL, Bewick AJ, Harrison TL et al (2015) Pan-African phylogeography of a model organism, the African clawed frog 'Xenopus laevis'. Mol Ecol 24(4):909–925

Greenbaum E, Meece J, Reed KD et al (2014) Amphibian chytrid infections in non-forested habitats of Katanga, Democratic Republic of the Congo. Herpetol Rev 45:610–614

Günther ACLG (1865 '1864') Descriptions of new species of batrachians from West Africa. Proc Zool Soc London 3:479–482

Guttman SI (1967) Transferrin and hemoglobin polymorphism, hybridization and introgression in two African toads, *Bufo regularis* and *Bufo rangeri*. Comp Biochem Physiol 23(3):871–877

Haacke WD (1999) Geographical distribution: *Ptychadena mapacha* Channing, 1993 – Mapacha Grass Frog. Afr Herp News 30:35

Hall BP (1960) The faunistic importance of the scarp of Angola. Ibis 102(3):420–442

Heinicke MP, Ceríaco LM, Moore IM et al (2017) *Tomopterna damarensis* (Anura: Pixicephalidae) is broadly distributed in Namibia and Angola. Salamandra 53(3):461–465

Hellmich W (1957a) Die reptilienausbeute der Hamburgischen Angola Expedition. Mitteilungen aus dem Hamburgischen Zoologischen Museum und Institut 55:39–80

Hellmich W (1957b) Herpetologische Ergebnisse einer Forschungsreise in Angola. Veröffentlichungen der Zoologischen Staatssammlung München 5:1–92

Herrmann HW, Branch WR (2013) Fifty years of herpetological research in the Namib Desert and Namibia with an updated and annotated species checklist. J Arid Environ 93:94–115

Huntley BJ (1974) Outlines of wildlife conservation in Angola. J S Afr Wildl Manag Assoc

4:157–166

Huntley BJ (2019) Angola in outline: physiography, climate and patterns of biodiversity. In: Huntley BJ, Russo V, Lages F, Ferrand N (eds) Biodiversity of Angola. Science & conservation: a modern synthesis. Springer, Cham

IUCN Red List of Threatened Species Version 2017-2. www.iucnredlist.org

IUCN Red List of Threatened Species. Version 2017-3. www.iucnredlist.org

Jongsma CF, Barej MF, Barratt CD et al (2018) Diversity and biogeography of frogs in the genus Amnirana (Anura: Ranidae) across sub-Saharan Africa. Mol Phylogenet Evol 120:274–285

Jordan K (1936) Dr Karl Jordan's expedition to South-West Africa and Angola. Narrative. Novitates Zooligicae 40:17–62, 2 maps, 5 pls

Jürgens N, Strohbach B, Lages F et al (2018) Biodiversity observation – an overview of the current state and first results of biodiversity monitoring studies. In: Revermann R, Krewenka KM, Schmiedel U et al (eds) Climate change and adaptive land management in southern Africa – assessments, changes, challenges, and solutions, Biodiversity & ecology, vol 6, pp 382–396

Köhler J, Vieites DR, Bonett RM et al (2005) New amphibians and global conservation: a boost in species discoveries in a highly endangered vertebrate group. AIBS Bull 55(8):693–696

Lamotte M, Perret JL (1961) Les formes larvaires de quelques espèces de *Leptopelis: L. aubryi, L. viridis, L. anchietae, L. ocellatus et L. calcaratus*. Bulletin de l'Institute fondamental d'Afrique noire, Sér. A 23:855–885

Laurent RF (1950) Reptiles et Batraciens de la region de Dundo (Angola du Nord-Est). Publicações culturais da Companhia de Diamantes de Angola 6:126–136

Laurent RF (1954) Reptiles et Batraciens de la région de Dundo (Angola) (Deuxième Note). Publicações culturais da Companhia de Diamantes de Angola 23:35–84

Laurent RF (1964) Reptiles et Amphibiens de l'Angola (Troisième contribution). Publicações culturais da Companhia de Diamantes de Angola 67:11–165

Marques MP, Ceríaco LMP, Bauer AM et al (2014) Geographic distribution of amphibians & reptiles of Angola: towards an Atlas of the Angolan Herpetofauna. 12th conference of the Herpetological Association of Africa, Gobabeb, Namibia

Marques MP, Ceríaco LMP, Blackburn DC et al (2018) Diversity and distribution of the amphibians and terrestrial reptiles of Angola atlas of historical and bibliographic records (1840¬–2017). Proceedings of the California academy of sciences, Series 4, Volume 65, Supplement II: 1-501

Mertens R (1938a) Amphibien und Reptilien aus Angola, gesammelt von W. Schack. Senckenbergiana 20:425–443

Mertens R (1938b) Herpetologische Ergebnisse einer Reise nach Kamerun. Abhandlungen der Senckenbergischen Naturforschenden Gesellschaft, Frankfurt am Main 442:1–52

MHNG – Muséum d'histoire naturelle de la Ville de Genève (2017) Partial Amphibians Collection. Occurrence Dataset https://doi.org/10.15468/iftvxc. Accessed via GBIF.org

Minter LR, Netherlands EC, Du Preez LH (2017) Uncovering a hidden diversity: two new species of Breviceps (Anura: Brevicipitidae) from northern KwaZulu-Natal, South Africa. Zootaxa 4300:195–216

Monard A (1937) Contribuition à la Batrachologie d'Angola. Bulletin de la Société Neuchâteloise des Sciences Naturelles 62:1–59

Nagy ZT, Kusamba C, Collet M et al (2013) Notes on the herpetofauna of Western Bas-Congo, Democratic Republic of the Congo. Herpetology Notes 6:413–419

Ohler A (1996) Systematics, morphometrics and biogeography of the genus *Aubria* (Ranidae, Pyxicephalinae). Alytes 13:141–166

Parker HW (1936) Dr. Karl Jordan's expedition to South West Africa and Angola: herpetological collection. Novitates Zoologicae 40:115–146

Passmore NI (1972) Intergrading between members of the "regularis group" of toads in South Africa. J Zool 167(2):143–151

Perret JL (1975) Les sous-espèces *d'Hyperolius ocellatus* Günther (Amphibia, Salientia). Annales de la Faculté des Sciences du Cameroun 20:23–31

Perret JL (1976) Révision des amphibiens africains et principalement des types, conservés au Musée Bocage de Lisbonne. Arquivos do Museu Bocage, Segunda Série 6(2):15–34

Perret JL (1977) Les *Hylarana* (Amphibiens, Ranidés) du Cameroun. Rev Suisse Zool 84:841–868

Perret JL (1996) Sur un énigmatique batracien d'Angola. Societé Neuchâteloise des Sciences Naturelles 119:95–100

Peters WCH (1877) Übersicht der Amphibien aus Chinchoxo (Westafrika), welche von der Afrikanischen Gesellschaft dem Berliner zoologischen Museum übergeben sind. Monatsberichte der Königlichen Preussische Akademie des Wissenschaften zu Berlin 1877:611–621

Peters WCH (1882) Neue Batrachier (*Amblystoma Krausei, Nyctibatrachus sinensis, Bufo buchneri*). Sitzungsberichte der Gesellschaft Naturforschender Freunde zu Berlin 1882:145–148

Pickersgill M (2007) Frog search. Results of expeditions to Southern and Eastern Africa from 1993–1999. Frankfurt contributions to natural history 28. Edition Chimaira, Frankfurt

Pietersen DW, Pietersen EW, Conradie W (2017) Preliminary herpetological survey of Ngonye Falls and surrounding regions in southwestern Zambia. Amphibian Reptile Conserv 11(1) [Special Section]:24–43 (e148)

Poynton JC (1964) The Amphibia of Southern Africa: a faunal study. Ann. Natal Mus 17:1–334

Poynton JC (1970) Guide to the *Ptychadena* (Amphibia: Ranidae) of the southern third of Africa. Ann. Natal Mus 20(2):365–375

Poynton JC, Broadley DG (1985a) Amphibia Zambesiaca 1. Scolecomorphidae, Pipidae, Microhylidae, Hemisidae, Arthroleptidae. Ann. Natal Mus 26(2):503–553

Poynton JC, Broadley DG (1985b) Amphibia Zambesiaca 2. Ranidae. Ann. Natal Mus 27(1):115–181

Poynton JC, Broadley DG (1987) Amphibia Zambesiaca 3. Rhacophoridae and Hyperoliidae. Ann. Natal Mus 28(1):161–229

Poynton JC, Broadley DG (1988) Amphibia Zambesiaca, 4. Bufonidae. Ann. Natal Mus 29(2):447–490

Poynton JC, Broadley DG (1991) Amphibia Zambesiaca 5. Zoogeography. Ann. Natal Mus 32:221–277

Poynton JC, Haacke WD (1993) On a collection of amphibians from Angola, including a new species of Bufo Laurenti. Ann Transv Mus 36(2):9–16

Poynton JC, Loader SP, Conradie W et al (2016) Designation and description of a neotype of *Sclerophrys maculata* (Hallowell, 1854), and reinstatement of *S. pusilla* (Mertens, 1937) (Amphibia: Anura: Bufonidae). Zootaxa 4098(1):73–94

Regester KJ, Lips KR, Whiles MR (2006) Energy flow and subsidies associated with the complex life cycle of ambystomatid salamanders in ponds and adjacent forest in southern Illinois. Oecologia 147(2):303–314

Rochebrune AT (1885) Vertebratorum novorum vel minus cognitorum orae Africae occidentalis incolarum. Diagnoses (1). Bulletin de la Société Philomathique de Paris 7(9):86–99

Royal Belgian Institute of Natural Sciences (2017). RBINS DaRWIN. Occurrence Dataset https://doi.org/10.15468/qxy4mc. Accessed via GBIF.org

Royal Museum for Central Africa, Belgium (2017). RMCA-HERP. Occurrence Dataset https://doi.org/10.15468/0inmlf. Accessed via GBIF.org

Ruas C (1996) Contribuição para o conhecimento da fauna de batráquios de Angola. Garcia de Orta, Série de Zoologia 21(1):19–41

Ruas C (2002) Batráquios de Angola em colecção no Centro de Zoologia. Garcia de Orta, Série de Zoologia 24(1–2):139–154

SASSCAL ObservationNet (2017) http://www.sasscalobservationnet.org/ consulted on March 6th 2018

Scheinberg L, Fong J (2017) CAS herpetology (HERP). Version 33.10. California Academy of Sciences. Occurrence dataset https://doi.org/10.15468/bvoyqy. Accessed via GBIF.org

Schick S, Kielgast J, Rödder D et al (2010) New species of reed frog from the Congo basin with discussion of paraphyly in Cinnamon-belly reed frogs. Zootaxa 2501:23–36

Schiøtz A (1999) Treefrogs of Africa. Editions Chimaira, Frankfurt am Main, 350 pp

Schiøtz A, Van Daele P (2003) Notes on the treefrogs (Hyperoliidae) of North-Western Province, Zambia. Alytes 20:137–149

Schmidt KP (1936) The amphibians of the Pulitzer-Angola expedition. Ann Carnegie Mus

25:127–133

Steindachner F (1867) Reise der österreichischen Fregatte Novara um die Erde in den Jahren 1857, 1858, 1859 unter den Bafehlen des Commodore B. von Wüllerstorf-Urbair. Zologischer Theil. 1. Amphibien. Wien: K. K. Hof- und Staatsdruckerei. 70 p + 5pls

Tandy M, Keith R (1972) *Bufo* of Africa. In: Blair WF (ed) Evolution in the Genus *Bufo*. University of Texas Press, Austin/London, pp 119–170

Turner AA, Channing A (2017) Three new species of *Arthroleptella* Hewitt, 1926 (Anura: Pyxicephalidae) from the Cape Fold Mountains, South Africa. Afr J Herpetol 66:53–78

Uyeda JC, Drewes RC, Zimkus BM (2007) The California Academy of Sciences Gulf of Guinea expeditions (2001, 2006) VI. Proc Calif Acad Sci 58(13–22):367

Van Dijk DE (1966) Systematic and field keys to the families, genera and described species of southern African anuran tadpoles. Ann Natal Mus 18:231–286

Van Dijk DE (1971) A further contribution to the systematics of Southern African anuran tadpoles—the genus *Bufo*. Ann Natal Mus 21:71–76

Waddle JH (2006) Use of amphibians as ecosystem indicator species. PhD Thesis. University of Florida, Gainesville

Whiles MR, Lips KR, Pringle CM et al (2006) The effects of amphibian population declines on the structure and function of Neotropical stream ecosystems. Front Ecol Environ 4(1):27–34

Zimkus BM, Rödel MO, Hillers A (2010) Complex patterns of continental speciation: molecular phylogenetics and biogeography of sub-Saharan puddle frogs (*Phrynobatrachus*). Mol Phylogenet Evol 55(3):883–900

Zimkus BM, Lawson LP, Barej MF et al (2017) Leapfrogging into new territory: how Mascarene ridged frogs diversified across Africa and Madagascar to maintain their ecological niche. Mol Phylogenet Evol 106:254–269

Permissions

All chapters in this book were first published by Springer; hereby published with permission under the Creative Commons Attribution License or equivalent. Every chapter published in this book has been scrutinized by our experts. Their significance has been extensively debated. The topics covered herein carry significant information for a comprehensive understanding. They may even be implemented as practical applications or may be referred to as a beginning point for further studies.

The contributors of this book come from diverse backgrounds, making this book a truly international effort. We would like to thank all the contributing authors for lending their expertise to make the book truly unique. They have played a crucial role in the development of this book. Without their invaluable contributions this book wouldn't have been possible. They have made vital efforts to compile up to date information on the varied aspects of this subject to make this book a valuable addition to the collection of many professionals and students.

This book was conceptualized with the vision of imparting up-to-date and integrated information in this field. To ensure the same, a matchless editorial board was set up. Every individual on the board went through rigorous rounds of assessment to prove their worth. After which they invested a large part of their time researching and compiling the most relevant data for our readers.

The editorial board has been involved in producing this book since its inception. They have spent rigorous hours researching and exploring the diverse topics which have resulted in the successful publishing of this book. They have passed on their knowledge of decades through this book. To expedite this challenging task, the publisher supported the team at every step. A small team of assistant editors was also appointed to further simplify the editing procedure and attain best results for the readers.

Apart from the editorial board, the designing team has also invested a significant amount of their time in understanding the subject and creating the most relevant covers. They scrutinized every image to scout for the most suitable representation of the subject and create an appropriate cover for the book.

The publishing team has been an ardent support to the editorial, designing and production team. Their endless efforts to recruit the best for this project, has resulted in the accomplishment of this book. They are a veteran in the field of academics and their pool of knowledge is as vast as their experience in printing. Their expertise and guidance has proved useful at every step. Their uncompromising quality standards have made this book an exceptional effort. Their encouragement from time to time has been an inspiration for everyone.

The publisher and the editorial board hope that this book will prove to be a valuable piece of knowledge for students, practitioners and scholars across the globe.

Index

A
Agricultural Zones, 97-98

Ammonites, 52-53, 59-61, 63, 68-70, 74

Amphibian, 9, 235-237, 239-240, 242-243, 253, 255-259, 266-269, 271-272

B
Biodiversity, 3-8, 10-15, 18, 32, 35, 37-42, 44, 46-47, 49-53, 116-118, 132, 135-136, 147-148, 158-159, 161, 194, 199, 205, 233-234, 266-268, 270

Biodiversity Surveys, 5, 7, 147

Biogeography, 5, 8-9, 12-14, 17-18, 29, 31, 37, 39, 41-42, 44, 71-72, 82, 135, 158, 213-214, 235, 238-239, 241, 254, 270, 272

Biomes, 3-4, 8, 10, 14, 29, 31-32, 34, 37, 113, 117, 161

Botanical Collections, 4, 78, 83

Botanical Collectors, 77, 83, 93

Botanical Diversity, 6, 12, 39, 77, 93, 132

C
Cenozoic, 17, 22, 53, 56, 58, 63, 66, 68, 71-72, 230, 232

Charcoal, 102, 118-121, 127, 131

Climate Change, 5, 7, 13-14, 27-28, 38, 48, 50, 63, 84, 102-105, 139, 148, 241, 243, 259, 266, 270

Clypeobarbus Bellcrossi, 212, 223

Coastal Belt, 18-19, 22, 25

Coastal Plain, 26, 80, 82, 87, 102, 205, 213

Coastal Rivers, 21, 200, 202, 204

Collaborative Research, 3, 7

Conservation, 3, 5-7, 10-13, 15, 22, 31-32, 37-43, 47-48, 77, 115-116, 131-132, 135, 147-148, 158, 160, 167, 169, 199, 216, 233, 235, 241-242, 253, 258-259, 266, 268-270

Conservation Value, 115

Cretaceous, 8-9, 17-18, 52-53, 55-56, 58-63, 68-72, 74, 194, 213-214, 230

D
Deforestation, 8, 18, 102-103, 118-119, 132, 147-148, 199

Dragonflies, 9, 12, 39, 133, 135, 139-140, 143, 148, 158-159

Drainages, 199-200, 202-203, 211, 213

E
Ecology, 5, 12-13, 16, 18, 39, 41, 97, 99, 106, 116, 148, 160, 201, 231, 250, 266, 270

Ecoregions, 3, 8, 14, 18, 29, 31-34, 36-38, 40, 46, 50, 98, 101, 104, 161, 202-203, 209, 233

Ecosystems, 3-6, 10, 13, 15, 29, 31, 37, 41, 46, 49-50, 53, 98-99, 105, 107, 113, 115-117, 132, 199, 272

Endemism, 9, 11, 29-30, 35, 37-39, 46, 77, 80-82, 86, 92-94, 103, 106, 108, 116-117, 135, 142, 146-148, 160, 162, 166, 235, 239, 253, 255-257

Environmental Conditions, 9, 108, 110, 116, 167

Escarpment Zone, 19-20, 24, 35, 131, 206, 255

Evolutionary History, 13, 31, 38, 41, 106, 116

F
Ferralitic Soils, 107, 110

Fishery, 44, 50, 216, 229

Flooded Savannas, 108, 111, 114-115

Floodplains, 24, 29, 42, 46, 115, 117, 126, 129, 138, 155, 158, 203, 210-211

Forest Loss, 119-120

Fossil Record, 8, 52-56, 61, 68

Fossiliferous Rocks, 8, 53-54

Freshwater Fishes, 9, 13, 17, 29, 40, 199-203, 206, 213-216, 218, 221-222, 230-233

G

Geology, 15, 18-19, 22, 62, 71, 96, 98, 102, 230, 232

Geomorphology, 15-16, 18, 213

Geoxyles, 18, 106, 108

Grasslands, 14, 20, 25, 29-31, 33-34, 36, 65, 83-84, 94, 106-108, 110, 112-119, 128-129, 140

H

Headwaters, 20, 82, 84, 89, 92, 107, 141, 146, 203, 207-210, 212, 231, 241

Hemichromis Elongatus, 210, 215, 221, 228

Hepsetus Cuvieri, 207-208, 210, 212, 215, 225, 233

Herbarium, 5, 11, 77-78, 83-84, 86, 88-92, 94

Herpetology, 235, 237, 241, 243, 258, 268-271

Hybridisation, 216, 252

Hydrocynus Vittatus, 210, 212, 215, 225

Hydrology, 21, 69, 84, 96

I

Inundation, 68, 113-114

Invasive Species, 5, 77, 97, 216

Invertebrates, 42, 45, 53, 55, 57-58, 60, 64-68, 240, 268

L

Land Transformation, 5, 28, 118

Landscape Change, 119, 127, 129, 146

M

Marine Areas, 42-43, 47, 49

Marine Protected Areas, 42-43, 47, 49-50

Marine Spatial Planning, 43, 48-49

Modern Synthesis, 3-4, 11-13, 37-41, 93, 103-104, 116, 132, 270

Mollusca, 53, 55, 72

Mosasaur, 53, 62, 72-73

N

Natural Resources, 71, 95, 99, 102, 118, 127, 233, 259

O

Oreochromis Niloticus, 204, 216, 221, 229, 233

P

Paleobiodiversity, 52-54, 56, 61, 68-69

Phytochorion, 106, 108-110, 115

Plant Communities, 8, 95, 97, 99, 101, 103, 107

Pleistocene, 17-18, 45, 53, 65-67, 72-73, 115, 230

Plesiosaurs, 56, 60, 69, 71-72

Pollimyrus Castelnaui, 210, 212, 222, 231

Pollution, 48, 131, 199, 216

Pseudocrenilabrus Philander, 210, 212-213, 228

R

Regionalisation, 30-31

Remote Sensing, 8, 95, 98-99, 102, 104-105, 131-132, 212

Reptiles, 9, 11, 30-31, 35, 37, 60-64, 236-238, 240-241, 266-268, 270

River Flows, 118, 126

Riverine Forest, 119, 170

S

Seaweed, 43, 45, 50, 82

Shifting Cultivation, 84, 92, 94, 102-103, 105, 118, 132

Suffrutices, 84, 106-109, 113-117, 128

Systematic Conservation Planning, 43, 47

T

Terminology, 30-31

Topography, 15-16, 18, 96, 110

U

Underground Forests, 8, 18, 39, 41, 97, 105-106, 117

Urbanisation, 47, 118, 148, 199, 216

V

Vegetation, 5-6, 8, 11, 13, 15-16, 18-19, 23-26, 29-32, 34, 37, 39-41, 75, 79, 82-84, 86, 91-107, 116-118, 123, 128-129, 161

Vegetation Communities, 16, 101

Vegetation Mapping, 99-100

Vegetation Types, 8, 23, 31-32, 34, 82, 86, 95, 97-99, 101-102, 107

Z

Zoogeography, 29, 201, 271

Printed in the USA
CPSIA information can be obtained
at www.ICGtesting.com
LVHW022132041023
760082LV00002B/25